Climate Investing

Climate Investing

New Strategies and Implementation Challenges

Edited by

Emmanuel Jurczenko

iSTE

WILEY

First published 2022 in Great Britain and the United States by ISTE Ltd and John Wiley & Sons, Inc.

ISTE Ltd
27-37 St George's Road
London SW19 4EU
UK

www.iste.co.uk

John Wiley & Sons, Inc.
111 River Street
Hoboken, NJ 07030
USA

www.wiley.com

Library of Congress Control Number: 2022945579

British Library Cataloguing-in-Publication Data
A CIP record for this book is available from the British Library
ISBN 978-1-78630-806-1

Contents

Foreword

Climate risk is one of the biggest challenges to humanity in the 21st century. As a consequence, we stand on the cusp of the next great economic revolution. One where sustainability will be the focus of all aspects of life, from consumption, to production of goods, commercial services and business practices.

But what will this change mean for the asset management industry? Asset managers are a key link between the providers of funding and those who need funding, and are therefore in a privileged position to help our society transition to exercise more sustainable investment practices. We will need to integrate in our asset allocation ways to assess the systemic risk posed by climate change. Climate risks still may be mis-priced by financial markets and climate change may expose investors to tail risks, which need to be actively managed.

This ground-breaking book on climate investing clarifies some practical considerations that asset managers should look at when assessing the implications that a net-zero carbon world might have on their portfolios. It assesses how to incorporate the risks and opportunities that exist as the world transitions to a lower-carbon future.

Such books are a necessity. While investors are increasingly concerned about the environmental footprint of their portfolios, they are puzzled by the plethora of climate integration methods currently out there. Although there is now a framework in place for how investors can meet the Paris climate goals, many challenges remain. This book helps address these challenges and provides research-led insights on how to tackle them.

We live in a world where climate and sustainability are at the forefront of the investment decision-making process. As such, asset managers will increasingly be assessed on the wider role they play in improving society – not just the financial returns they deliver. In other words, “social alpha” will be assessed alongside “performance alpha” and while “doing good” will hopefully translate into “doing well” over time, it is not a directly correlated relationship. The valuable insights of this book will help us understand the implication between social and performance alpha.

Fiona Frick
CEO, Ungestion
September 2022

1

The Financial Materiality of Climate Change: Evidence from a Global Survey

This chapter presents evidence from a global survey of nearly 700 investors and companies on the materiality of climate risk for financial reporting. There are growing calls by investors and other stakeholders to expand the scope of materiality judgments and mandate the disclosure of climate-related financial risks. Field evidence shows that investors believe climate risk to be financially material and to represent heightened regulatory and litigation risk. The study further finds a misalignment of the materiality perception of investors compared with companies. Far fewer companies believe that they are exposed to climate risks and consequently do not make any disclosures about it. Investors state difficulties with identifying and quantifying risks due to the lack of disclosures as the main challenge in assessing the impact of climate change. The findings suggest that current disclosure practices do not provide investors with adequate information about climate-related financial risks.

1.1. Introduction

What constitutes material information for investors evolves over time. In recent years, various organizations have advocated for enhanced corporate disclosures in

Chapter written by Amir AMEL-ZADEH.
This study benefited from discussions with Sandra Carlisle, David Chambers, Howard Covington, Philippe Desfosses, Elroy Dimson, Robert Eccles, Jessica Fries, Guy Gresham, Jeffrey Hales, Andrew Karolyi, Sam Knox, Philipp Krueger, Harry Markowitz, Martina McPherson, Geoff Meeks, Anthony Miller, George Serafeim, Martin Skancke, Geoff Whittington and participants at the 3rd Stranded Assets Forum, the Berkeley Sustainable Business & Investment Forum and the Judge Business School Conference on Long-Term Investing at the Newton Centre for Endowment Asset Management. The author would like to thank Emmanuel Jurczenko (editor), and Bank of New York Mellon, IPREO and Institutional Investor Research Group for support in survey distribution and administration. This research was funded as a gift from BNY Mellon.

financial reports beyond traditional financial metrics to include material environmental, social and governance (ESG) information[1]. Specifically, the potential financial risks associated with climate change have attracted increasing attention. The World Economic Forum in their Global Risk Report rates risks related to climate change among the most likely and with the most damaging impact on economies and businesses highlighting their potential financial materiality (World Economic Forum 2020). Although the economic costs of climate change are increasingly being recognized (OECD 2015; TCFD 2016), it is unclear whether investors are being provided with sufficient disclosures by companies to assess the financial risks from climate change.

Over the past several years, momentum has been building to recognize the financial materiality of climate change and thus to require companies to disclose their exposure to and management of climate change risks. For example, in 2014, a group of large institutional investors, the Carbon Disclosure Project (CDP) and the UNEP Finance Initiative formed a coalition to raise awareness of climate risks among institutional investors. In 2015, the G20 Finance Ministers and Central Bank Governors mandated the Financial Stability Board (FSB) to review companies' disclosures of risks related to climate change. The FSB established the Task Force on Climate-related Financial Disclosures (TCFD), which in December 2016 published recommendations on voluntary climate-related financial disclosures[2]. The TCFD recommendations are being adopted in the United Kingdom and the European Union as the de facto standards for the disclosure of climate change-related risks[3].

1 These international bodies include initiatives such as the Carbon Disclosure Project (CDP), Sustainability Accounting Standards Board (SASB), International Integrated Reporting Council (IIRC) to name but a few.

2 The TCFD recommends forward-looking disclosures on the financial impacts of risks and opportunities related to the physical impacts of climate change, as well as of those related to the transition to a lower-carbon global economy as specified under the United Nations Framework Convention on Climate Change (see United Nations Framework Convention on Climate Change. "The Paris Agreement", December 2015, http://unfccc.int/files/essential_background/convention/application/pdf/english_paris_agreement.pdf).

3 See *Wall Street Journal*, "U.K. Requires Companies to Report on Climate Change by 2025", 9 November 2020, https://www.wsj.com/articles/u-k-requires-companies-to-report-on-climate-change-by-2025-11604964183 (last accessed 10 January 2021); EU Technical Expert Group on Sustainable Finance, "Report on Climate-related Disclosures, January 2019, https://ec.europa.eu/info/sites/info/files/business_economy_euro/banking_and_finance/documents/190110-sustainable-finance-teg-report-climate-related-disclosures_en.pdf (last accessed 10 January 2021).

In the United States, the SEC in 2016 issued a concept release on the existing disclosure requirements in Regulation S-K asking for public comments. The concept release drew responses calling for improved disclosures of ESG information, particularly on climate change[4]. Moreover, in March 2022, the SEC approved a proposed rule to amend the Securities Acts requiring companies to disclose certain risks and impacts related to climate change in their annual report. Pressure is also coming from investors. The CEOs of the largest asset managers have warned their investee companies they would vote against management and directors if they did not improve their climate-related disclosures[5].

A growing academic literature examines the financial impact of environmental risks and climate change and how companies disclose these risks and opportunities (Cho et al. 2010, 2012; Flammer 2013; Bernstein et al. 2019). Yet, it is still unclear whether current disclosure regulation provides investors with the necessary means to assess the risks of climate change and whether and how investors consider these risks in asset allocation decisions.

This chapter provides evidence from a global survey of nearly 700 institutional investors and listed companies on their views of the financial materiality of climate change. The study documents whether and why investors view climate risks as material for investment decisions and whether their views are consistent with the companies' own views and their disclosure practices. The study identifies the diverging views of investors and companies on the materiality of climate risk, and the types of risks climate change poses to companies, as one potential impediment to the efficient pricing of climate risks in capital markets. The findings in this chapter lend support to call on the IFRS and the SEC to require broader applications of materiality judgments and to expand current disclosure requirements to climate-related risks in financial reports.

---

4 See SEC, Comments on Concept Release: Business and Financial Disclosure Required by Regulation S-K available on https://www.sec.gov/comments/s7-06-16/s70616.htm (last accessed 22 March 2020); SASB (2016), Business and Financial Disclosure Required by Regulation S-K – the SEC's Concept Release and Its Implications, available at https://www.sasb.org/wp-content/uploads/2016/09/Reg-SK-Comment-Bulletin-091416.pdf (last accessed 22 March 2020).

5 Larry Fink, CEO of BlackRock, CEO letter: A fundamental reshaping of finance, January 14, 2020; Cyrus Taraporevala, CEO, State Street Global Advisors, CEO's Letter on our 2020 Proxy Voting Agenda, January 28, 2020.

This chapter contributes to the literature on the financial materiality of climate risk and to the large literature on environmental disclosures and financial performance more broadly. A growing number of studies find that climate risks might be mis-priced by financial markets and that climate change might expose investors to tail risks that are largely avoidable (Anderson et al. 2016; Khan et al. 2016; Bansal et al. 2019). The academic evidence also seems to support that mandatory climate-related disclosures are positively valued by investors (Krüger 2015). This study contributes to this line of research by providing field evidence on institutional investors' perceptions of climate risks and whether they believe climate risk to be financially material and why[6]. In contrast to concurrent surveys that focus on investor perceptions and the tools they employ to manage climate risks (Krüger et al. 2020), this chapter compares the climate risk perceptions and climate disclosure requirements of investors with the views and disclosure practices of companies. The field evidence in this chapter thus informs the academic, policy and practitioner debate on the information disclosure environment with regard to climate change.

1.2. Survey design and demographic data

1.2.1. *Survey design*

The survey questions in this study were developed based on a review of the academic literature and feedback from a group of six academic researchers in finance and accounting as well as a group of institutional investors and financial market organizations. An Internet-based survey instrument was then developed with the help of Institutional Investor Research Group, a financial market's survey design and execution specialist with the aim to reduce biases introduced by the wording and tone of the questions. The penultimate version was beta-tested with a small number of investors and financial market experts.

6 Although the survey in this chapter was initially conducted in mid-2016, the survey results remain relevant for several reasons. First, the disclosure environment on climate-related risks has not materially changed since 2016 suggesting that the investor demands and perceptions as well as company's disclosure practices are likely relatively stable over time. Second, given the heightened attention on the changing climate in the subsequent years and the therefore increasing relevance of climate-related risks to investors, the results of this survey likely underestimate investors' and companies' perceptions about their financial materiality.

Although the survey was conducted anonymously, it allowed respondents to voluntarily reveal their names and institutions. About 40% of respondents chose to do so. Respondents could skip questions if they chose not to answer them.

The final version of the questionnaire was sent via email to a list of senior investment professionals at 4,523 asset-managing and asset-owning institutions and 2,171 listed companies compiled by Bank of New York Mellon and IPREO, a financial services data company. A total of 652 investors and 364 issuers responded, for an overall response rate of 15.2% (or 14.4% among investors and 16.7% among companies). However, across the survey questions included in the analysis for this study, the average response rate ranges between 7 and 16% for a total of 683 responses as some respondents skipped the questions related to this study or did not fully complete the survey.

1.2.2. *Demographic data*

Table 1.1 reports demographic information about the responding person and institution, summarizing the responses of investors in Panel A and of issuers in Panel B. Panel A shows that the survey captures a large distribution of investors in terms of AUM. On the one hand, 35% of respondents report AUM below 1 billion dollars, and, on the other hand, 15% report AUM of more than 100 billion dollars. The sample has a higher proportion of large investors from the United States and Europe and a higher proportion of smaller investors from other parts of the world, i.e. Asia, the Middle East and Africa. The larger institutions are also more likely to be asset managers than asset owners. Overall, the respondents' total AUM is approximately US $31 trillion[7].

7 Untabulated results reveal that 70% of the responding organizations report that less than 10% of their AUM is allocated to ESG investments and half of those report no ESG allocation at all. In addition, none of the respondents report a 100% ESG allocation.

Panel A: Demographic characteristics of investors						
Assets under management (in US $ billion)		Region			Type of organization	
	All	US	Europe	RoW	Asset Manager	Asset owner
< 1	35%	30%	34%	36%	70%	30%
1–5	20%	25%	41%	34%	55%	45%
5–10	11%	33%	39%	28%	64%	36%
10–50	12%	21%	39%	39%	56%	44%
50–100	6%	28%	55%	17%	78%	23%
> 100	15%	34%	47%	19%	79%	21%
Respondent titles/positions		Region			Size	
	Total	US	Europe	RoW	Large	Small
Fund/portfolio manager	28%	23%	49%	28%	48%	52%
Chief executive officer	13%	12%	37%	51%	12%	88%
Investment analyst/strategist	13%	44%	26%	29%	60%	40%
Executive/managing director	11%	20%	47%	32%	67%	33%
Chief investment officer	8%	35%	26%	40%	44%	56%
ESG/responsible investment specialist	8%	44%	40%	16%	56%	44%
CFO/COO/chairman/other executive	5%	30%	33%	37%	49%	51%
Other	13%	40%	29%	31%	32%	68%
Respondent organizations		Region			Size	
	Total	US	Europe	RoW	Large	Small
Asset manager	65%	28%	41%	31%	47%	53%
Asset owner	35%	31%	32%	37%	41%	59%

Panel B: Demographic characteristics of companies				
Revenues (in US $ billion)			Sectors	
	%			%
< 1	32%		Financial services	22%
1–5	30%		Consumer and retail	16%
5–10	13%		Technology, media and telco	16%
10–50	18%		Industrials/capital good	11%
50–100	2%		Basic materials	10%
> 100	4%		Transportation	7%
Geography			Energy	6%
	%		Health care	5%
Asia	48%		Business and professional services	4%
Europe	26%		Utilities	4%
North and South America	21%			
Middle East and Africa	5%			
Respondent titles/positions				
	%			%
Chief executive officer/chairman	4%		Investor relations officer/manager	65%
Chief financial officer	7%		Business manager	4%
Chief operating officer	1%		Other	14%
Chief sustainability officer/manager	6%			

Table 1.1. *Respondent characteristics*

Among the individuals responding to the investor survey, only 8% are specialists on ESG investing. About a quarter of the respondents are senior executives (CEO, CIO, CFO or similar), and another quarter are senior fund or portfolio managers (28%). Conditioning on the region shows a larger proportion of senior executives coming from Europe and other parts of the world compared to the United States[8]. Overall, based on the sample composition, this research reflects the views of a diverse group of largely mainstream investors.

Panel B shows the size, sector and geographical distribution of the issuer respondents as well as their roles within their organizations. Of the issuers in the sample, 62% report revenues up to 5 billion dollars, another 31% report between 5 and 50 billion dollars and 6% report above 50 billion dollars. In terms of sectors, 22% belong to financial services, followed by consumer and retail companies (16%) and technology, media and telecoms (16%). Overall, the sector composition is roughly in line with sector rankings of the MSCI World Index: 48% of the companies are headquartered in Asia, 26% in Europe, 21% in the Americas and 5% in the Middle East and Africa. The majority of the respondents to the issuer survey are investor relations officers (65%), and around 12% of the responses come from CEOs, CFOs and COOs.

In the subsequent analyses, survey results are also cut by investor type (asset manager or asset owner), whether they are above the median of 5 billion dollars in AUM (large versus small), based on their ESG allocation at the median cut-off point of 5% of AUM (high versus low) and based on their geography (United States, Europe and the rest of the world)[9]. Issuer survey results are also cut by size and sector to understand differences in views on climate change between large international and resource-intensive companies that are more likely impacted by climate-related matters compared to smaller companies from less resource-intensive sectors.

1.3. Survey results

The survey addresses three key topics to shed light on the materiality of climate change. I will start by examining the importance of climate change for investment

8 Untabulated results also show that among asset owners, there are public and private pension funds (19%), financial institutions (4%), charities and endowments (4%), sovereign wealth funds (3%) and family offices (3%).

9 The size cut-off provides a value-weighted rather than an equal-weighted view of investor beliefs. I condition on their ESG allocation to distinguish between ESG-aware investors that naturally have stronger views on climate change-related matters. The cut by geography gives a view on the geographical distribution in climate risk awareness.

decisions generally and whether investors' views have changed over time. I will then focus more specifically on whether investors believe climate change to pose a material risk to companies in their portfolios and whether companies' executives themselves consider climate change a material risk. I will then follow up by investigating in more detail the kind of risk investors and companies that I expect climate change to pose. Thirdly, I will examine the disclosure environment of climate risk and ask investors (issuers) whether they analyze (disclose) information on climate change and the challenges they face when evaluating (disclosing) this information.

1.3.1. *Importance of climate change for investment decisions*

A growing number of investors are beginning to consider ESG information in their investment allocations, suggesting that such information has become important to investment decisions over the past few years (Amel-Zadeh and Serafeim 2018). Yet, we know little about whether investors consider information related to climate change in asset allocation decisions despite the growing attention to climate-related investment risks by policy makers and academic research (e.g. Carney 2015; Bansal et al. 2019). Equally, little is known about how investors' views on the materiality of climate change for investment decisions have evolved over time. Among a few, in a recent, concurrent study, Krüger et al. (2020) find that institutional investors believe that climate change is financially material to their investment portfolios, in particular stemming from regulatory risks.

Table 1.2 presents results concerning the question of how important climate change has been to investors in the past five years (Panel A) and will be in the next 5–10 years (Panel B). The respondents were asked to rank the importance of climate change to their investment decisions on a scale: 1 (not important) to 2 (somewhat important) to 3 (very important). The table shows the percentage of responses for each rank in Column (1), and also by type of investor, whether asset manager or asset owner (Columns 2–4); by AUM size, i.e. whether above the median of 5 billion dollars in AUM (Columns 5–7); by ESG awareness, i.e. whether above the median of 5% of AUM allocated to ESG (Columns 8–10) and by region, i.e. whether based in the United States, Europe or the Middle East, Asia and Africa (Columns 11–15).

Table 1.2 Panel A Column (1) shows that about 22% of the respondents considered climate change to be very important to investment decisions in the past five years, while 46% considered it somewhat important. Nevertheless, about a third responded that climate change was not important to their investment decisions in the past five years. The average rating is 1.89, which is statistically significantly smaller than 2 ($p\text{-value} < 0.01$), suggesting that on average investors did not consider climate change to be important to investment decisions in the past five years. The results in the next columns in Panel A reveal that somewhat surprisingly, a higher proportion of asset managers consider climate change to be very important compared to asset owners. This is also reflected in a significantly higher average rating (1.94 versus 1.76, p-value of difference <0.01). This is surprising as the effects of climate change are generally thought to materialize over the longer term and thus are more likely to be important to long-term asset owners. However, this result might be due to the fact that my sample has a larger fraction of European asset managers, who – as Columns (11) and (12) reveal – are significantly more likely to consider climate change as very important.

The results in Panel A also reveal that larger investors and more ESG-aware investors have considered climate change as important in the past five years. The average rating for ESG-aware investors is significantly higher than 2 and higher than for those with a low awareness (p-value<0.001).

Panel B reports the results to the question of how important climate change will be to investment decisions in the next 5–10 years. Column (1) shows that the proportion of investors that deem climate change as very important to investment decisions in the near future is about 45%, which is more than double from the proportion reported in Panel A that reflected the importance of climate change in the past. The average rating is 2.29, which is statistically significantly higher than 2 (p-value<0.001), suggesting that on average investors in my sample consider climate change as becoming important to investment decisions in the near future. I also found that a higher fraction of asset managers, large investors, ESG-aware investors and European investors consider climate change as very important for investment decisions in the future. Almost 60% of ESG-aware investors (compared to 35% of those with a low awareness) and 56% of European investors (compared to 41% from the United States and 33% from the rest of the world) have responded that climate change will be very important to their investment decisions.

Panel C in Table 1.2 reports the change in percentage of respondents that consider climate change to be very important (from the past five years to the next 5–10 years) and the change in average rating. The panel documents an increase in percentages and average ratings across all groups suggesting that all respondents believe that climate change is becoming more important for investment decisions. I found the largest increase overall for large investors and European investors and a significantly higher increase in importance for large compared to small and European compared to U.S. investors.

Panel A: How important has climate change been to your investment decisions in the past 5 years?

	ALL	Type			AUM Size			ESG Awareness			Region				
N = 309	(1)	(2)	(3)	(4)	(5)	(6)	(7)	(8)	(9)	(10)	(11)	(12)	(13)	(14)	(15)
	%	Asset Manager	Asset Owner	Diff	Large	Small	Diff	High	Low	Diff	US	Europe	RoW	Diff (11)-(12)	Diff (11)-(13)
3 Very important	21.7%	24.5%	15.1%	**	24.8%	18.9%		35.2%	12.5%	***	18.2%	29.5%	14.9%	**	
2 Somewhat important	45.6%	45.4%	46.2%		50.3%	41.5%		46.4%	45.1%		48.5%	41.8%	48.3%		
1 Not important	32.7%	30.1%	38.7%		24.8%	39.6%	***	18.4%	42.4%	***	33.3%	28.7%	36.8%		
Average rating	1.89	1.94	1.76	**	2.00	1.79	***	2.17	1.70	***	1.85	2.01	1.78	*	
H0: Average rating = 2	p-value<0.01														

Panel B: How important will climate change be to your investment decisions in the next 5-10 years?

	ALL	Type			AUM Size			ESG Awareness			Region				
N = 309	(1)	(2)	(3)	(4)	(5)	(6)	(7)	(8)	(9)	(10)	(11)	(12)	(13)	(14)	(15)
	%	Asset Manager	Asset Owner	Diff	Large	Small	Diff	High	Low	Diff	US	Europe	RoW	Diff (11)-(12)	Diff (11)-(13)
3 Very important	44.5%	47.9%	36.6%	*	55.5%	34.6%	***	58.9%	34.8%	***	40.8%	56.1%	32.6%	**	
2 Somewhat important	40.3%	38.1%	45.2%		34.2%	45.7%	**	33.1%	45.1%	**	36.7%	34.1%	53.5%		**
1 Not important	15.3%	14.0%	18.3%		10.3%	19.8%	**	8.1%	20.1%	***	22.4%	9.8%	14.0%	**	
Average rating	2.29	2.34	2.18	*	2.45	2.15		2.51	2.15	***	2.18	2.46	2.19	***	
H0: Average rating = 2	p-value<0.001														

Panel C: Differences in means of past 5 years to next 5-10 years?

	ALL	Type			AUM Size			ESG Awareness			Region				
N = 309	(1)	(2)	(3)	(4)	(5)	(6)	(7)	(8)	(9)	(10)	(11)	(12)	(13)	(14)	(15)
	%	Asset Manager	Asset Owner	Diff	Large	Small	Diff	High	Low	Diff	US	Europe	RoW	Diff (11)-(12)	Diff (11)-(13)
Δ Very important	22.8%	23.4%	21.5%		30.7%	15.7%	***	23.7%	22.3%		22.6%	26.6%	17.6%		
p-value	<0.001	<0.001	<0.001		<0.001	<0.01		<0.001	<0.001		<0.001	<0.001	<0.01		
Δ Average rating	0.40	0.40	0.42		0.45	0.36	*	0.34	0.45		0.34	0.46	0.40	*	
p-value	<0.001	<0.001	<0.001		<0.001	<0.001		<0.001	<0.001		<0.01	<0.001	<0.001		

Column (1) presents the percent of respondents choosing the response in a given row. Columns (2)–(3) show the results by investor type. Column (4) reports the results of a test of the null hypothesis that the percentages in Columns (2) and (3) are equal to each other. Columns (5) and (7) report the percentages for investors with AUM >$ 5 billion (large) and AUM <$ 5 billion (small) and the results of a test of the null hypothesis that the percentages are equal, respectively. Columns (8)–(10) report the percentages for investors with ESG allocation > 5% of AUM (high ESG awareness) and ESG allocation < 5% of AUM (low ESG awareness) and the results of a test of the null hypothesis that the percentages are equal, respectively. Columns (11–(15) report the percentages by geographical regions and the respective tests of the null hypothesis that the percentages in the specified columns are equal to each other. ***, ** and * denote significance at the 1%, 5% and 10% levels, respectively.

Table 1.2. *Importance of climate change for investment decisions – past and present*

Overall, the results in this section suggest that there is heterogeneity of opinions among different types of investors and different regions about the importance of climate change for investment decisions. My findings show that although overall climate change was not considered important in the past, investor views have changed significantly. The average investor, irrespective of size, ESG awareness or region, seems to consider climate change to be important to investment decisions and to become more so in the next 5–10 years. The results also suggest that European investors generally are more likely to find climate change important for investment decisions.

1.3.2. *Financial materiality of climate risk*

The previous section has documented a general increase in importance of climate change for investment decisions. In this section, I therefore further investigate whether climate change is financially material in terms of posing a risk to companies and investment portfolios. Several contemporaneous archival studies suggest that climate risks are financially material to companies (Bernstein et al. 2019; Giglio et al. 2021) and practitioner literature recommends businesses to prepare for the potential risks that are emerging because of climate change and the policy responses that aim to address it (e.g. International Finance Corporation 2010; Oliver Wyman 2018).

Table 1.3 Panel A reports the response to the question to investors of whether climate change poses a material risk to companies in their portfolio. The vast majority, 73.5% of investors, responded "yes". This proportion, statistically, is significantly different from a random draw ($p\text{-value}<0.001$)[10].

Panel B summarizes investors' beliefs about what type of risks climate change poses to companies they follow for investment purposes. Respondents were asked to rate the risks on a scale from 1 to 5, where 1 is the most important and 5 is the least important risk. The results are ordered by the risk that received the highest proportion of first and second rankings (Column 1). The panel reveals that the majority of investors believe climate change to pose regulatory and litigation risks to their portfolio companies and a risk to their operational reliability and business continuity (i.e. physical risk). Overall, 64% and 59% of investors ranked these two risks among the top two, respectively. The average ranking of these two risks is significantly higher than the average (and median) of 3 and compared to the other three risks ($p\text{-value}<0.01$). The third-ranked risk is the risk to customer demand (37%), then the risk of stranded assets (37%) and lastly the risk to employee safety (15%).

10 To assess a potential bias I examine whether those that elected to identify themselves are more inclined to respond "yes" due to reputational or image concerns. I do not find a significant difference between the two groups. On the contrary, I find a slightly lower proportion of positive answers by investors that identified themselves.

Panel A: Does Climate change pose a material risk to companies in your investment portfolio?		
YES	73.5%	H0: yes = 0.5
NO	26.5%	p-value<0.001
N = 310		

Panel B: What kind of risk does climate change pose?	ALL				Type		
	(1)	(2)	(3)	(4)	(5)	(6)	(7)
For companies in our portfolio, climate change poses material…	% ranked first/second	% ranked fourth/fifth	Average Ranking	Significant difference in average ranking vs rows	Asset Manager	Asset Owner	Diff
1 Regulatory and litigation risk	63.8%	22.9%	2.37***	3-5	2.44***	2.18***	
2 Risk to operational reliability/ business continuity/ supply chain	59.3%	21.6%	2.40***	3-5	2.31***	2.58***	
3 Risk to customer demand	37.3%	38.2%	3.08	1-2, 5	3.16	2.90	
4 Risk of stranded assets	36.9%	42.3%	3.09	1-2, 5	3.04	3.19	
5 Risk to employee safety	14.9%	60.6%	3.81***	1-4	3.76***	3.93***	

Panel C: Responses by size, ESG awareness and Region

	AUM Size			ESG Awareness			Region				
	(1)	(2)	(3)	(4)	(5)	(6)	(7)	(8)	(9)	(10)	(11)
For companies in our portfolio, climate change poses material…	Large	Small	Diff	High	Low	Diff	US	Europe	RoW	Diff (7)-(8)	Diff (7)-(9)
Regulatory and litigation risk	2.19***	2.56***	**	2.35***	2.39***		2.29***	2.33***	2.57**		
Risk to operational reliability/ business continuity/ supply chain	2.47***	2.31***		2.59***	2.23***	**	2.38***	2.54***	2.13***		
Risk to customer demand	3.17	2.99		3.15	3.02		3.11	3.00	3.15		
Risk of stranded assets	2.84	3.32***	***	2.91	3.22*	*	3.11	2.92	3.33		
Risk to employee safety	4.00***	3.63***	**	3.73***	3.87***		3.79***	3.91***	3.67***		
Climate change poses no risk	18.7%	32.7%	***	16.0%	33.0%	***	25.3%	22.8%	31.0%		

Panel A presents the percentage of respondents choosing the response to the question in the panel heading in a given row and the test of the null hypothesis that the percentage of "yes" responses is equal to 50%. Panel B Column (1) presents the percentage of respondents that ranked the risk in a given row first or second by importance and Column (2) shows the percentage of respondents that ranked the risk fourth or fifth on a scale from 1 (highest importance) to 5 (lowest importance). Column (3) shows the average ranking. Column (4) shows the results of a t-test of the null hypothesis that the average ranking for a given row is equal to the ranking of each of the other rows, where only significant differences at the 5% level are reported. For example, for row 2, the recorded "3–5" in the column signifies that the average ranking for the response in row 2 is significantly different from the average ranking for the responses in rows 3 to 5. Columns (5)–(6) show the average ranking by investor type where the asterisks denote the significance level of the rejection a test of the null hypothesis that the average ranking is equal to 3. Column (7) reports the results of a test of the null hypothesis that the rankings in Columns (5) and (6) are equal to each other. Panel C Columns (1)–(2) show the average ranking by investor size where the asterisks denote the significance level of the rejection a test of the null hypothesis that the average ranking is equal to 3. Column (3) reports the results of a test of the null hypothesis that the rankings in Columns (1) and (2) are equal to each other. Columns (4)–(6) report the average rankings for investors by ESG awareness and the results of a test of the null hypothesis that the rankings are equal, respectively. Columns (7)–(11) report the average rankings by geographical regions and the respective tests of the null hypothesis that the rankings in the specified columns are equal to each other. ***, ** and * denote significance at the 1%, 5% and 10% levels, respectively.

Table 1.3. *Financial materiality of climate risk – investor views*

Panel C reports the average ranking conditioned on size, ESG awareness and region. Columns 1–3 show that large investors rank regulatory and litigation risk and the risk of stranded assets significantly more highly than small investors (p-values of the difference in average ranking <0.05 and <0.01, respectively) and the risk to employee safety significantly less highly (p-value of the difference in average ranking <0.05). A significantly larger proportion of small (compared to large) investors also responded that climate change poses no risk (33% versus 19%, p-value of difference <0.01). Similarly, ESG-aware investors rank the risk of stranded assets more highly and the risk to operational reliability less highly compared to investors with low ESG awareness. I do not find any statistically significant differences between the regions with respect to their rankings of climate risks.

Overall, the results in Table 1.3 show that a large majority of investors consider climate change to be financially material for their portfolio companies and in particular to pose regulatory, litigation and physical risks. The risk of stranded assets does not rank very prominently, although it seems to be more important to large investors.

Table 1.4 presents the results of issuers' responses whether they believe climate change is posing a risk to them and how they rank the different types of risks. Panel A reports that about 53% of companies responded that they believe climate risk to pose a material risk, while 47% responded it does not. This proportion is not statistically different from a random draw and is significantly lower than the proportion of investors that believe climate risk to be material.

Panel B reports how company respondents have ranked the different risks on a scale from 1 to 5 as before, where 1 is the most important and 5 is the least important risk. The risks are ordered according to the order in Table 1.3 for ease of comparison. The panel reveals in Column (1) that risk to operational reliability and risk to customer demand are the two highest ranked risks by companies with 67% and 57% of respondents ranking these first and second, respectively. The least important is the risk of stranded assets having been ranked first or second by only 18% of respondents. This is consistent with the view of major oil companies that argue that there is little risk of being left with stranded assets[11].

11 See, for example, ExxonMobil report "Energy and Carbon: Managing the Risks", March 2014. Available at https://cdn.exxonmobil.com/~/media/global/files/energy-and-environment/report---energy-and-carbon---managing-the-risks.pdf (last accessed 24 Sep 2018).

Panel A: Does Climate change pose a material risk to your company?		
YES	52.6%	H0: yes = 0.5
NO	47.4%	p-value=0.47
N = 364		

Panel B: What kind of risk does climate change pose?	ALL				Industry		
	(1)	(2)	(3)	(4)	(5)	(6)	(7)
To our companiy climate change poses material…	% ranked first/second	% ranked fourth/fifth	Average Ranking	Difference in average ranking vs investors	Resource Intensive	Knowledge Intensive	Diff
1 Regulatory and litigation risk	42.9%	39.8%	2.94	0.58***	2.77	3.11	
2 Risk to operational reliability/ business continuity/ supply chain	66.8%	12.8%	2.10***	-0.29**	2.23	1.98	
3 Risk to customer demand	56.8%	25.4%	2.52***	-0.56***	2.73	2.31	
4 Risk of stranded assets	17.9%	58.4%	3.69***	0.61***	3.70	3.68	
5 Risk to employee safety	20.6%	58.9%	3.55***	-0.25*	3.44	3.65	
Climate change poses no risk					35.4%	55.7%	***

Panel A presents the percentage of respondents choosing the response to the question in the panel heading in a given row and the test of the null hypothesis that the percentage of "yes" responses is equal to 50%. Panel B Column (1) presents the percentage of respondents that ranked the risk in a given row first or second by importance and Column (2) shows the percentage of respondents that ranked the risk fourth or fifth on a scale from 1 (highest importance) to 5 (lowest importance). Column (3) presents the average ranking where the asterisks denote the significance level of the rejection a test of the null hypothesis that the average rating is equal to 3. Column (4) shows the difference in average rankings of a given row minus the average ranking of the same row in Table 1.3, Panel B and the results of a test of the null hypothesis that the difference is zero. Columns (5)–(6) show the average ranking by industry type (see discussion in the main text). Column (7) reports the results of a test of the null hypothesis that the rankings in Columns (5) and (6) are equal to each other. ***, ** and * denote significance at the 1%, 5% and 10% levels, respectively.

Table 1.4. *Financial materiality of climate risk – company views*

Column (4) shows the difference in average ranking of the different risks compared to responses by investors. The results show that regulatory and litigation risk as well as the risk of stranded assets are considered significantly more important by investors (p-value<0.01), while risk to operations, customer demand and employee safety are significantly more important to companies.

The panel further divides the responses by sectors. I label those sectors related to energy, materials, utilities, transportation and industrials as "resource-intensive" and those related to consumer and retail, financials, health care, technology and media and professional services as "knowledge-intensive". I do not find any significant differences in rankings of the various risks between these two sector categories. However, a significantly higher proportion of respondents of knowledge-intensive sectors responded that climate change poses no risk to their company (56%). This finding is consistent with the finding in Panel A that almost half of all the respondents did not believe climate change to pose a risk to their company as 63% of the respondents come from knowledge-intensive sectors (see Table 1.1, Panel B).

The results in Table 1.4 lead to two conclusions. First, investors and companies rank different climate risks as important, which to some extent aligns with a portfolio view by investors (regulatory risk and risk of stranded assets) versus a company-specific view by company executives (operational risk and risk to customer demand). Secondly, a significantly smaller proportion of companies believe climate risk is material to them compared to investors, and this view is dependent on the particular sector affiliation. This suggests that climate change will have different materiality considerations for different sectors.

I investigated this notion of sector materiality of climate risk further by asking investors for which sectors they believe climate change to be the most material. Table 1.5 presents the results. The table is ordered by the percentage of responses for each sector shown in Column (1). Companies in the energy sector are ranked first with 68% of investors responding that they believe the impact of climate change to be material for companies in this sector. This is followed by companies in basic materials (62%), utilities (60%), transportation (56%) and industrials (56%). The proportion of responses for all of these sectors is significantly higher than 50%. Not surprisingly, these sectors are all resource-intensive. Among them, climate change is considered the most material for the energy sector as the proportion of responses is statistically significantly larger than that for any other sector (see Column 2). All other, i.e. knowledge-intensive sectors, received a significantly lower than 50% proportion of responses from investors suggesting that climate change is not considered material for knowledge-intensive sectors.

For which sectors is the impact of climate change material?	ALL			Region				
	(1)	(2)	(3)	(4)	(5)	(6)	(7)	(8)
	%	Significant difference in proportion vs rows	H0: p = 0.5	US	Europe	RoW	Diff (4)-(5)	Diff (4)-(6)
1 Energy	68.3%	2-10	***	71.3%	77.6%	55.2%		***
2 Basic materials	62.0%	1, 4-10	***	63.2%	69.7%	52.4%		*
3 Utilities	60.4%	1, 6-10	***	66.9%	67.3%	46.9%		***
4 Transportation	56.2%	1-3, 5-10	***	65.4%	63.0%	39.9%		***
5 Industrials	56.0%	1, 6-10	**	58.8%	63.0%	45.5%		**
6 Consumer & Retail	36.9%	1-5	***	43.4%	39.4%	28.0%		***
7 Financial	35.7%	1-5	***	45.6%	37.0%	25.2%		***
8 Health Care	35.7%	1-5	***	42.6%	38.2%	26.6%		***
9 Tech, Media & Telecoms	33.5%	1-5	***	40.4%	35.8%	24.5%		***
10 Professional Services	31.5%	1-5	***	39.0%	33.3%	22.4%		***
The impact of climate change is not material for any sector	5.8%			6.6%	1.8%	9.1%	*	

Column (1) presents the percent of respondents choosing the response in a given row. Column (2) shows the results of a t-test of the null hypothesis that the percentage for a given alternative is equal to the percentage for each of the other alternative responses, where only significant differences at the 5% level are reported. For example, for response 1 the recorded "2–10" in the column signifies that the percentage for the response in row 1 is significantly different from the percentages for the responses in rows 2 to 10. Column (3) shows the results of a test of the null hypothesis that the percentages in Column (1) are equal to 50%. Columns (4)–(8) report the percentages by geographical regions and the respective tests of the null hypothesis that the percentages in the specified columns are equal to each other. ***, ** and * denote significance at the 1%, 5% and 10% levels, respectively.

Table 1.5. *Sector materiality of climate risk*

Interestingly, it seems that climate change is not considered as material for the financial services sector, which is contrary to the recommendations by the TCFD to make banks and insurance companies disclose climate-related risks of their lending and investment portfolios (TCFD 2016). This might be because the respondents only considered the impact of climate change directly related to the business of the companies (and not the indirect exposure through lending or investment portfolios) or because they do not believe these risks to be material to diversified portfolios held by banks and insurance companies.

The table further reveals that a significantly higher proportion of U.S. and European investors considers climate change to be material for all sectors compared to investors from the rest of the world. Particularly, only 1.8% of investors from Europe believe that climate change is not material for any sector.

Overall, the findings in this section suggest that the impact of climate change is considered to be material particularly for companies in resource-intensive sectors and that investors are most concerned about the regulatory and litigation risks associated with climate change. On the other hand, consistent with the sector distribution in the sample, fewer companies believe climate change to pose a material risk and those that do are mostly concerned with the risk to their business operations and to customer demand.

1.3.3. *Challenges for the disclosure and use of climate change information*

The results in the previous section have implications for how investors use climate-related information for investment decisions and the challenges they face given the disclosure of climate-related information by companies.

Table 1.6 presents results to the question whether investors routinely analyze information on climate change for investment decisions and which challenges they face doing so. Panel A shows that about half of the investors in the sample responded that they analyze information about climate change for investment decisions[12].

12 This percentage is likely somewhat upward biased as I also find in untabulated results that this percentage is significantly lower (38%) among those investors that do not identify themselves (p-value<0.001) suggesting that social expectations and image concerns might have played a role in choosing the positive response. These results suggest, however, that the percentage of investors analyzing climate risk information is somewhere between 40 and 50%.

Panel A: Do you analyze information on climate change for investment decisions?		
YES	50.5%	H0: yes = 0.5
NO	49.5%	p-value=0.89
N = 319		

Panel B: What challenges do you face when evaluating the impact of climate change?	ALL				Type		
	(1)	(2)	(3)	(4)	(5)	(6)	(7)
	% ranked first/second	% ranked fourth/fifth	Average Ranking	Significant difference in average ranking vs rows	Asset Manager	Asset Owner	Diff
1 Quantifying opportunities and risks posed by climate change	58.5%	20.3%	2.39***	4-5	2.34***	2.57**	
2 Identifying opportunities and risks posed by climate change	50.7%	27.0%	2.62***	4-5	2.57***	2.81	
3 Lack of disclosures from companies about climate change	49.6%	25.6%	2.58***	4-5	2.58***	2.56**	
4 Risk of sacrificing investment returns and/or diversification benefits	31.9%	52.0%	3.25**	1-3, 5	3.29**	3.10	
5 Lack of interest from our clients/constituents	20.4%	64.0%	3.77***	1-4	3.82***	3.56**	
We do not analyze information on climate change for investment decisions					31.3%	86.7%	***

Panel C: Responses by size, ESG awareness and Region

	AUM Size			ESG Awareness			Region				
	(1)	(2)	(3)	(4)	(5)	(6)	(7)	(8)	(9)	(10)	(11)
Which challenges do you face when evaluating the impact of climate change?	Large	Small	Diff	High	Low	Diff	US	Europe	RoW	Diff (7)-(8)	Diff (7)-(9)
Quantifying opportunities and risks posed by climate change	2.25***	2.54***	*	2.33***	2.44***		2.48***	2.25***	2.47***		
Identifying opportunities and risks posed by climate change	2.74*	2.49***		2.54***	2.70*		2.65*	2.41***	2.88		
Lack of disclosures from companies about climate change	2.29***	2.87	***	2.55***	2.60***		2.50***	2.66*	2.53***		
Risk of sacrificing investment returns and/or diversification benefits	3.39***	3.11		3.33**	3.18		2.79	3.59***	3.25	***	*
Lack of interest from our clients/constituents	3.80***	3.73***		3.98***	3.59***	**	3.88***	3.79***	3.64***		
We do not analyze information on climate change for investment decisions	40.3%	57.1%	**	31.0%	61.7%	***	59.0%	41.2%	49.5%	**	

Panel A presents the percentage of respondents choosing the response to the question in the panel heading in a given row and the test of the null hypothesis that the percentage of "yes" responses is equal to 50%. Panel B Column (1) presents the percentage of respondents that ranked the challenges in a given row first or second by importance and Column (2) shows the percentage of respondents that ranked the challenges fourth or fifth on a scale from 1 (highest importance) to 5 (lowest importance). Column (3) shows the average ranking where the asterisks denote the significance level of the rejection a test of the null hypothesis that the average ranking is equal to 3. Column (4) shows the results of a t-test of the null hypothesis that the average ranking for a given row is equal to the ranking of each of the other rows, where only significant differences at the 5% level are reported. Columns (5)–(6) show the average ranking by investor type where the asterisks denote the significance level of the rejection a test of the null hypothesis that the average ranking is equal to 3. Column (7) reports the results of a test of the null hypothesis that the rankings in Columns (5) and (6) are equal to each other. Panel C Columns (1)–(2) show the average ranking by investor size where the asterisks denote the significance level of the rejection a test of the null hypothesis that the average ranking is equal to 3. Column (3) reports the results of a test of the null hypothesis that the rankings in Columns (1) and (2) are equal to each other. Columns (4)–(6) report the average rankings for investors by ESG awareness and the results of a test of the null hypothesis that the rankings are equal, respectively. Columns (7)–(11) report the average rankings by geographical regions and the respective tests of the null hypothesis that the rankings in the specified columns are equal to each other. ***, ** and * denote significance at the 1%, 5% and 10% levels, respectively.

Table 1.6. *Assessing climate risk*

Panel B summarizes the challenges investors face in doing so. Investors were asked to rank the challenges on a scale from 1 to 5, where 1 is the most important and 5 is the least important. The results in Panel B are ordered by the proportion of responses that ranked the particular challenge first or second. The most important challenge that investors face when evaluating the impact of climate change is to *quantify* the opportunities and risks. Almost 59% of respondents ranked this first or second. This is followed by challenges in *identifying* the opportunities and risks (51%) and by lack of disclosures from companies (50%): 31% responded that they believe they risk sacrificing investment returns if they consider climate-related information and 20% responded that there is no interest from clients for considering climate risk. The average rankings in Columns (3) and (4) show that the differences in importance between the top 3 challenges (quantifying risks, identifying risks and lack of disclosure) and the bottom two (sacrificing investment returns and lack of client interest) are statistically significant[13].

Panel C reports the responses by size, ESG awareness and region. A significantly higher proportion of small investors (57% versus 40%) acknowledge that they do not analyze climate-related information for investment decision (p-value of difference<0.05) and also rank identifying climate risks as their biggest challenge. Larger investors find the lack of disclosures from companies as a significantly more important challenge (p-value of difference<0.01). Unsurprisingly, a higher proportion of ESG-aware investors respond that they analyze climate-related information (69% compared to 38%, p-value of difference<0.01) and also state that the lack of interest from clients is significantly less important. Regional differences reveal that a significantly larger proportion of U.S. investors respond that they do not analyze climate-related information compared to European investors (59% versus 41%, p-value<0.05). U.S. investors compared to European investors also seem to be more concerned that considering climate risk in investment decision might sacrifice investment returns (p-value<0.01). Overall, the investor responses seem to suggest that they face challenges when identifying and quantifying climate risk in part due to a lack of disclosures by companies. Furthermore, U.S. investors in particular seem to be concerned about sacrificing investment returns if they considered climate related-information for investment purposes.

13 The results in Panel A further reveal that a significantly larger proportion of asset managers analyze climate risks for investment decisions compared to asset owners (68% versus 13%). As before, this might partly be explained by the geographical distribution of asset managers and asset owners as well as by the possibility that asset owners often delegate the investment analysis to their asset managers.

Panel A: Do you disclose information about climate risk (in your CSR report, annual report or other outlets)?

	ALL		Industry		
			Resource Intensive	Knowledge Intensive	Diff
YES	37.5%	H0: yes = 0.5	48.0%	31.7%	***
NO	62.5%	p-value<0.001	52.0%	68.3%	
N = 264					

Panel B: Climate risk disclosure by sector

Sector				Region		
Energy	56.3%	Consumer/retail	35.6%	US	Europe	RoW
Basic materials	55.6%	Industrials	31.0%	10.0%	32.1%	40.5%
Transportation	55.6%	Financial services	27.8%			
Utilities	50.0%	Health Care	16.7%	US - region:	p<0.1	p<0.05
TMT	40.5%	Business Services	12.5%			

Panel C: What challeneges do you face with respect to your dislcosures on climate change?

	ALL		Industry		
	(1)	(2)	(3)	(4)	(5)
	%	Significant difference vs. rows	Resource Intensive	Knowledge Intensive	Diff
1 Quantifying opportunities and risks posed by climate change	42.4%	2-6	55.2%	35.1%	***
2 Identifying opportunities and risks posed by climate change	26.3%	1, 4-6	35.6%	20.9%	***
3 Lack of regulatory standards requiring disclosure	21.4%	1, 5-6	25.3%	19.4%	
4 Lack of interest from investors	19.6%	1-2, 5-6	28.7%	14.2%	**
5 Risk of revealing sensitive proprietary information	12.5%	1-4	16.1%	10.4%	
6 Opening ourselves to potential legal risk	8.9%	1-4	16.1%	4.5%	***

Panel A presents the percentage of respondents choosing the response to the question in the panel heading in a given row and the test of the null hypothesis that the percentage of "yes" responses is equal to 50%. The panel also shows the percentages by industry as well as the result of a test of the hypothesis that the industry percentages in the respective rows are equal to each other. Panel B shows the percentage responses of companies in the respective sectors that disclose climate risks as well as the responses by region and the p-values of their difference in proportions. Panel C Column (1) presents the percent of respondents choosing the response in a given row. Column (2) shows the results of a t-test of the null hypothesis that the percentage for a given alternative is equal to the percentage for each of the other alternative responses, where only significant differences at the 5% level are reported. For example, for response 1, the recorded "2–6" in the column signifies that the percentage for the response in row 1 is significantly different from the percentages for the responses in rows 2 to 16. Columns (3)–(5) report the percentages by industry type and the respective tests of the null hypothesis that the percentages in the specified columns are equal to each other. ***, ** and * denote significance at the 1%, 5% and 10% levels, respectively.

Table 1.7. *Disclosing climate risk*

Table 1.7 reports issuers' views and practices on climate-related disclosures. Panel A shows that only about 38% of the companies in the sample say that they disclose information about climate risks in their annual report, CSR report or other outlets. Considering that this figure might be somewhat upward biased due to image concerns, the actual percentage might be somewhat lower. In untabulated results, I found that the percentage is significantly lower for those respondents that do not identify themselves (17%)[14]. This is also significantly lower than the proportion of companies, almost half, that responded in Table 1.4 that climate risk is not material to them. Panel A further reports that a larger proportion of companies in resource-intensive sectors disclose information about climate-risk compared to those in knowledge-intensive sectors (48% versus 32%). However, the percentage of resource-intensive companies that disclose information about their exposure to climate-risk seems rather low given that the results in Table 1.5 show that the majority of investors believe climate risk to be material to companies in these sectors.

Panel B summarizes the proportions by sector and region. About half or little over half of the companies in the energy, basic materials, transportation and utilities sectors disclose climate risks. Of the companies in the other more resource-intensive sector, i.e. industrials, less than a third disclose climate-related risks in their reports to stakeholders. Health care and business services companies have the lowest percentage of climate risk disclosures with 17% and 13%, respectively. Panel B also shows that a significantly lower percentage of U.S. companies (10%) disclose climate risks compared to European companies (32%) and those from other parts of the world (41%) (p-values of the difference = 0.1 and 0.05, respectively).

Panel C summarizes the responses to the question of which challenges companies face with respect to their disclosures on climate change. The results in Panel B are ordered by the proportion of responses associated with each challenge. Similar to investors, the top two challenges faced by companies are quantifying and identifying the opportunities and risks posed by climate change (42% and 26% of responses). These are followed by the lack of standards (21%) and the lack of interest from investors (20%). Companies state that they are least concerned about revealing proprietary information and potential litigation risk which are commonly stated in the academic literature as the main obstacles of voluntary disclosure. However, a significantly higher proportion of respondents from resource-intensive sectors state the litigation risk as a challenge (p-value<0.01). Significantly more

14 The percentages are also much smaller in studies that examine the actual disclosure of climate-related risks in annual reports. For example, a 2016 study on sustainability practices of S&P Global 1200 companies reveals that only 19% disclose business risks related to climate change in their annual report (The Conference Board 2016). However, the study does not examine other disclosure outlets such as sustainability reports or others.

respondents from these sectors also chose the quantification and identification of climate risk and the lack of interest from investors as a challenge.

Overall, the results in this section suggest that although investors seem to consider climate risk to be material, particularly for firms in resource-intensive industries, only about half or less actually analyze information on climate risks for investment decisions. They also state that identifying and quantifying climate risks and lack of disclosures by companies are major hurdles in using climate-related information. This is consistent with responses by issuers that suggest that only a third overall and perhaps a half of companies in the resource-intensive sectors disclose climate-related information. The percentages of U.S. companies disclosing this information are also significantly lower. Companies also state identifying and quantifying climate risks as major hurdles and, perhaps surprisingly, a fifth also suggest that the lack of standards might impede their efforts to disclose. The findings further suggest that few companies, even in those sectors for which climate risk is perceived material, actually disclose any information about it and that this lack of disclosure is one factor impeding the investors' ability in incorporating climate-related information in their investment decisions.

1.4. Summary and conclusion

While mandatory disclosures requirements have generally been introduced to overcome economic frictions that inhibit less than socially efficient levels of disclosures, at the same time, materiality judgments have been applied to allow managers to focus their mandatory reporting on the most relevant items specific to their firm and to balance the economic benefits and costs of disclosure. By providing evidence from a large global survey of institutional investors and companies, this chapter contributes to the academic, policy and practitioner debate on the financial materiality of climate risk.

The field evidence reveals that even though investors have largely not considered climate risk as important for investment decisions in the past, their views have changed significantly. A large majority of investors believe that climate change poses a material risk to companies. Investors seem concerned about heightened regulatory and litigation risk as well as physical risks to productive assets. As one respondent commented:

> As the impacts of climate change (CC) have become more widely known and understood, with high profile government responses, the business case for paying attention to CC has increased, and data tools have surfaced to allow us to better identify and track CC risk. However, physical impact CC risk and regulatory CC risk along with

related stranded asset CC risk are separate and distinct buckets – highly complex, market specific, and dynamic. Their importance cannot be over emphasized.

Moreover, the survey results also reveal that there is a misalignment of the risk perception of investors compared with companies with potential consequences for the usefulness of climate-related disclosures. Far fewer issuers believe that they are exposed to climate risks and also regard the risks to manifest themselves mainly through changes in consumer demand and impacts on their supply chain. The findings also suggest that large heterogeneity exists across sectors and that climate risks are mainly financially material to companies in "resource-intensive" sectors. As an impediment to the use of climate change information, investors state difficulties in identifying and quantifying climate risks due in part to a lack of disclosures. The findings further suggest that not enough companies in sectors, for which investors believe climate risks to be material, actually disclose information on climate change, citing challenges with the identification and quantification of these risks.

The field evidence in this study thus informs the academic, policy and practitioner debate on the materiality of climate change supporting recent calls on policy makers and standard setters to require broader applications of materiality judgments and to expand current disclosure requirements to climate-related risks in financial reports.

1.5. References

Amel-Zadeh, A. and Serafeim, G. (2018). Why and how investors use ESG information: Evidence from a global survey. *Financial Analysts Journal*, 74, 1–17.

Anderson, M., Bolton, P., Samama, F. (2016). Hedging climate risk. *Financial Analysts Journal*, 72(3), 13–20.

Bansal, R., Kiku, D., Ochoa, M. (2019). Climate change risk. Working paper, Duke University.

Bernstein, A., Gustafson, M., Lewis, R. (2019). Disaster on the horizon: The price effect of sea level rise. *Journal of Financial Economics*, 134, 253–272.

Carney, M. (2015). Breaking the tragedy of the horizon: Climate change and financial stability. Bank of England, Speech on 29 September at Lloyds of London.

Cho, C., Roberts, R., Patten, D. (2010). The language of US corporate environmental disclosure. *Accounting, Organizations and Society*, 35, 431–443.

Cho, C., Guidry, R., Hageman, A., Patten, D. (2012). Do actions speak louder than words? An empirical investigation of corporate environmental reputation. *Accounting, Organizations and Society*, 37, 14–25.

Flammer, C. (2013). Corporate social responsibility and shareholder reaction: The environmental awareness of investors. *Academy of Management Journal*, 56(3), 758–781.

Giglio, S., Maggiori, M., Rao, K., Stroebel, J., Weber, A. (2021). Climate change and long-run discount rates: Evidence from real estate. *The Review of Financial Studies*, 34 (8), 3527–3571.

International Finance Corporation (2010). Climate risk and business: Practical methods for assessing risk. Report, IFC, Washington D.C.

Khan, M., Serafeim, G., Yoon, A. (2016). Corporate sustainability: First evidence on materiality. *The Accounting Review*, 91, 1697–1724.

Krüger, P. (2015). Climate change and firm valuation: Evidence from a quasi-natural experiment. Working paper, Geneva School of Economics and Management, Geneva.

Krüger, P., Sautner, Z., Starks, L. (2020). The importance of climate risks for institutional investors. *The Review of Financial Studies*, 33, 1067–1111.

OECD (2015). *The Economic Consequences of Climate Change*. OECD Publishing, Paris. DOI: 10.1787/9789264235410-en.

Oliver Wyman (2018). Reporting climate resilience: The challenges ahead. Paper, Marsh & McLennan Companies, London.

Securities Exchange Commission (2010). Commission guidance regarding disclosure related to climate change, 17 CFR Parts 211, 231 and 241. SEC, Release nos 33-9106; 34-61469; FR-82 [Online]. Available at: http://www.sec.gov/rules/interp/2010/33-9106.pdf [Accessed 3 June 2015].

TCFD (2016). Recommendations of the Task Force on Climate-related Financial Disclosures. Report, Task Force on Climate-Related Financial Disclosures, Basel.

World Economic Forum (2020). The Global Risks Report 2020, 15th edition. Report, World Economic Forum, Geneva.

2

Looking Forward with Historical Carbon Data

Increasingly many allocators are interested in computing the carbon footprint of their portfolios. The usual way of doing this relies on historical greenhouse gas emissions data, which typically lag by one to two years relative to when investment portfolios are built. In this chapter, we show that historical emissions data are useful despite this significant lag. First, we use point-in-time data to show that the relative greenness of portfolio companies has remained very stable over time. Perhaps unsurprisingly, companies that were historically green (or brown) remain green (or brown) today and for several years to come, overall and compared to same-sector peers. Second, and more surprisingly, we show that the portfolio carbon footprint measured using historical data is remarkably informative about the current portfolio carbon footprint (using current or same-fiscal-year emissions that the portfolio finances). We believe this observation should give allocators comfort that historical carbon data still provides important insights into their portfolio's current, or even future, climate exposure.

2.1. Introduction

Climate-aware investing is one of the most important recent trends in asset management. In a survey of institutional investors, Krueger et al. (2020) found that 93% of respondents had incorporated climate risk into their investment process. Asset owners who publicly committed to net zero targets jointly manage over

Chapter written by Steffen BIXBY, Alfie BRIXTON and Lukasz POMORSKI.

For a color version of all figures in this chapter, see www.iste.co.uk/jurczenko/climate.zip.

We thank Emmanuel Jurczenko (the Editor), Jeff Dunn, Bradley Jones, Thomas Maloney and Daniel Villalon for helpful discussion and comments. AQR Capital Management is a global investment management firm, which may or may not apply similar investment techniques or methods of analysis as described herein. The views expressed here are those of the authors and not necessarily those of AQR.

$10 trillion as of the end of 2021[1], and many more have at least partial commitments to decarbonization, even if not net zero per se. These investors face a major challenge, and frequently complain about the lack of relevant climate data to help position their portfolios. The investor community generally agrees that the key variable is the carbon emissions of portfolio companies (and perhaps also countries), but unfortunately, these data are only available to investors with a significant lag. An investor building a low-carbon portfolio today is likely using data that reflects emissions as of a year or two earlier, leading to an understandable concern that such staleness limits the usefulness of such data in a portfolio context[2]. Using historical data would only make sense if such data are informative about their portfolio companies' present, or even future emissions. We take this important concern to the data to gauge if "standard" data on emissions, historical in nature, still can be useful for investors. Our answer is a resounding "yes!"

Our first set of tests uses point-in-time data, meaning that we use the exact observations that are actually available to an investor on a given date. We ask a simple question: how often would investors change their mind about which companies are relatively greener, and which are relatively browner? Unsurprisingly, we show that historical data is very persistent and thus informative about companies' relative emissions for many years after it is released. While this finding will be consistent with many investors' priors, the specific magnitudes we document will likely be novel to many: for example, we show that over 90% of companies in the brownest decile (i.e. the 10% of companies with the highest emissions) remain in the brownest decile for at least the next three years; the vast majority of those companies that move categories only move to an adjacent decile.

Our second set of findings is based on a more challenging test, looking at a portfolio's absolute carbon footprint[3], rather than the relative emissions of portfolio companies. We ask whether portfolios built using stale carbon data, available at the

1 As reported by the United Nations Environment Programme Finance Initiative in "The Net-Zero Asset Owner Alliance New Year Countdown: A Review of 2021 and a Vision for 2022".

2 This view is increasingly voiced by climate scientists, investors and regulators alike. For example, Markwat (2021) notes that the backward-looking nature of carbon footprint data is a "fundamental problem"; Bocquet et al. (2021) state that these lagging indicators "provide limited insights for the future" and a technical document from the Network for Greening the Financial System (NGFS 2021) points out that such metrics "only provide a snapshot of an evolving problem".

3 Portfolio carbon footprint, or financed emissions, refers to the quantity of emissions "financed" by allocating the carbon footprint of underlying companies to a portfolio based on the percentage ownership of the companies' capital. We focus on this metric here, acknowledging that there are other ways of measuring climate exposure, for example, carbon intensity or NLP measures; for example, see Engle et al. (2020) and Brixton et al. (2021).

time, preserve their carbon characteristics when assessed using contemporaneous carbon data, only available ex post. For example, we ask whether portfolios that target reductions in historical carbon footprint still show a similar reduction in their current carbon footprint (i.e. the current, same fiscal year emissions they finance). This is a crystal ball-type exercise because such contemporaneous emissions data will not be available to investors for perhaps two years after portfolio formation. We find that a portfolio built to a carbon reduction target using historical data shows a similar reduction when measured using data contemporaneous with portfolio formation. The portfolios we investigate (a broad market index and a hypothetical low carbon portfolio) have realized financed emissions that are within 5% of the estimates based on historical data. We believe this is an important message for many allocators. While the experiment we conduct to show this is intuitive and straightforward, remarkably, we do not believe this simple exercise has been conducted by earlier studies on this topic[4].

Empirically at least, historical data are a powerful predictor of future emissions in the short term. Is this reasonable from first principles? We think so, because, as shown below, companies that have been heavy polluters historically remain so going forward, even within sector. We do see examples of specific companies that dramatically lower their emissions, usually as a result of a corporate action. However, such companies are outliers and their impact on a diversified portfolio tends to be limited over a horizon of a few years. So, relative brownness can be expected to be persistent. Of course, portfolio-level emissions could still fall if the market as a whole decarbonizes – which is the real aim of global initiatives such as the Paris Agreement. Unfortunately, our results suggest this trend is not material over short horizons. We realize that this is not a comforting insight, but it is consistent with macro-level data. Even a massive Covid-related intervention only decreased economy-wide greenhouse gas emissions by 6-7% from 2019 to 2020 (Le Quéré et al. 2020; IEA 2021). Thus, it should not be surprising that historical data are informative about current and possibly even future emissions in a diversified portfolio.

Of course, the usefulness of emissions data, historical or not, for reporting and even portfolio construction does not mean that such data is flawless. It clearly is not, and in our discussion below, we point out the noise and outliers in the data. While the issue may be greater for third-party estimates, even company-reported emissions are by no means perfect. There is no common or binding accounting standard: for

4 There are of course many relevant studies that use carbon data, for example, Kalesnik et al. (2020), who find that emissions data is very persistent at a one-year horizon. However, we are not aware of studies that would gauge whether portfolios based on historical data meet their carbon targets when assessed with emissions data contemporaneous with portfolio construction.

example, the Centre for Audit Quality (CAQ) found that just over half of S&P 500 companies had some form of assurance or verification over ESG metrics, with just 6% receiving assurance from a public company auditing firm (CAQ 2021). Despite these limitations, such data are our best estimate of actual company emissions.

Our article also contributes to the heated debate about "forward-looking carbon data", or data measuring a company's emission trajectory over a longer, potentially decades-long, future horizon. We show that historical data is a good proxy for emissions today or even in the near future. Forecasts longer than a few years out are potentially interesting and useful, even if there are obvious challenges inherent in them[5]. A 2021 consultation of financial firms carried out by the Task Force on Climate Related Disclosures (TCFD 2021) found that three-quarters of respondents admit to using some form of forward-looking metrics – for example, implied temperature rise, climate value-at-risk and portfolio alignment estimates – but around 80% of respondents had concerns around the reliance on assumptions required.

In our view, the forward-looking data is not a replacement for the obvious question about the emissions an investment portfolio finances *today*. For this question, historical emissions data provide quite a precise answer, and while some analysts may be able to incrementally improve the estimate, our evidence here shows that even the simplest direct approach is surprisingly worthwhile.

2.2. Data

For our discussion below, we use data obtained from Trucost, one of the leading providers of greenhouse gas emissions data for asset owners and managers, consultants and academia. In our empirical tests, we use scope 1+2 emissions data because such data is most often used in investment practice, unlike scope 3 where a lack of consistency in measuring the broad scope of related indirect emissions leads to lower availability and quality of data. We would expect similar results using data from alternative providers because, as shown by Busch et al. (2020), emissions data across major data providers when using scope 1+2 emissions combined is very consistent. For scope 3 data, they find very low levels of consistency for both reported and estimated data. Moreover, we focus on large-cap stocks in developed

5 Predicting a firm's emissions 10, 20 or 30 years from now is extremely challenging – for example, few analysts would be comfortable predicting company earnings over such horizons. Moreover, there is no discounting of carbon emissions, so estimated 2050 emissions count as much as actual emissions today. This can greatly amplify the noise in the data – in comparison, company earnings 30 years from now would typically be discounted, which would attenuate any data errors and their impact on, say, the implied enterprise value.

markets, generally similar to the MSCI World universe. This allows us to sidestep another concern about climate data: coverage. Our chosen universe has coverage throughout time that exceeds 95%.

The key data in our study are firms' greenhouse gas emissions, expressed in tons of CO2-equivalents. We use this data to compute portfolio carbon footprint (financed emissions). In addition to emissions data, we leverage Trucost descriptions to identify the source of the emission data (reported by the company or estimated by Trucost). Trucost also provides a data field, effective date, indicating when the data became available to their clients. We use this variable before 2017; from 2017 onward, the data we use is collected on a point-in-time basis. Finally, to identify the period in which the emissions occurred, we use the companies' fiscal year, or the fiscal year end, as reported by Trucost.

2.3. How stale is historical carbon data?

To assess how stale carbon data may be, it is useful to review how providers of carbon emissions data obtain it in the first place. At a very high level, such data comes from two sources: from portfolio companies themselves (we loosely refer to this as "reported data") and from data vendors' and/or third-party models ("estimated data"). Between reported and estimated data, the former is likely to be more reliable. Using the data descriptors available from Trucost, we assess that reported data accounts for roughly two-thirds of large cap developed market issuers, but only for one-third of large-cap emerging markets firms. But even for emerging markets, company reports are going to be a very important source, and the availability of such reports will determine how stale historical data is and when it is updated.

When do companies report their carbon emissions? There are two important channels: one is company-issued reports such as CSR (Corporate Social Responsibility) reports, the other is data aggregators such as CDP (formerly known as Carbon Disclosure Project). CSR reports are released on a company-determined schedule with many companies doing so at some point in the second or third quarter. CDP aggregates the questionnaires it receives back from the companies and releases the data to its clients and supporting organizations, usually in the fourth quarter. Importantly, in both cases, the reporting period is the prior fiscal year – for example, CSR or CDP data released in late 2021 generally covers the emissions reporting companies had in their fiscal year 2020.

When reported data is not available, many data vendors provide their own estimates instead. Such estimates could in principle be released and revised throughout the year, but they are generally provided on a fixed schedule for a given company, and often aligned with when data reported by other companies is released. This is not surprising given that peer groups usually play a central role in the estimation models.

All in all, this means that historical data, both reported and estimated, will be one to two years stale, meaning it describes the greenhouse gases emitted one to two years before the data is available to investors. Figure 2.1 shows how this lag evolves over time, taking four snapshots of Trucost data, as of the end of each quarter 2021, and assessing what fraction of the data refers to emissions portfolio companies had in their 2020, 2019 and 2018 fiscal years.

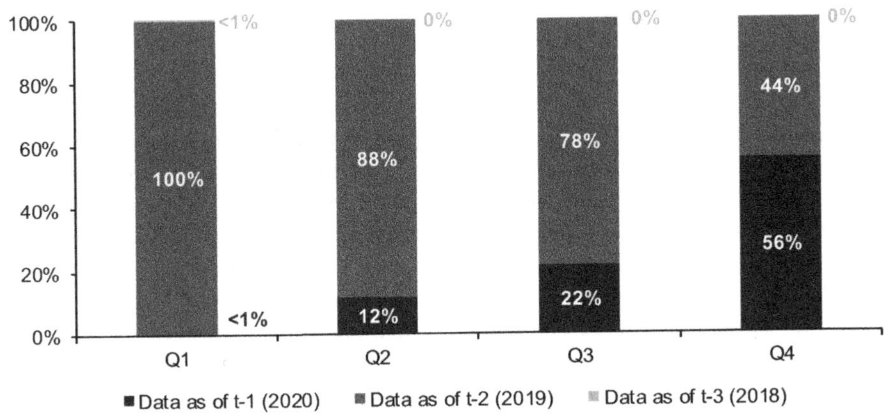

Figure 2.1. *Lag in carbon emissions data, as of 2021 quarter ends. The chart shows, for MSCI World index constituents and as of each quarter end in 2021, the fraction of carbon data corresponding to the emissions firms had in the prior year (2020), two years earlier (2019) or three years earlier (2018). Source: AQR, S&P Global Trucost, MSCI*

It is clear from Figure 2.1 that carbon data is indeed stale. This is most directly visible from the leftmost bar in the chart. On March 31, 2021, only a tiny fraction of the data reflects 2020 emissions (i.e. the immediately preceding year). The vast majority of the data refers to companies' 2019 fiscal year; data earlier than that (2018 and before) are only a rounding error. It is worth noting that while for some companies, the 2019 fiscal year-end may be December 31, for others, it may be at any other point in the 2019 calendar year, meaning that this data could lag by as much as two years relative to March 31, 2021. The average delay is 1.5 years, while 10% of the data is delayed by two years or more. The coverage of fiscal year 2020 emissions gradually increases over time, jumping to about 20% at the end of Q3 and over 50% at the end of 2021, corresponding to the reporting milestones we mentioned above. It is important to note that the reported data typically is even more stale than the estimated data. For example, at the end of 2021, only 45% of reported emissions covered the fiscal year 2020. In contrast, 71% of the estimated emissions were from this fiscal year.

2.4. Are historically brown firms getting greener? Might green firms become browner?

How useful is historical carbon data for understanding today's or even future emissions? This is an empirical question, and we turn to the data to address it.

We start by checking how often the relative greenness, or brownness, of companies in a developed equity portfolio changes. We do this using point-in-time data, meaning we capture the actual numbers investors saw at each point in time, and gauge how often these numbers change. It is natural to expect persistence in emissions, if only driven by industry membership: an oil and gas company last year is likely to still be an oil and gas company today or going forward, and as such can be expected to have meaningfully higher emissions than, say, an IT Services firm. Nonetheless, there is at least a theoretical possibility that firms within a given industry may meaningfully change their emissions versus peers – to test this theory, we ran the same analysis for sector-adjusted emissions and found comparable results which we present in Figure 2.3. In Figure 2.2, we present results for emissions without controlling for industry membership because the objective of most investors is to lower their overall portfolio emissions, and in most cases, they do so by divesting from heavy emitters overall rather than relative to industry or sector peers. We acknowledge that neutralizing the impact of industry exposure could improve risk-return trade-offs when building carbon-based factors, but this is less common in the context of overall decarbonization of portfolios. For example, Cheema-Fox et al. (2021) found that minimizing constraints on portfolio construction (i.e. not controlling for sector or industry membership) led to both greater portfolio decarbonization and higher decarbonization alpha.

Figures 2.2 and 2.3 quantify changes in relative greenness by looking at how often firms in different emissions bins jump to other bins over time. Such "transition matrices" are an intuitive way of describing how stocks move between categories. In Figure 2.2, we show how regularly firms transition from one year to the next, comparing emissions using the data that was available to investors at the end of 2020 and 2021, respectively.

	Low Emitters in 2021									High Emitters in 2021
	89%	10%	0%	0%	1%	0%	0%	0%	0%	0%
Low Emitters in 2020	6%	77%	12%	2%	1%	1%	0%	1%	0%	0%
	1%	10%	75%	12%	1%	0%	0%	0%	0%	1%
	1%	1%	9%	76%	12%	1%	0%	0%	0%	0%
	1%	1%	3%	8%	77%	9%	0%	0%	0%	1%
	0%	1%	1%	1%	6%	85%	6%	0%	0%	0%
	0%	0%	0%	0%	3%	4%	87%	4%	1%	1%
High Emitters in 2020	1%	0%	0%	0%	0%	0%	7%	87%	6%	0%
	0%	0%	0%	0%	0%	0%	1%	6%	91%	2%
	0%	0%	0%	0%	0%	0%	1%	0%	4%	96%

Figure 2.2. *2020–2021 Carbon transition matrix: MSCI World stocks are sorted into deciles by carbon emissions using data available to investors at the end of 2020 (with 10% of firms with the lowest emitters in the top row; 10% of the highest emitters in the bottom row) and again using data available to investors at the end of 2021 (with 10% of the lowest emitters in the leftmost column; 10% of the highest emitters in the rightmost column). Each cell depicts the fraction of firms in the corresponding row and column. Source: AQR, S&P Global Trucost, MSCI*

Figure 2.2 gives striking evidence of persistence in emissions. The color code in the figure indicates the importance of the diagonal in our transition matrix, meaning that by far the most usual outcome is for firms not to change their emissions decile. For example, the bottom right corner of the figure shows that almost all firms in the highest emissions decile (i.e. the 10% of firms with the highest emissions) in 2020 are again in the highest emissions decile in 2021. Only 4% of such firms "migrate" to the adjacent cell, meaning that only a small minority of top-decile emitters in 2020 ended up in the second highest decile in 2021 and virtually no firms became any greener than that.

This remarkable persistence of the brownest companies is practically relevant for investors. These companies account for the vast majority of emissions in the investible universe: the emissions of the brownest decile of companies (largest 10% of emitters) add up to 82% of MSCI World emissions in 2021.

Similarly, there is a strong persistence in the green firms: the top left corner indicates that 89% of firms in the lowest emissions decile in 2020 are again in the greenest decile in 2021. As we noted above, this pattern persists even when bucketing within industries. For example, 97% of companies in the brownest decile of sector-adjusted emissions in 2020 remain in the same or the second-highest decile a year later. As we noted when discussing Figure 2.1, there could be differences in the emissions data disclosed directly by portfolio companies, as compared to the emissions data estimated by the data provider. To test this conjecture, we reproduced Figure 2.2 using only disclosed emissions (Figure 2.7 in section 2.8) and using only estimated emissions (Figure 2.8 in section 2.8). Our overall conclusion is unchanged: emissions data is remarkably persistent. Figures 2.7 and 2.8 suggest that the estimated data is somewhat less persistent than disclosed data, which could be explained by the noise inherent in the former. But, even here, 80–90% of the greenest, or the brownest, companies remain in the same decile year-to-year or at most move to an adjacent one.

To illustrate the persistence even more forcefully, we present an extremely conservative analysis in Figure 2.3. First, we assume the exceptional lag of three years. As we saw in Figure 2.1, investors can be expected to have data less stale than three years. Second, we perform the analysis within sector to look at the persistence of company greenness relative to their sector peers. This is more stringent than what most investors would do: ultimately, we care about overall emissions by portfolio companies and the resulting portfolio carbon footprint, and not necessarily about sector-adjusted emissions.

	Low Emitters in 2021									*High Emitters in 2021*
Low Emitters in 2018	71%	19%	7%	1%	2%	0%	0%	0%	0%	1%
	18%	47%	19%	8%	2%	2%	1%	2%	0%	2%
	4%	18%	38%	29%	6%	2%	2%	1%	0%	0%
	1%	9%	19%	39%	20%	8%	4%	0%	0%	0%
	2%	3%	7%	15%	45%	21%	5%	2%	0%	0%
	0%	3%	5%	3%	19%	41%	20%	7%	1%	2%
	2%	2%	1%	3%	4%	19%	47%	22%	2%	0%
	0%	0%	2%	1%	0%	4%	16%	54%	22%	1%
	0%	0%	0%	1%	2%	2%	4%	12%	64%	17%
High Emitters in 2018	0%	1%	0%	0%	1%	0%	1%	2%	14%	81%

Figure 2.3. *2018–2021 Carbon transition matrix: the chart is built similarly to Figure 2.2, except decile groupings are formed within sector, and groupings in rows are based on emissions data available at the end of 2018 (in Figure 2.2, groupings were for the whole universe and rows were based on data available at the end of 2020). Source: AQR, S&P Global Trucost, MSCI*

Figure 2.3 demonstrates that even if carbon data is lagged by three years relative to when a portfolio decision is made, and even if it is adjusted by sector, it is still a valuable indicator of which companies are greener or browner within their sector. Over 81% of companies in the brownest decile in their sector in 2018 remain in the highest emissions decile in 2021; most of the remaining firms only move to the second-brownest decile.

There are a few outliers, some of which are modeling issues, such as for a company for which the emissions data was derived from firm disclosures in 2018 but modeled in 2021, leading to a difference of about three orders of magnitude. In other cases, a dramatic change may be driven by a corporate action. For example, we have observed cases of acquisitions that increased a company's emissions dramatically. Interestingly, the impact of M&A activity tends to be more muted for carbon intensity rather than raw carbon emissions, as acquisitions tend to change both emissions and revenue in the same direction. Finally, in the results shown in section 2.8 (Figure 2.9), we found similar levels of persistence for carbon intensity, a complementary measure of greenhouse gas emissions. If we sort firms on their carbon intensity, over 90% of firms in either the greenest or the brownest decile remain there in the following year; the firms that move tend to move to the adjacent deciles only. Formulae for the measures of portfolio greenhouse gas emissions discussed are included in section 2.8.

2.5. Nowcasting financed emissions using historical data

Having established strong persistence in relative carbon emissions, we now ask a more challenging question: can we describe a portfolio's financed emissions (carbon footprint) today, if all we have is historical data?

In principle, one could build a regression model to address this question, formally modeling how past financed emissions may predict today's or future financed emissions in a portfolio. But even so, the intuitive first step would be to simply compare a portfolio's carbon footprint computed using historical data to what the carbon footprint actually is using the data on emissions contemporaneous with when the portfolio is built. Remarkably, this very simple approach gets us most of the way there.

We used terms "historical", "current" and "future" somewhat loosely above, but the timeline in Figure 2.4 makes them precise for our subsequent empirical tests. The timeline is meant to illustrate the quandary low-carbon investors are in. Consider such an investor today (at the "year *t*" point on the timeline). As we discussed, such investors only have access to data that lags by one to two years relative to when they make portfolio decisions (in the timeline, we label that historical period at "year *t*-1"). The investor builds a low carbon portfolio at time *t*. This portfolio, by construction, will meet its carbon target when assessed using year *t*-1 data. The question we ask is whether this portfolio is still low carbon, and whether its carbon characteristics are similar, when assessed using data on emissions from year *t*, i.e. contemporaneous with the investor's decision. Such data will not be known until one to two years into the future ("year *t*+1" in the timeline), so obviously cannot be used for portfolio construction – but can be used for a portfolio assessment ex post. We believe this simple experiment will be informative and of interest to many allocators interested in low carbon investing. Remarkably, we do not believe this simple exercise has been conducted by earlier studies on this topic.

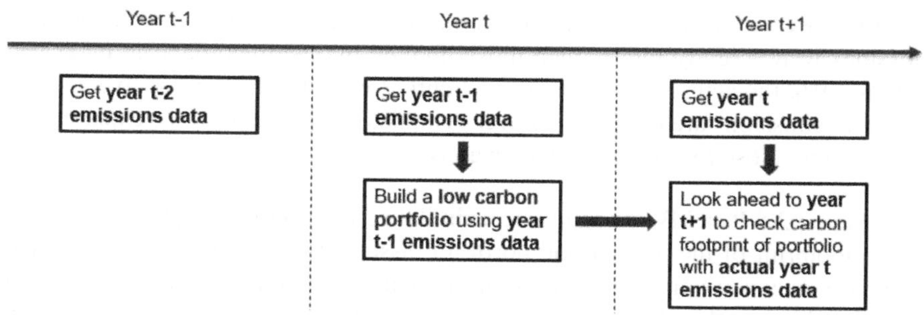

Figure 2.4. *Stylized description of how to assess the usefulness of historical data for portfolios with climate targets. Such portfolios are necessarily built using historical data, but can still be evaluated ex post, using data on emissions in the year contemporaneous with when the portfolio was built. Source: AQR*

Figure 2.5, the key exhibit in our study, takes the timeline from Figure 2.4 to the data. For the ease of presentation, the bar charts in Figure 2.5 show the carbon footprint using the actual emissions data covering a given year (e.g. using firms' emissions in fiscal year 2018 to compute a portfolio's carbon footprint as of 12/31/2018), as a fraction of the carbon footprint assessed using the latest historical data that were available to investors at that time (as of year t-1). For example, if the carbon footprint is identical in both cases, we record this as a value of 100%; if actual emissions are 20% higher than those estimated using historical data, the value is 120%, etc.

In the two panels of Figure 2.5, we consider two realistic portfolios. In Panel A, we look at one of the most important benchmark indexes, the MSCI World index. While not designed with climate investing in mind, this broad benchmark is a good approximation of many investors' equity portfolios. These investors in turn may, for a variety of reasons, produce climate reporting, and may be naturally curious as to whether their portfolio companies are low carbon not just in the past but also on an ongoing basis. In Panel B, we consider a hypothetical portfolio, annually rebalanced, that minimizes the tracking error versus MSCI World while at the same time seeking to deliver a carbon footprint that is at most half of that of MSCI World. While the data coverage in Figure 2.5 is excellent for most years, we note that the 2020 fiscal year emissions are only partially reported as of December 31, 2021 (as Figure 2.1 shows, we know 2020 emissions only for about half of the MSCI World constituents, which translates to about 2/3 of the index by weight).

Panel A shows that the carbon footprint of the MSCI World index, assessed using actual emissions data for a given year, is remarkably close to the carbon footprint assessed using historical data. In some years such as 2012, the actual emissions were slightly higher than those implied by historical data; in other years such as 2016, the actual emissions were slightly lower. Overall, in the average year, the actual emissions are only 1.2% lower than historical, which as we explain below is remarkably close for this type of an exercise.

In Figure 2.5 Panel B, we again see a similar pattern to that in Panel A: the actual carbon footprint of the low-carbon portfolio (assessed using emissions in the same fiscal year as construction) is very similar to that assessed using historical data. The portfolio's historical carbon footprint is only 4.1% higher than the contemporaneous carbon footprint on average.

These patterns are striking, for three reasons. First, while our analysis in Figures 2.2 and 2.3 only dealt with firms' relative emissions, Figure 2.5 is also affected by the overall emissions in the market. For example, if all firms produce exactly twice as many emissions as they did last year, the patterns in Figures 2.2 and 2.3 would not change: they only reflect a firm's greenness relative to peers. But such doubling of emissions would double the carbon footprint of the portfolio, relative to what historical data would have said, which would clearly affect Figure 2.5. For example, historical emissions overstate the actual carbon footprint in 2020, when COVID-related policies reduced global energy demand and hence emissions by 6–7% (Le Quéré et al. 2020; IEA 2021). Thus, if (or when) the world economy de-carbonizes, we can expect that the actual carbon footprint will be consistently lower than that assessed using historical data. This may be good news for allocators, in that historical data will tend to give them a conservative view of their financed emissions.

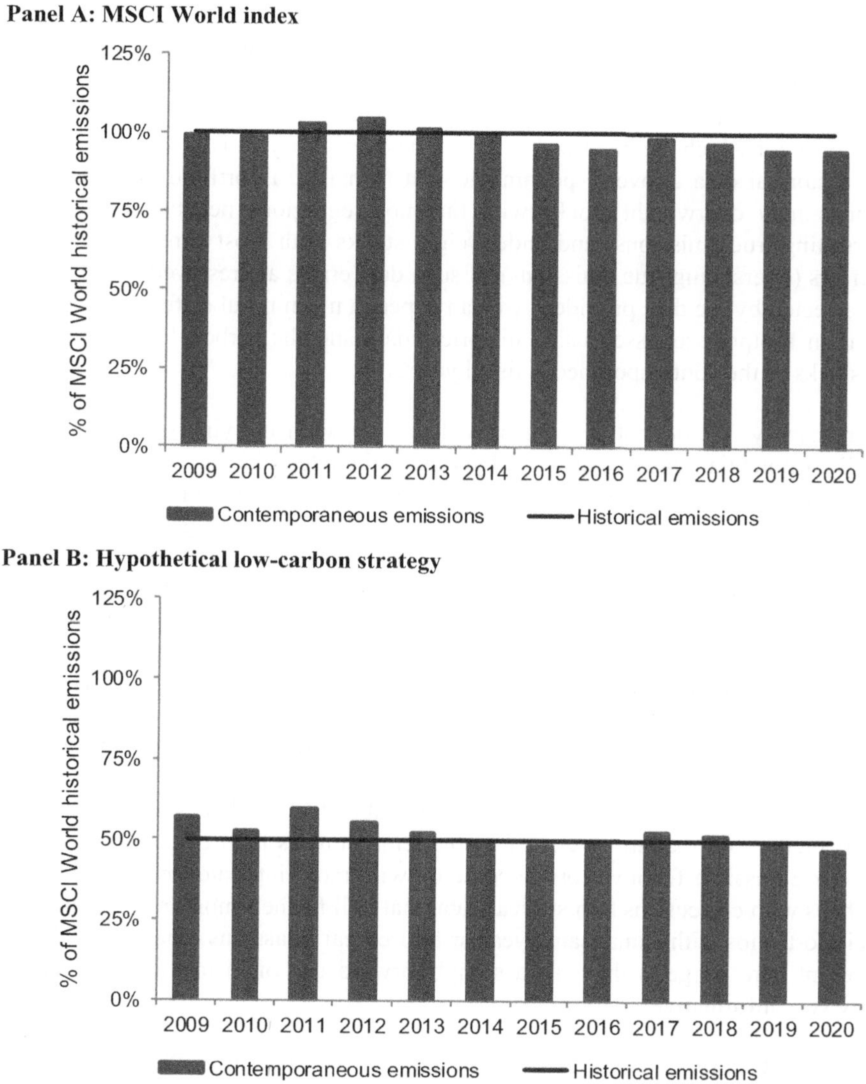

Figure 2.5. *Assessing the usefulness of historical data for portfolios with climate objectives: the charts show the carbon footprint in tons of CO2e per $M investment assessed two different ways: using historical data (the snapshot available at portfolio construction) and using data on emissions portfolio companies produced in the year of portfolio construction (only available at a lag). The top panel shows the MSCI World index. The bottom panel shows a low-carbon portfolio based on the MSCI World index targeting a 50% reduction in carbon emissions while maintaining a low tracking error. Source: AQR, S&P Global Trucost, MSCI. Construction of the low-carbon portfolio is described in the text below*

The second reason why Figure 2.5 is striking relates to Panel B, showing a portfolio managed to deliver a low carbon footprint. When optimizing any portfolio for a characteristic of choice, there is always a concern that a human analyst or a quantitative optimizer will fall prey to some mistakes in the data, overweighting those stocks for which the historical data may be overly rosy and avoiding stocks for which historical data is overly pessimistic. Put formally, a portfolio optimized for emissions may overweight stocks with the most egregious negative data errors (understating true emissions) and underweight stocks with most egregious positive data errors (overstating true emissions). If such data errors are resolved in the future (e.g. corrected by the data provider), we can expect a meaningful difference between the carbon footprint assessed using historical data and the carbon footprint of the same stocks in the contemporaneous fiscal year[6].

Surprisingly, we do not see much evidence for such outlying observations in Panel B. A potential reason is that our hypothetical strategy is very well diversified, given its objective to minimize the tracking error. Consequently, a few outliers may not influence its carbon footprint very dramatically. We saw in Figures 2.2 and 2.3 that such outliers do occur – some firms meaningfully change their emissions category year to year – but this happens in a minority of the data. Thus, we conjecture that a well-diversified portfolio is unlikely to be so concentrated in such companies to see meaningful changes in its carbon footprint. With this diversification caveat, we expect the broad result to hold for other strategies as well.

Finally, Figure 2.5 is interesting in that it uses the simplest possible forecast (or "nowcast") of emissions: it implicitly assumes that current fiscal year emissions will be equal to those historical ones. It is remarkable that such a simple approach delivers predictions that are as close to actual emissions as Figure 2.5 documents. It is of course possible for investors to come up with more intricate forecasting tools, or perhaps with corrections to historical data, that will further improve the alignment of their portfolios with actual same-year or future-year emissions – but the evidence we present here suggests that just a straightforward use of historical data as-is is already very informative.

In our last exhibit, Figure 2.6, we present the distribution of "forecast errors" implied by Figure 2.5, showing the difference in carbon footprint based on same-year emissions compared to that based on historical emissions. This is simply to provide another perspective, while Figure 2.5 scaled emissions in each year to the historical estimate (showing current emissions as a percentage of what we thought

6 To the extent that changes are driven by updates in emissions estimated by the data provider, the issue might be less that of historical data being stale and more of the precision of the estimates in the first place. This is because seemingly small changes in estimation methodology may lead to dramatic changes in the final estimate.

they would be using data available at the time), Figure 2.6 shows these differences in terms of overall tons of CO2-equivalents per $M invested.

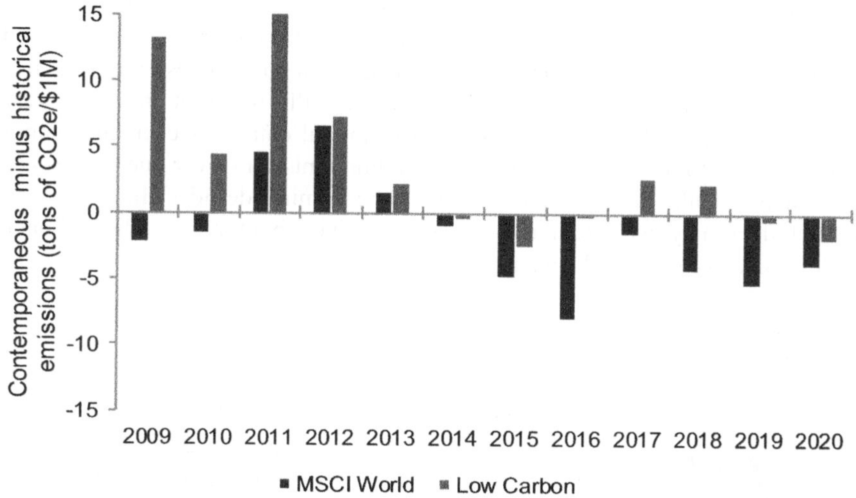

Figure 2.6. *Assessing the usefulness of historical data for portfolios with climate objectives: the chart shows the difference in the portfolio carbon footprint based on same-year emissions and based on historical emissions in terms of overall tons of CO2-equivalents per $M invested. The left bars show the MSCI World index, and the right bars show the low-carbon portfolio described in Figure 2.5. Source: AQR, S&P Global Trucost, MSCI*

2.6. Conclusion

In this chapter, we have presented evidence that historical carbon data is a valuable predictor of how green a stock may be today or in the future, or of what a portfolio's current financed emissions may be. This evidence suggests that historical data can be very important for allocators, contrary to recent commentary about the (lack of) value of such data. Of course, historical data is not a panacea for all relevant questions. It is probably going to be meaningfully less useful in predicting emissions in, say, 2030 or 2040. Such long-term predictions present formidable research challenges and may call for a variety of additional data, for example, company decarbonization targets (ideally suitably adjusted given that many such targets may have very low quality; reduction commitments are not audited, and some companies have set public goals with no clear plan on how to achieve them

(Vincent 2020)). But, for most obvious use cases such as carbon reporting, we conclude that historical carbon data are meaningful and actionable.

Could historical data become less informative going forward? It is possible: ironically, we could only verify this after future emissions are realized and included in historical data. As of right now, the answer depends on an investor's prior. If you expect that next year's emissions will be meaningfully different from this year's, then you would probably be cautious about historical data. However, the economy may be unlikely to dramatically change its carbon emissions over one or two years, even if it eventually reaches net zero over the coming decades. In such a case, historical data may well remain very informative about a portfolio's financed emissions today or even in the near future.

2.7. Appendix

	Low Emitters in 2021									*High Emitters in 2021*
Low Emitters in 2020	94%	6%	0%	0%	0%	0%	0%	0%	0%	0%
	9%	87%	4%	0%	0%	0%	0%	0%	0%	0%
	0%	12%	83%	3%	2%	0%	0%	0%	0%	0%
	0%	0%	19%	78%	3%	0%	0%	0%	0%	0%
	0%	0%	0%	14%	83%	3%	0%	0%	0%	0%
	0%	0%	0%	0%	11%	88%	2%	0%	0%	0%
	0%	0%	0%	0%	0%	15%	85%	0%	0%	0%
	0%	0%	0%	0%	0%	0%	11%	89%	0%	0%
	0%	0%	0%	0%	0%	0%	0%	10%	90%	0%
High Emitters in 2020	0%	0%	0%	0%	0%	0%	0%	0%	11%	89%

Figure 2.7. *2020–2021 Carbon transition matrix based on reported data only: MSCI World stocks are sorted into deciles by carbon emissions using reported (company-disclosed) data available to investors at the end of 2020 (with 10% of firms with the lowest emitters in the top row; 10% of the highest emitters in the bottom row) and again using reported data available to investors at the end of 2021 (with 10% of the lowest emitters in the leftmost column; 10% of the highest emitters in the rightmost column). Each cell depicts the fraction of firms in the corresponding row and column. Source: AQR, S&P Global Trucost, MSCI*

	Low Emitters in 2021						*High Emitters in 2021*			
	89%	5%	4%	0%	0%	2%	0%	0%	0%	0%
Low Emitters in 2020	4%	**84%**	11%	2%	0%	0%	0%	0%	0%	0%
	0%	4%	**69%**	16%	4%	5%	0%	2%	0%	0%
	0%	2%	14%	**63%**	20%	0%	0%	0%	0%	2%
	0%	0%	2%	10%	**76%**	12%	0%	0%	0%	0%
	0%	2%	0%	4%	11%	**66%**	18%	0%	0%	0%
	0%	0%	4%	4%	0%	13%	**75%**	4%	0%	0%
High Emitters in 2020	0%	0%	0%	0%	0%	4%	10%	**77%**	6%	4%
	2%	0%	0%	0%	0%	0%	2%	12%	**76%**	8%
	0%	0%	0%	0%	0%	0%	0%	4%	12%	**84%**

Figure 2.8. *2020–2021 Carbon transition matrix using estimated data only: MSCI World stocks are sorted into deciles by carbon emissions using estimated (non-company-disclosed) data available to investors at the end of 2020 (with 10% of firms with the lowest emitters in the top row; 10% of the highest emitters in the bottom row) and again using estimated data available to investors at the end of 2021 (with 10% of the lowest emitters in the leftmost column; 10% of the highest emitters in the rightmost column). Each cell depicts the fraction of firms in the corresponding row and column. Source: AQR, S&P Global Trucost, MSCI*

	Low Intensity Emitters in 2021						*High Intensity Emitters in 2021*			
	90%	6%	1%	1%	0%	1%	0%	1%	0%	0%
Low Intensity Emitters in 2020	7%	**82%**	9%	1%	0%	0%	0%	0%	1%	1%
	1%	9%	**75%**	12%	1%	1%	0%	0%	0%	0%
	1%	2%	10%	**75%**	11%	0%	0%	0%	0%	1%
	1%	1%	2%	7%	**74%**	12%	3%	0%	0%	0%
	0%	0%	0%	1%	12%	**74%**	12%	1%	0%	1%
	1%	1%	0%	2%	2%	9%	**80%**	5%	0%	0%
High Intensity Emitters in 2020	0%	0%	0%	1%	1%	1%	5%	**85%**	5%	1%
	0%	0%	0%	0%	0%	0%	0%	7%	**89%**	4%
	0%	0%	0%	0%	0%	0%	1%	1%	6%	**93%**

Figure 2.9. *2020–2021 Carbon intensity transition matrix: MSCI World stocks are sorted into deciles by carbon intensity using data available to investors at the end of 2020 (with 10% of firms with the lowest-intensity emitters in the top row; 10% of the highest-intensity emitters in the bottom row) and again using data available to investors at the end of 2021 (with 10% of the lowest-intensity emitters in the leftmost column; 10% of the highest-intensity emitters in the rightmost column). Each cell depicts the fraction of firms in the corresponding row and column. Source: AQR, S&P Global Trucost, MSCI*

2.7.1. *Measures of portfolio greenhouse gas emissions*

Carbon Footprint: The carbon footprint, also known as **financed emissions**, is the portion of the underlying assets' GHG emissions that the investor owns per $M invested. Emissions of individual investors are assigned to investors based on a proportional ownership logic:

$$\frac{\sum_{n}^{i}\left(\frac{current\ value\ of\ investment_j}{issuer's\ market\ cap_j} x\ issuer's\ Scope\ 1\ and\ 2\ emissions_j\right)}{current\ portfolio\ value\ (\$M)}$$

Carbon Intensity: Carbon intensity is the volume of emissions per $M of revenue. This allows for optimizing portfolios towards more carbon-efficient holdings:

$$\frac{\sum_{n}^{i}\left(\frac{current\ value\ of\ investment_j}{issuer's\ market\ cap_j} x\ issuer's\ Scope\ 1\ and\ 2\ emissions_j\right)}{\sum_{n}^{i}\left(\frac{current\ value\ of\ investment_j}{issuer's\ market\ cap_j} x\ issuer's\ \$M\ revenue_j\right)}$$

Weighted average carbon intensity (WACI): WACI is a portfolio-weight-based variation of carbon intensity that ignores ownership ratios. It is the preferred metric of several frameworks such as the Task Force for Climate Related Disclosures (TCFD):

$$\sum_{n}^{i}\left(\frac{current\ value\ of\ investment_j}{current\ portfolio\ value} x \frac{issuer's\ Scope\ 1\ and\ 2\ emissions_j}{issuer's\ \$M\ revenue_j}\right)$$

2.8. References

Bocquet, R., Georgieva, A., Mehrotra, S. (2021). Untangling the confusing landscape of forward-looking climate metrics. *Responsible Investor*, 20 July.

Brixton, A., Hecht, P., Kelly, B. (2021). Can machine learning help manage climate risks? *4Q2021 AQR Alternative Thinking*, 1–16.

Busch, T., Johnson, M., Pioch, T. (2020). Corporate carbon performance data: Quo vadis? *Journal of Industrial Ecology*, 26(1), 1–14.

CAQ (2021). S&P 500 and ESG Reporting. Report, Center For Audit Quality.

Cheema-Fox, A., LaPerla, B., Serafeim, G., Turkington, D., Wang, H. (Stacie) (2021). Decarbonization factors. *The Journal of Impact & ESG Investing*, 2(1), 47–73.

Engle, G.K. and Stroebel, L. (2020). Hedging climate change news. *Review of Financial Studies*, 33(3), 1184–1216.

IEA (2021). Global energy review: CO2 emissions in 2020. Report, IEA.

Kalesnik, V., Wilkens, M., Zink, J. (2020). Green data or greenwashing? Do corporate carbon emissions data enable investors to mitigate climate change? Working paper.

Krueger, P., Sautner, Z., Starks, L. (2020). The importance of climate risks for institutional investors. *Review of Financial Studies*, 33(3), 1067–1111.

Le Quéré, C., Jackson, R.B., Jones, M.W. (2020). Temporary reduction in daily global CO2 emissions during the COVID-19 forced confinement. *Nature Climate Change*, 10, 647–653.

Markwat, T. (2021). The data dilemma in tackling climate change. *Robeco Insights*, 19 April.

NGFS (2021). *Progress report on Bridging Data Gaps*. Technical document, Network for Greening the Financial System.

TCFD (2021). Summary of forward looking financial metrics consultation. Report, TCFD.

Vincent (2020). The problem with zero carbon pledges. *Financial Times Wealth*, 30 November.

3

Portfolio Construction with Climate Risk Measures

With the 2015 Paris Agreement, the development of ESG investing and the emergence of net zero emission policies, climate risk is certainly the most important topic and challenge for asset owners and managers today, and will remain so for the next five years. It considerably changes portfolio allocation and the investment framework of both passive and active investors. The goal of this chapter is to conduct a survey of the various climate risk measures that are available in the asset management industry and the practices of portfolio construction that use these metrics. Therefore, the first part of this chapter lists the different climate risk metrics – for example, carbon footprint, carbon transition pathway, carbon transition and physical risks. The second part is dedicated to portfolio optimization, in particular portfolio decarbonization and portfolio alignment (Paris-based benchmarks and net zero carbon objective). Among the different findings, two are of great importance to investors. First, portfolio decarbonization is more difficult when we include scope 3 carbon emissions. In fact, optimizing using the sum of scopes 1, 2 and 3 emissions leads to a portfolio with a higher risk of tracking error than using direct plus first tier indirect carbon emissions. Second, portfolio alignment is more complex than portfolio decarbonization. Since aligning portfolios with scope 3 is becoming the standard approach to climate portfolio construction, the impact on portfolio management may be significant, increasing the divergence between carbon investing and traditional investing.

3.1. Introduction

Climate risk is the biggest challenge to humanity of the 21st century. Indeed, climate change implies rises in temperature that increase the likelihood of extreme weather events and that impact living patterns. Similarly, the latest IPCC assessment report paints a grim picture of the future of the world (IPCC 2021). Beyond the

---

Chapter written by Théo LE GUENEDAL and Thierry RONCALLI.
For a color version of all figures in this chapter, see www.iste.co.uk/jurczenko/climate.zip.

direct effect on natural hazards, biodiversity, health or migration of human and animal populations, climate change may also result in a new economic order because of the transition to a low-carbon economy. In this context, the resilience of the financial system to climate-related risks is being called into question. This is why climate change has become one of the largest systemic risks (FSB 2020) and is the top priority on the agenda of financial institutions, supervisors and policymakers. Since all stakeholders play a key role in tackling climate risk, it also concerns the asset management industry. Therefore, portfolio decarbonization, temperature alignment, net zero carbon investment and Paris-aligned benchmarks are the everyday reality of both asset owners and managers.

Since 2014, interest in climate-related financial risks has been boosted by the development of ESG investing in Europe (Bennani et al. 2018; Drei et al. 2019). While environmental issues have been lagging behind social issues during the Covid-19 crisis, the net zero carbon race and the Glasgow COP 26 event have recently changed the equation, and climate risk is now the hottest topic in asset management. Climate risk is definitively back on the agenda of investors and goes beyond ESG ratings. However, climate risk assessment methodologies have not reached maturity. We are in an early stage, and there are many initiatives for proposing climate risk measures that are different from traditional scope $1 + 2$ carbon emissions. In particular, taking into account scope 3 changes the way we measure the risk to portfolio construction. Indeed, the dispersion of carbon intensities with scope 3 is dramatically reduced, implying that it is more difficult to decarbonize a portfolio. Similarly, investors are increasingly using carbon trajectories rather than just current carbon measures. Therefore, the introduction of trend or target dynamics means a new approach to portfolio allocation. More generally, the development of climate risk measures and their sophistication have an impact on portfolio optimization. We note that these approaches were first developed on the side of passive management and low-carbon equity indices but are now gaining a lot of traction in active management. As such, the goal of our research is to conduct a survey of the current practices in terms of climate risk measures and portfolio construction, which are now widely used by asset owners and managers.

This chapter is organized as follows. In section 3.2, we review the different climate risk measures. First, we consider the carbon footprint and explain the concept of carbon intensity. Then, we focus on carbon transition pathways that are a dynamic extension of carbon emissions. These metrics are particularly important for defining a net zero carbon emission policy. We also study carbon transition and physical risks and show their differences from an investment point of view. Finally, this section ends with several key performance indicators that are derived from ESG scoring, green revenues. Section 3.3 is dedicated to portfolio optimization when imposing climate risk constraints. First, we present several portfolio decarbonization methodologies. In particular, we show how using carbon emissions differs from using carbon intensity. Then, we investigate the portfolio alignment approach. In particular, we focus on two methods. The first one corresponds to the approach of Paris-aligned benchmarks. We highlight the impact of using scope 3 instead of scopes 1 and 2. The second method defines the net zero objective. We show how the dynamics of carbon intensity change the portfolio allocation with respect to a pure decarbonization approach. Finally, section 3.4 offers some concluding remarks.

3.2. Climate risk measures

Traditional portfolio optimization requires input data such as the vector μ of expected returns, the vector σ of asset volatilities and the correlation matrix ρ of asset returns[1] (Roncalli 2013). Let $x = (x_1, \ldots, x_n)$ be the vector of portfolio weights. We can then compute the first and second moments of the stochastic portfolio return[2]. In particular, the portfolio risk corresponds to the portfolio volatility $\sigma(x)$. When introducing climate-related risks, we have to define another measure that assesses the risk of climate change. For that, we generally consider a climate metric $\mathcal{C}_i$ associated with asset i and assume that the climate risk measure is linear:

$$\mathcal{C}(x) = \sum_{i=1}^{n} x_i \mathcal{C}_i \tag{3.1}$$

The nature of the climate risk measure is then different from the volatility risk measure since the latter satisfies the subadditivity property:

$$\sigma(x) \leq \sum_{i=1}^{n} x_i \sigma_i \tag{3.2}$$

Even if $\mathcal{C}(x)$ is a convex risk measure, it is an abuse of language because there is no way to diversify the climate risk:

$$\mathcal{C}(x) \not< \sum_{i=1}^{n} x_i \mathcal{C}_i \tag{3.3}$$

Therefore, $\mathcal{C}(x)$ plays more the role of an expected loss than a risk measure.

3.2.1. *Carbon footprint*

Carbon footprint is a generic term used to define the total greenhouse gas (GHG) emissions caused by a given system, activity, company, country or region. Greenhouse

---

1 Using the parameters σ and ρ, we can form the covariance matrix Σ, which is defined by $\Sigma_{i,j} = \rho_{i,j}\sigma_i\sigma_j$.

2 The first moment is equal to the expected return of the portfolio $\mu(x) = x^\top \mu$, while the second moment is equal to the variance of the portfolio return $\sigma^2(x) = x^\top \Sigma x$.

gases are made up of water vapor (H_2O), carbon dioxide (CO_2), methane (CH_4), nitrous oxide (N_2O) and ozone (O_3). They absorb and emit radiation energy, causing the greenhouse effect. The greenhouse effect was a crucial factor for the development of human life on Earth. Indeed, without the greenhouse effect, the average temperature of the Earth's surface would be about $-18°C$. With the greenhouse effect, the current temperature of the Earth's surface is about $+15°C$. Nevertheless, the increasing concentration of some GHGs is an issue because it is a factor in global warming. It mainly concerns carbon dioxide, and to a lesser extent, methane and nitrous oxide.

Carbon footprint is generally measured in carbon dioxide equivalent (CO_2e), which is a term for describing different GHGs in a common unit. In this framework, a quantity of GHG is expressed as CO_2e by multiplying the GHG amount by its global warming potential (GWP). The GWP of a gas is the amount of CO_2 that would warm the Earth equally. For instance, the IPCC's 4th Assessment Report has used the following rules[3]: 1 kg of methane corresponds to 25 kg of CO_2 and 1 kg of nitrous oxide corresponds to 298 kg of CO_2. The definition of a common unit allows two companies to be compared properly. However, carbon emissions cannot be compared fairly if the size of the two companies differs. Therefore, it can be useful to transform carbon emissions into normalized metrics, which are called carbon intensities.

3.2.1.1. *Carbon emissions*

It is now possible to access the carbon emissions of an increasing number of companies. However, since GHG are not physical materials, even this raw indicator is subject to uncertainty at the company level. To provide a common accounting basis that can be used by states and companies, and to limit the measurement gap, the greenhouse gas protocol (GHGP) classifies a company's greenhouse gas emissions in three scopes[4]:

– Scope 1 denotes direct GHG emissions occurring from sources that are owned or controlled by the company.

– Scope 2 corresponds to the indirect greenhouse gas emissions from the consumption of purchased electricity, heat or steam. Scope 2 emissions can be computed using the energy mix of the country (location-based) or the energy mix of the utility company supplying the electricity.

– Finally, scope 3 are other indirect emissions[5], such as the extraction and production of purchased materials and fuels, transport-related activities in vehicles not

3 GWP estimates depend on several factors such as the atmospheric lifetime of the gas. This explains that estimates are different from one study to another (Derwent 2020).

4 The standards can be found at www.ghgprotocol.org.

5 See Appendix 3.5.1 about the issues of scope 3 emission measurement.

owned or controlled by the reporting entity, electricity-related activities not covered in scope 2. Scope 3 upstream emissions include the indirect emissions that come from the supply side, while scope 3 downstream emissions are mostly associated with the product sold by the entity.

In what follows, we distinguish these three carbon emission measures by introducing the notations $\mathcal{CE}_1$, $\mathcal{CE}_2$ and $\mathcal{CE}_3$. In carbon emission databases, we also observe that carbon scopes can be self-reported or estimated. Moreover, they are generally expressed in tons of carbon dioxide equivalent or tCO_2e.

REMARK 3.1.– *Except when it explicitly refers to downstream items, the scope 3 emissions that we use in this chapter are defined in the Appendix (section 3.5.1) on page 84 and mainly concern the upstream items.*

Data on GHG emissions are easily and freely available when they concern countries and regions[6]. For corporations, two main sources of GHG emissions are generally used: the CDP database and the S&P Trucost data[7]. In what follows, we use the latter since it seems to be more comprehensive. Indeed, it contains the self-reported values of companies from CDP. When they are not available, estimated values are provided. For the year 2019, we have about 15 700 corporations. If we sum up all the observations, we obtain the following figures (expressed in $GtCO_2e$): $\mathcal{CE}_1 = 15.57$, $\mathcal{CE}_2 = 2.45$ and $\mathcal{CE}_3 = 10.17$. It follows that the total emissions of these corporate firms are about 28.2 $GtCO_2e$, which represents more than 75% of the 36 $GtCO_2e$ global emissions.

Figure 3.1 shows the breakdown of these GHG emissions by sector. We note that the direct emissions are concentrated in a few sectors: Utilities, Materials, Energy and to some extent Industrials. Scope 2 emissions are more equally distributed. On the other hand, every sector's scope 2 emissions are already accounted for in the scope 1 emissions of the Utilities sector. This *double-counting* issue becomes particularly problematic when trying to build the carbon footprint of a portfolio. Finally, scope 3 emissions are discriminant for Information Technology, Health Care, Consumer Staples, Consumer Discretionary and Financials. However, estimating scope 3 emissions is still subject to uncertainty. The distribution of the absolute emissions within GICS sectors is also an issue. Contrary to the commonly used centered scores, the emissions are concentrated on very few actors. Filtering the

6 They can be retrieved from the World Bank (data.worldbank.org/topic/climate-change), Climate Watch Data (www.climatewatchdata.org/ghg-emissions), Global Carbon Project (www.globalcarbonproject.org).

7 A description of these two databases can be found at www.cdp.net/en/data and www.spglobal.com/esg/trucost.

outliers[8], we obtain the distribution of the values given in Figure 3.2. Although the filtration allows us to visualize the distribution, we are faced with multiple scaling issues (sector, size, country), making absolute emissions difficult to use in portfolio construction.

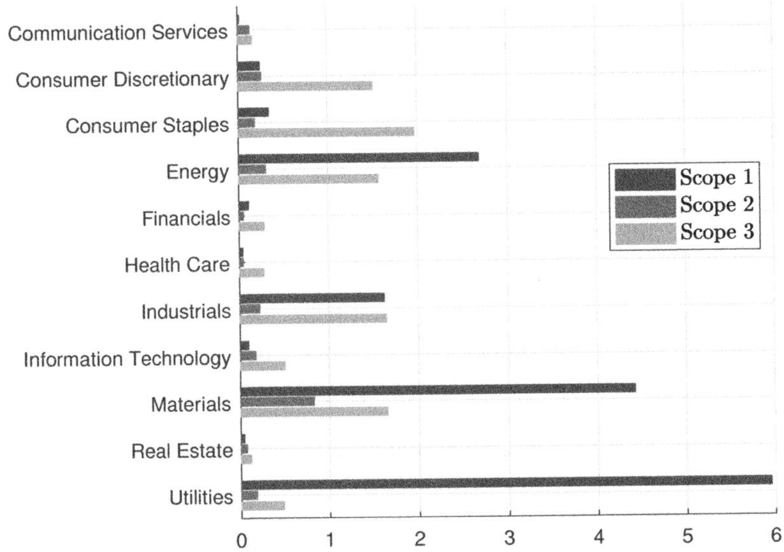

Source: Trucost reporting year 2019 and authors' calculations.

Figure 3.1. *Total absolute scopes per GICS sector in GtCO$_2$e*

3.2.1.2. *Carbon metrics*

3.2.1.2.1. Definition

The carbon intensity of company i with respect to scope j is a normalization of the carbon emissions:

$$\mathcal{CI}_{i,j} = \frac{\mathcal{CE}_{i,j}}{Y_i} \qquad [3.4]$$

where $\mathcal{CE}_{i,j}$ is the company's absolute scope j emissions and Y_i is an output indicator measuring its activity. In general, revenues (expressed in \$) are used to compute carbon intensities. For some major sectors, it is possible to find intensities per production unit. For instance, for a company from the Utilities sector, a carbon intensity in CO_2e/kWh is more informative in terms of carbon efficiency than a carbon intensity in CO_2e/\$. In what follows, we use revenue as the normalization variable when one is not defined. Some examples are provided in Table 3.1. Comparing the carbon emissions of a football club like Juventus with a petroleum company

8 We exclude the companies below the 20th percentile and above the 80th percentile.

like BP does not make sense, because their economic size is not comparable. This explains why the carbon intensity measure is more popular than the carbon emission measure in portfolio management. These examples also illustrate how interconnected the three scopes are. Thus, if a company extensively outsources the manufacturing of its products, it reduces its scope 1 but increases its scope 3 emissions. It is also possible to deal with more disaggregated data, for instance by making the distinction between upstream and downstream emissions resulting from the entire supply chain. However, it is not straightforward since the accuracy of scope 3 is not obvious[9].

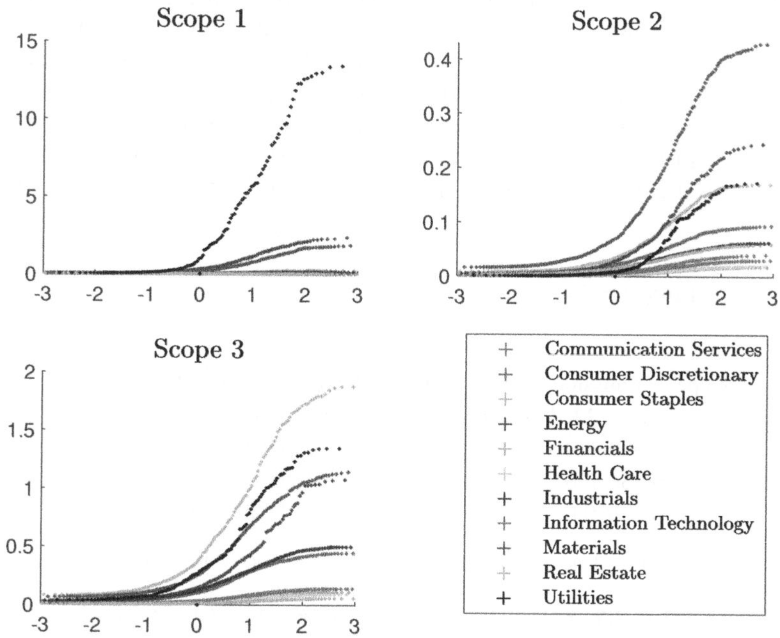

Source: Trucost reporting year 2019 and authors' calculations.

Figure 3.2. *QQ-plot of carbon scopes per GICS sector in $MtCO_2e$*

Another advantage of carbon intensity is reducing the skewness of the distribution compared to absolute emissions. In Table 3.2, we report some statistics (average, median, 95% percentile and maximum). We verify the skewness reduction. For instance, if we compute the ratio between $Q\,(95\%)$ and the median, it takes the value 233, 43 and 37 for the three scopes when we consider carbon emissions. These figures become 69, 9 and 5 when the ratio is computed with the carbon intensity.

9 For example, the figures of BP, Enel and Total are very different and may be difficult to understand.

Company	Emission (in tCO_2e)			Revenue	Intensity (in tCO_2e/$ mn)		
	Scope 1	Scope 2	Scope 3	(in $ mn)	Scope 1	Scope 2	Scope 3
Alphabet	74 462	5 116 949	7 166 240	161 857	0.460	31.614	44.275
Amazon	5 760 000	5 500 000	20 054 722	280 522	20.533	19.606	71.491
Apple	50 463	862 127	27 618 943	260 174	0.194	3.314	106.156
BP	49 199 999	5 200 000	103 840 194	276 850	177.714	18.783	375.077
Danone	722 122	944 877	28 969 780	28 308	25.509	33.378	1023.365
Enel	69 981,891	5 365 386	8 726 973	86 610	808.016	61.949	100.762
Juventus	6 665	15 739	35 842	709	9.401	22.198	50.553
LVMH	67 613	262 609	11 853 749	60 083	1.125	4.371	197.291
Microsoft	113 414	3 556 553	5 977 488	125 843	0.901	28.262	47.500
Nestle	3 291 303	3 206 495	61 262 078	93 153	35.332	34.422	657.647
Netflix	38 481	145 443	1 900 283	20 156	1.909	7.216	94.277
Total	40 909 135	3 596 127	49 831 487	200 316	204.223	17.952	248.764
Volkswagen	4 494 066	5 973 894	65 335 372	282 817	15.890	21.123	231.016

Source: Trucost reporting year 2019.

Table 3.1. *Examples of carbon emissions and intensity*

Scope	Emission (in $10^6 \cdot tCO_2e$)				Intensity (in $10^3 \cdot tCO_2e$/$ mn)			
	Avg.	Med.	Q (95%)	Max.	Avg.	Med.	Q (95%)	Max.
1	0.992	0.010	2.28	587.1	0.277	0.016	1.14	207.4
2	0.156	0.012	0.53	99.1	0.053	0.021	0.19	11.9
3	0.648	0.067	2.50	137.5	0.170	0.099	0.51	2.0

Source: Trucost reporting year 2019 &and authors' calculations.

Table 3.2. *Statistics of carbon emissions and intensity*

3.2.1.2.2. Additivity property

Carbon intensity is additive when we consider a given issuer. For instance, the carbon intensity of scopes 1, 2 and 3 is the sum of the different scopes:

$$\begin{aligned} \mathcal{CI}_{i,1+2+3} &= \frac{\mathcal{CE}_{i,1} + \mathcal{CE}_{i,2} + \mathcal{CE}_{i,3}}{Y_i} \\ &= \mathcal{CI}_{i,1} + \mathcal{CI}_{i,2} + \mathcal{CI}_{i,3} \end{aligned} \quad [3.5]$$

Nevertheless, the additivity property is not satisfied when we consider a set of issuers. Let us consider a portfolio x investing in n assets (stocks or bonds). Its carbon emissions are equal to:

$$\mathcal{CE}_j(x) = \sum_{i=1}^{n} \frac{W_i}{\mathcal{MV}_i} \mathcal{CE}_{i,j} = \sum_{i=1}^{n} \varpi_i \mathcal{CE}_{i,j} \quad [3.6]$$

where $\mathcal{MV}_i$ is the market value of the company i, W_i is the dollar value invested in the company i and ϖ_i is the ownership ratio:

$$\varpi_i = \frac{W_i}{\mathcal{MV}_i} \tag{3.7}$$

Let x_i be the weight of the asset i in the portfolio x. By definition, the wealth invested in this asset is equal to $W_i = x_i W$, where $W = \sum_{i=1}^n W_i$ is the dollar value of the portfolio. We deduce that:

$$\varpi_i = x_i \frac{W}{\mathcal{MV}_i} \tag{3.8}$$

and:

$$\mathcal{CE}_j(x) = \sum_{i=1}^n \left(x_i \frac{W}{\mathcal{MV}_i} \right) \mathcal{CE}_{i,j} \tag{3.9}$$

If we consider the weighted-average carbon intensity (WACI), we obtain:

$$\begin{aligned} \mathcal{CI}_j(x) &= \sum_{i=1}^n x_i \mathcal{CI}_{i,j} \\ &= \sum_{i=1}^n x_i \frac{\mathcal{CE}_{i,j}}{Y_i} \end{aligned} \tag{3.10}$$

It is obvious that equations [3.9] and [3.10] are mutually satisfied if and only if the normalization metric Y_i of the carbon intensity is proportional to the market value $\mathcal{MV}_i$. Indeed, we can interpret the carbon emissions of a portfolio as a weighted-average carbon intensity:

$$\mathcal{CE}_j(x) = W \left(\sum_{i=1}^n x_i \mathcal{CI}_{i,j}^{\mathcal{MV}} \right) \tag{3.11}$$

where $\mathcal{CI}_{i,j}^{\mathcal{MV}}$ is a carbon intensity measure, where the normalization factor is the market value of the company:

$$\mathcal{CI}_{i,j}^{\mathcal{MV}} = \frac{\mathcal{CE}_{i,j}}{\mathcal{MV}_i} \tag{3.12}$$

Let us assume that equation [3.6] holds. We can show that:

$$\mathcal{CI}_j(x) = \sum_{i=1}^n \omega_i \mathcal{CI}_{i,j} \tag{3.13}$$

where the weights ω_i are equal to:

$$\omega_i = \frac{x_i \dfrac{Y_i}{\mathcal{MV}_i}}{\sum_{k=1}^n x_k \dfrac{Y_k}{\mathcal{MV}_k}} = \frac{x_i \mathcal{SR}_i}{\sum_{k=1}^n x_k \mathcal{SR}_k} \quad [3.14]$$

and $\mathcal{SR}_i = Y_i / \mathcal{MV}_i$ is the revenue-to-market value (or sales-to-price) ratio. Contrary to common belief, analyzing the carbon risk of a portfolio using carbon intensity or carbon emissions is then not equivalent.

REMARK 3.2.– *$\mathcal{CE}_j(x)$ is generally expressed in tCO2e per 1\$ mn invested ($W = \$10^6$).*

3.2.2. *Carbon transition pathway*

3.2.2.1. *Carbon reduction scenario*

The most recent IPCC report urges action to drastically reduce carbon emissions to net zero by 2050 (IPCC 2021). The carbon reduction trajectory imposes that we need to reduce total emissions by at least 7% every year between 2019 and 2050. IEA (2021) has also published its net zero emission (NZE) scenario (see Table 3.3). It implies a 40.11% reduction of carbon emissions by 2030 and 61.84% by 2035. By 2050, gross emissions would be 1.94 $GtCO_2e$ offset by the carbon capture and storage (CCS) technology. These two scenarios are reported in Figure 3.3.

Year	2019	2025	2030	2035	2040	2045	2050
Gross emission	35.90	30.30	21.50	13.70	7.77	4.30	1.94
CCS	0.00	−0.06	−0.32	−0.96	−1.46	−1.80	−1.94
Net emission	35.90	30.24	21.18	12.74	6.31	2.50	0.00
Reduction (in %)	0.00	15.60	40.11	61.84	78.36	88.02	94.60

Source: IEA (2021, Chapter 2, Figure 2.3, page 55).

Table 3.3. *IEA NZE scenario (in $GtCO_2e$)*

Besides net zero carbon scenarios, we can find other carbon scenarios, whose goal is to limit global warming. For instance, the European Union has framed new climate benchmarks designed to align investors' portfolios with the Paris agreement. For instance, the EU Paris-aligned benchmark (PAB) label implies a year-on-year self-decarbonization of 7% and an initial carbon intensity reduction of 50% with respect to the investable universe in 2021. In the case of the EU climate transition benchmark (CTB) label, this last figure is equal to 30%. We note that these two labels use the carbon intensity[10] instead of the carbon emission measure.

REMARK 3.3.– *Carbon reduction scenarios can be applied to regions, countries, sectors. For example, IEA (2021, Chapter 3, pages 99–150) proposes sectoral pathways to net zero emissions for electricity, transport, buildings. Moreover, we observe the development of physical carbon intensity by production unit (CO_2/km, CO_2/ton, CO_2/km^2, CO_2/kWh, etc.). The issue is then how to compare and aggregate these different intensity measures.*

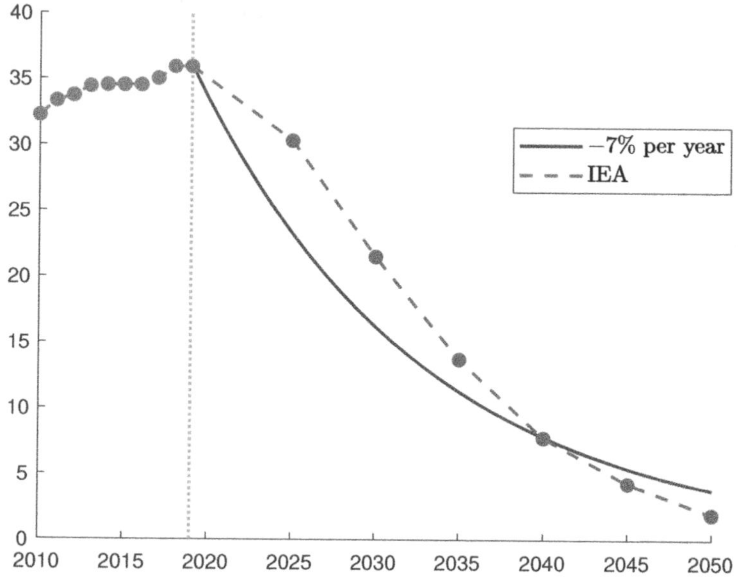

Source: IEA (2021) and authors' calculations.

Figure 3.3. *Two net zero emission scenarios*

3.2.2.2. *Carbon trajectory*

Defining indicators describing individual transitions is not trivial. For some issuers, we can retrieve declared emissions since 2005, but the coverage starts to become more widespread starting in 2010. Using the historical records, it is possible to estimate the trend and make some carbon budget projections. For instance, we can consider a linear regression model:

$$\mathcal{CE}_{i,j}(t) = \beta_0 + \beta_1 t + u(t) \qquad \text{for } t \leq t_0 \tag{3.15}$$

10 The normalization variable is then the enterprise value including cash and not the revenue.

where t_0 is the current date. Once the model is estimated, we can project the future carbon emissions:

$$\mathcal{CE}_{i,j}^{\text{trend}}(t_0+h)=\hat{\beta}_0+\hat{\beta}_1(t_0+h) \qquad \text{for } h=1,2,\ldots \tag{3.16}$$

Figure 3.4 illustrates this approach when we consider the company Λ and the direct carbon emissions (scope 1). We note that the projected value of direct carbon emissions is equal to zero by 2040. Using these trends, we can define the annual contribution of company i to reducing emissions:

$$\mathcal{R}_i^{\text{trend}}=\frac{1}{t-t_0}\frac{\sum_{j=1}^{3}\mathcal{CE}_{i,j}^{\text{trend}}(t_0)-\sum_{j=1}^{3}\mathcal{CE}_{i,j}^{\text{trend}}(t)}{\sum_{j=1}^{3}\mathcal{CE}_{i,j}^{\text{trend}}(t_0)} \tag{3.17}$$

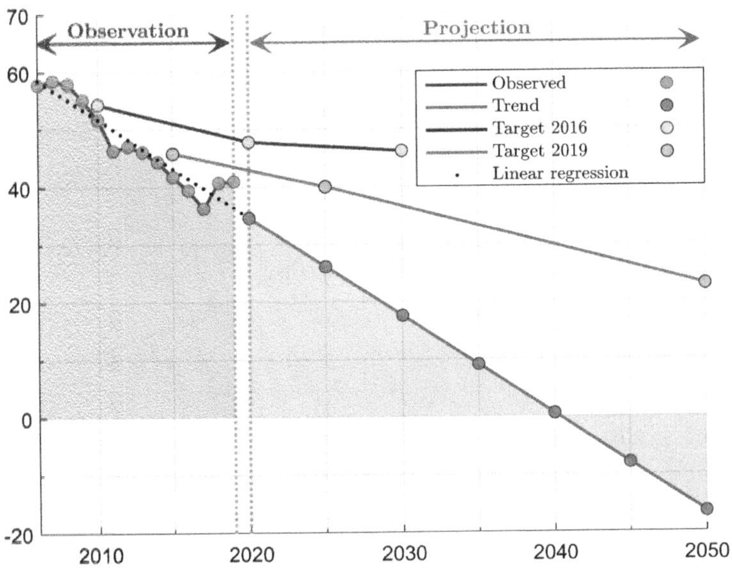

Source: Carbon Disclosure Project reporting year 2019 and authors' calculations.

Figure 3.4. *Scope 1 trajectory of the company Λ (in $MtCO_2e$)*

In addition, companies can directly disclose carbon targets $\mathcal{CE}_{i,j}^{\text{target}}(t)$ at a given horizon date t for a given scope j. The Carbon Disclosure Project (CDP) gathers information about reduction targets (base year, target year, target value, scope, percentage of the scope concerned, etc.) in its climate change emission management dataset. For example, we have reported the company Λ's targets announced in 2016 and 2019. We observe that the 2019 targets are more ambitious than the 2016 targets, and the targets are greater than the projections. As previously, we can compute the annual reduction using equation [3.17] and replacing $\mathcal{CE}_{i,j}^{\text{trend}}(t)$ with $\mathcal{CE}_{i,j}^{\text{target}}(t)$.

Therefore, the synthetic trajectory of the company can be derived from both projected and reported annual reductions.

3.2.2.3. *Temperature*

Temperature tags are indicators assigned to companies by mixing various metrics. The main ingredient is the carbon trajectory. Based on the carbon reduction required to achieve a climate scenario, and individual trajectories, it is possible to interpolate a temperature for each issuer. For instance, if we would like to achieve a 1.8°C scenario, this implies reducing absolute carbon emissions with respect to year 2005 by 51.5%. For each company, we can consider a linear projection of the carbon trajectory, and perform some corrections using additional information such as reduction targets, changes in the share of green revenues, investment in research and development associated with transition and green patents, etc.

Range	CDP				Iceberg
	Scope 1 + 2 + 3	Short-term	Mid-term	Long-term	
$\mathcal{T} \leq 1.0°\text{C}$	0.00	0.00	0.00	0.00	1.01
$1.0°\text{C} < \mathcal{T} \leq 1.5°\text{C}$	1.44	2.92	10.68	2.71	2.60
$1.5°\text{C} < \mathcal{T} \leq 2.0°\text{C}$	6.20	1.26	13.03	3.94	3.14
$2.0°\text{C} < \mathcal{T} \leq 2.5°\text{C}$	6.86	3.07	7.46	2.68	21.76
$2.5°\text{C} < \mathcal{T} \leq 3.0°\text{C}$	7.64	1.99	4.21	0.48	30.87
$3.0°\text{C} < \mathcal{T} \leq 3.5°\text{C}$	76.95	89.77	62.80	90.07	32.30
$3.5°\text{C} < \mathcal{T} \leq 4.0°\text{C}$	0.78	0.81	1.44	0.09	2.23
$4.0°\text{C} < \mathcal{T} \leq 4.5°\text{C}$	0.12	0.18	0.36	0.03	3.31
$4.5°\text{C} < \mathcal{T} \leq 5.0°\text{C}$	0.00	0.00	0.00	0.00	0.77
$\mathcal{T} = 3.2°\text{C}$	52.09	88.50	61.39	89.95	0.01

Source: CDP Temperature Ratings Dataset, version 1.1, February 2021 and Iceberg Data Lab (2021).

Table 3.4. *Frequency of temperature ratings (in %)*

However, there are multiple methodologies that do not always provide the same output in terms of aggregated temperature (Raynaud et al. 2021a,2021b). One of the main pitfalls is the aggregation of individual trajectories, or in other terms, the definition of the portfolio trajectory based on the trajectories of issuers (Le Guenedal et al. 2020). For example, the CDP provides temperature ratings that mix the trajectories with issuer targets[11]. Among the 8,500 companies in the database, about 3,300 have

11 The methodology is available at www.cdp.net/en/investor/temperature-ratings. The short-term measure corresponds to target years up to four years from the reporting year (e.g. 2021–2024), whereas the time horizon of mid-term and long-term categories is between 5–15 years and greater than 15 years (e.g. 2025–2035 and 2036–2050).

a CDP temperature rating. When carbon targets are used, they are based on the Science-Based Target (SBT) initiative; however, it only represents 12% of the sample for scope 1 + 2, compared to 1% for scope 3. In Table 3.4, we report the frequencies of the temperature ratings provided by CDP and Iceberg. For CDP data, we note that most of companies have a rating corresponding to a temperature of 3.2°C.

3.2.3. *Other metrics*

There are many other metrics, some of them are related to carbon transition risks, while others concern climate physical risks. Moreover, portfolio managers can also use variables from ESG scoring systems and key performance indicators. Some examples are avoided emissions, green revenues, energy mix, stranding risk indicators, or green R&D and capital expenditure.

While carbon footprint and pathway are measured by CO_2 emissions, which are based on fundamentals and physical data, we can consider carbon risk metrics, which are by nature based on market data. In other words, they are related to the financial market's perception of the potentially reduced impact of climate policies on securities issued by corporations. These metrics are related to carbon transition risks. Two main approaches have been developed. The first one assesses how an increase in carbon prices and taxes influences the credit risk of the issuer (Bouchet and Le Guenedal 2020), whereas the second method measures how sensitive the asset price is to a carbon market factor[12] (Harris 2015; Görgen et al. 2019; Roncalli et al. 2021). A related approach is to replace the factor mimicking portfolio by a climate risk news index based on text mining or machine learning techniques (Engle et al. 2020; Apel et al. 2021; Ardia et al. 2021).

Responsible investors have paid more attention to the transition risk than to the physical risk. However, recent events show that physical risk is also a big concern. It corresponds to the financial losses that really come from climate change, and not from the adaptation of the economy to prevent them. It includes droughts, floods, storms, etc. This risk is more difficult to quantify, and its evaluation requires multidisciplinary methodologies: climate modeling, physical asset geolocation, financial loss estimation, etc. For example, we refer to Le Guenedal et al. (2021) for the description of a full integrated methodology to measure cyclone-related physical risk.

12 In this case, the central measure is the carbon beta of the issuer, which can be used in portfolio optimization (Roncalli et al. 2020).

3.3. Portfolio optimization

The development of sustainable and impact investing strategies related to climate change implies including climate risk metrics in modern portfolio optimization. Generally, the solution consists of adding some management constraints. From a theoretical and mathematical viewpoint, this does not fundamentally change the traditional approach to portfolio construction. However, this can lead to high portfolio distortion from a practical viewpoint, especially when constraints are tight, which can be the case with net zero carbon portfolios.

3.3.1. *General framework*

Traditional portfolio optimization problems can be written as:

$$x^\star(\gamma) = \arg\min \frac{1}{2}(x-b)^\top \Sigma (x-b) - \gamma (x-b)^\top \mu \quad [3.18]$$

$$\text{s.t.} \begin{cases} \mathbf{1}_n^\top x = 1 \\ x \in \Omega \end{cases}$$

where $x = (x_1, \ldots, x_n)$ is the active portfolio, $b = (b_1, \ldots, b_n)$ is the benchmark portfolio, Σ is the covariance matrix of asset returns, μ is the vector of expected returns (or alpha tilts/scores) and Ω is the set of constraints. For instance, we can impose a long-only constraint[13], maximum sector deviation[14], etc. If there is no benchmark ($b = \mathbf{0}_n$), we retrieve the Markowitz framework with the trade-off between the portfolio variance $\sigma^2(x) = x^\top \Sigma x$ and the average return $\mu(x) = x^\top \mu$. The efficient frontier is then obtained by considering different values of $\gamma \in \mathbb{R}$. If we consider a benchmark ($b \neq \mathbf{0}_n$), the trade-off concerns the tracking error variance $\sigma^2(x \mid b) = (x-b)^\top \Sigma (x-b)$ and the average excess return of the portfolio $\mu(x \mid b) = (x-b)^\top \mu$.

A climate risk measure can be introduced into the portfolio optimization by adding a new constraint[15]:

$$\Omega = \left\{ x : \mathcal{C}(x) = \sum_{i=1}^{n} x_i \mathcal{C}_i \leq \mathcal{C}^+ \right\} \quad [3.19]$$

13 In this case, we have $\Omega = \{x : 0 \leq x_i \leq 1\}$.

14 Let $\omega^-(\mathcal{S}_j)$ and $\omega^+(\mathcal{S}_j)$ be the minimum and maximum weights for the sector $\mathcal{S}_j$, respectively. We can impose that $\Omega = \left\{ x : \omega^-(\mathcal{S}_j) \leq \sum_{i \in \mathcal{S}_j} x_i \leq \omega^+(\mathcal{S}_j) \right\}$.

15 We assume that the climate metric is an increasing function with respect to the climate risk. If we consider a decreasing function, we replace the constraint by $\mathcal{C}(x) = \sum_{i=1}^{n} x_i \mathcal{C}_i \geq \mathcal{C}^-$. In this case, we would like to achieve a minimum value $\mathcal{C}^-$ of the climate metric $\mathcal{C}(x)$.

For instance, we may want to limit the portfolio's carbon emissions – $\mathcal{CE}_j(x) \leq \mathcal{CE}_j^+$ – or its carbon intensity – $\mathcal{CI}_j(x) \leq \mathcal{CI}_j^+$. We may also want to reduce the portfolio's carbon emissions or intensity with respect to a benchmark: $\mathcal{CE}_j(x) \leq (1-\mathcal{R})\,\mathcal{CE}_j(b)$ or $\mathcal{CI}_j(x) \leq (1-\mathcal{R})\,\mathcal{CI}_j(b)$, where $\mathcal{R} > 0$ is the reduction rate. An alternative approach to problem [3.18-3.19] is to formulate the following objective function:

$$x^\star(\gamma,\delta) = \arg\min \frac{1}{2}(x-b)^\top \Sigma (x-b) - \gamma (x-b)^\top \mu + \gamma \mathcal{C}(x) \qquad [3.20]$$

$$\text{s.t.} \begin{cases} \mathbf{1}_n^\top x = 1 \\ x \in \Omega \end{cases}$$

In this case, we obtain a three-dimensional efficient frontier: mean return versus variance risk versus climate risk. As noted by Pedersen et al. (2020), optimizing over three characteristics can be challenging. However, these authors showed that the problem can be reduced to a trade-off between the Sharpe ratio and the climate risk measure[16]. In what follows, we consider another approach by fixing a minimum level of expected excess return – $\mu(x \mid b) \geq \mu^-$ – or a maximum level of tracking error volatility – $\sigma(x \mid b) \leq \sigma^+$. In this case, the efficient frontier is reduced to a two-dimensional trade-off: variance risk versus climate risk or mean return versus climate risk. This is the preferred approach of professionals when implementing active management. A third approach consists of defining the benchmark portfolio and finding the climate-optimal portfolio that minimizes the tracking error volatility. This is equivalent to imposing $\gamma = 0$ (or $\mu = \mathbf{0}_n$) in problem [3.20]. This approach is extensively used in passive management (Andersson et al. 2016).

3.3.2. *Portfolio decarbonization*

3.3.2.1. *Using carbon intensity*

One of the most important topics in climate finance is portfolio decarbonization. The underlying idea is to construct a portfolio x that tracks the benchmark portfolio b but with a lower carbon risk metric. From a mathematical point of view, the optimization problem is defined as[17]:

$$x^\star(\mathcal{R}) = \arg\min \frac{1}{2}(x-b)^\top \Sigma (x-b) \qquad [3.21]$$

$$\text{s.t.} \begin{cases} \mathbf{1}_n^\top x = 1 \\ x \geq \mathbf{0}_n \\ \sum_{i=1}^n x_i \mathcal{CI}_i \leq (1-\mathcal{R})\,\mathcal{CI}(b) \end{cases}$$

16 Although Pedersen et al. (2020) used ESG scores instead of climate scores, their analysis remains valid when considering climate risks and not ESG risks.

17 In the following, we omit the subscript j that defines the scope to simplify the notations.

where $\mathcal{R}$ is the reduction rate. Therefore, we minimize the tracking error volatility of the portfolio x with respect to the benchmark by imposing a long-only constraint and reducing the weighted average carbon intensity (WACI) of the benchmark. We note that the carbon intensity metric is used and not the carbon emission metric, which may cause problems when considering net zero carbon portfolios[18]. Since we both impose a constraint and minimize the tracking error volatility, the portfolio x has fewer stocks than the benchmark b. In fact, the number of stocks depends on several parameters: the reduction rate $\mathcal{R}$, the number n of stocks in the benchmark and the covariance matrix. This implies that the portfolio x is less diversified than the benchmark b. In order to explicitly control the number of removed stocks, Andersson et al. (2016) proposed a second portfolio decarbonization approach by eliminating the m worst-performing issuers in terms of carbon intensity. Let $\mathcal{CI}_{i:n}$ be the order statistics of $(\mathcal{CI}_1, \ldots, \mathcal{CI}_n)$ such that:

$$\min \mathcal{CI}_i = \mathcal{CI}_{1:n} \leq \mathcal{CI}_{2:n} \leq \cdots \leq \mathcal{CI}_{i:n} \leq \cdots \leq \mathcal{CI}_{n:n} = \max \mathcal{CI}_i \tag{3.22}$$

The carbon intensity bound $\mathcal{CI}^{(m,n)}$ is defined as $\mathcal{CI}^{(m,n)} = \mathcal{CI}_{n-m+1:n}$, where $\mathcal{CI}_{n-m+1:n}$ is the $(n-m+1)$-th order statistic of $(\mathcal{CI}_1, \ldots, \mathcal{CI}_n)$. Eliminating the m worst-performing assets is equivalent to imposing the following constraint: $\mathcal{CI}_i \geq \mathcal{CI}^{(m,n)} \Rightarrow x_i = 0$. Finally, we obtain the following optimization problem:

$$x^\star(m) = \arg\min \frac{1}{2}(x-b)^\top \Sigma (x-b) \tag{3.23}$$

$$\text{s.t.} \begin{cases} \mathbf{1}_n^\top x = 1 \\ x \geq \mathbf{0}_n \\ \mathcal{CI}_i \geq \mathcal{CI}^{(m,n)} \Rightarrow x_i = 0 \end{cases}$$

Problem [3.23] is called the "*order-statistic*" approach, whereas problem [3.21] is known as "*max-threshold*". Finally, a "*naive*" solution consists of re-weighting the remaining assets:

$$x_i = \frac{\mathbf{1}\left\{\mathcal{CI}_i < \mathcal{CI}^{(m,n)}\right\} \cdot b_i}{\sum_{k=1}^n \mathbf{1}\left\{\mathcal{CI}_k < \mathcal{CI}^{(m,n)}\right\} \cdot b_k} \tag{3.24}$$

REMARK 3.4.– *Portfolio decarbonization requires the covariance matrix Σ to be estimated. A first approach is to use the empirical covariance matrix of stock returns. A second solution is to postulate a multivariate risk factor model:*

$$R_i(t) = \alpha_i + B\mathcal{F}(t) + \varepsilon_i(t) \tag{3.25}$$

18 This issue is developed in section 3.3.3.

where $\mathcal{F}(t)$ is the vector of risk factors and $\varepsilon_i(t)$ is the specific risk. By denoting $\Omega = \text{var}(\mathcal{F}(t))$ and $D = \text{var}(\varepsilon_1(t), \ldots, \varepsilon_n(t)) = \text{diag}(\tilde{\sigma}_1^2, \ldots, \tilde{\sigma}_n^2)$, we obtain $\Sigma = B\Omega B^\top + D$.

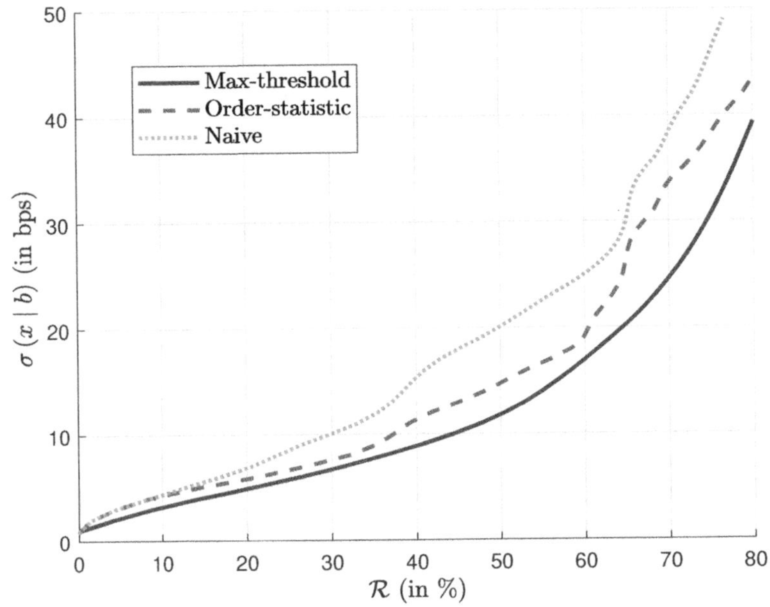

Source: Trucost reporting year 2020 and authors' calculations.

Figure 3.5. *Efficient frontier of optimal decarbonization portfolios (S&P 500 index, October 2021, scope $1+2$)*

Let us illustrate the difference between the three approaches. The benchmark corresponds to the S&P 500 index at the end of October 2021. To estimate the covariance matrix, we use the single-factor CAPM model and the historical prices of the last three years. In Figure 3.5, we report the tracking error volatility $\sigma(x \mid b)$ with respect to the reduction rate $\mathcal{R}$ when we consider scope $1+2$. For example, we note that reducing the carbon intensity of 50% leads to a tracking error volatility of 12 bps for the max-threshold solution. For the naive solution, the tracking error volatility is equal to 20 bps. We observe some discrepancies between the three approaches. In particular, the order-statistic and naive solutions are satisfying only when the reduction is small, for example, less than 30%. We also note that the slope of the tracking error risk is steeper when the reduction is high (see Figure 3.6). For scope 1 or $1+2$, the slope changes when $\mathcal{R} \approx 50\%$. Today, most asset managers use scope $1+2$ to decarbonize their portfolios because the data are more robust than scope $1+2+3$.

Nevertheless, there is pressure from ESG investors to increasingly consider scope $1 + 2 + 3$. In this case, the changing point is lower than and equal to $\mathcal{R} \approx 30\%$. The reason lies in the fact that the range of scope 3 is smaller than the range of scope 1 or 2. Therefore, it is more difficult to remove assets with high-carbon intensity and replace them by assets with low-carbon intensity. In Figure 3.7, we show the overlap[19] between optimal decarbonization portfolios. We note that the portfolio composition is very different if we consider scope $1+2$ or scope $1+2+3$. Indeed, when the reduction is greater than 50%, the overlap is less than 90%. Considering the preference of ESG investors for scope $1 + 2 + 3$, we can anticipate large portfolio rebalancing in the future.

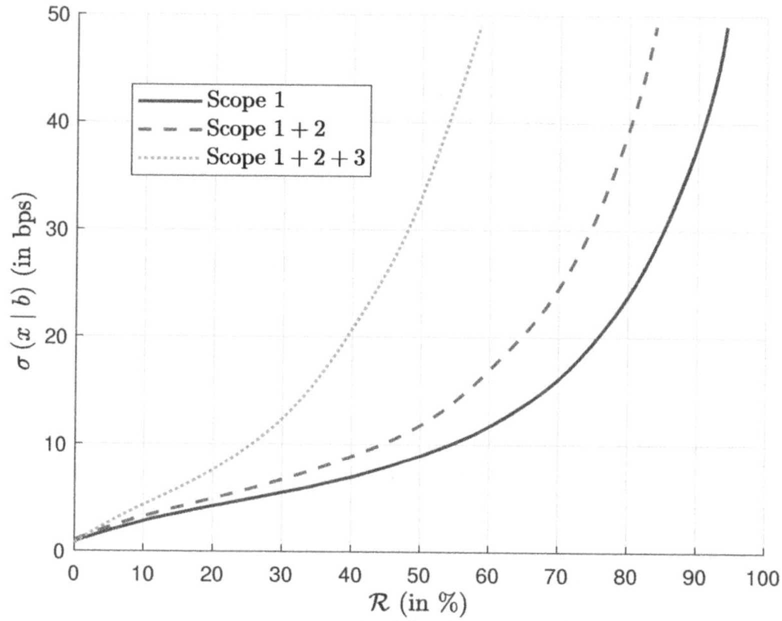

Source: Trucost reporting year 2020 and authors' calculations.

Figure 3.6. *Impact of the carbon scope on the tracking error volatility (S&P 500 index, October 2021)*

19 The overlap measure between two portfolios x and y is defined as:

$$\text{overlap}(x, y) = 1 - \frac{1}{2} \sum |x_i - y_i| \qquad [3.26]$$

It is equal to 100% if the two portfolios are the same and 0% if the two portfolios have no common trading positions.

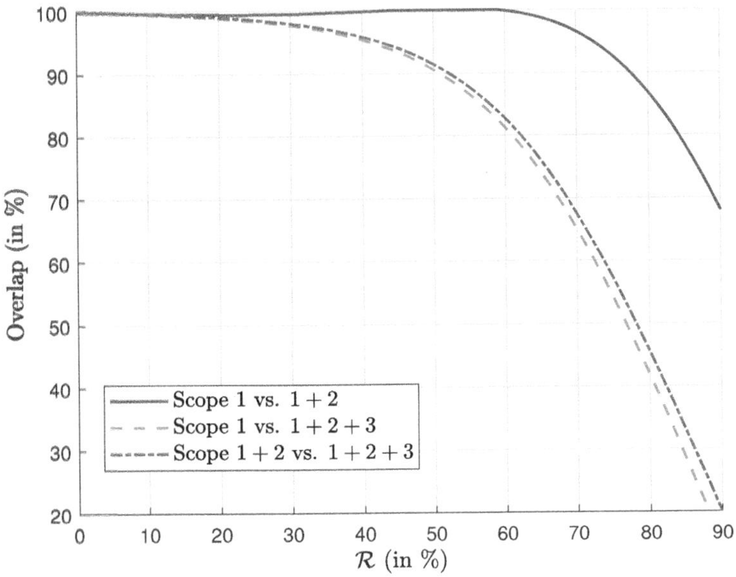

Source: Trucost reporting year 2020 and authors' calculations.

Figure 3.7. *Overlap of optimal decarbonization portfolios (S&P 500 index, October 2021)*

3.3.2.2. *Using carbon emissions*

Let us consider an investor who has a nominal exposure W_i to the stock i. Their carbon emission contribution is equal to:

$$\mathcal{CEC}_i(W_i) = \frac{W_i \mathcal{FP}_i}{\mathcal{MC}_i} \mathcal{CE}_i \qquad [3.27]$$

where $\mathcal{CE}_i$ is the carbon emissions of the company i, $\mathcal{FP}_i$ is the float percentage associated with the stock i and $\mathcal{MC}_i$ is the free-float market capitalization. If we consider a portfolio $x = (x_1, \ldots, x_n)$, we cannot directly compute its carbon emission contribution by averaging the carbon emissions:

$$\mathcal{CE}(x) \neq \sum_{i=1}^{n} x_i \mathcal{CE}_i \qquad [3.28]$$

Indeed, this formula does not depend on the nominal value W of the portfolio. Therefore, the correct formula is:

$$\begin{aligned} \mathcal{CE}(x; W) &= \mathcal{CEC}(x; W) \\ &= \sum_{i=1}^{n} \frac{W_i \mathcal{FP}_i}{\mathcal{MC}_i} \mathcal{CE}_i \end{aligned} \qquad [3.29]$$

where $W_i = x_i W$. We note that the carbon emission $\mathcal{CE}(x)$ of the portfolio is in fact the carbon emission contribution $\mathcal{CEC}(x)$.

As explained on page 56, we can compute the carbon intensity of this investment as:

$$\mathcal{CI}^{\text{exact}}(x) = \frac{\mathcal{CE}(x; W)}{Y(x; W)} \tag{3.30}$$

where:

$$Y(x; W) = \sum_{i=1}^{n} \frac{W_i \mathcal{FP}_i}{\mathcal{MC}_i} Y_i \tag{3.31}$$

Finally, we obtain[20]:

$$\mathcal{CI}^{\text{exact}}(x) = \frac{\sum_{i=1}^{n} \omega_i \mathcal{CE}_i}{\sum_{i=1}^{n} \omega_i Y_i} \tag{3.32}$$

where:

$$\omega_i = \frac{x_i \mathcal{FP}_i}{\mathcal{MC}_i} \tag{3.33}$$

Formula [3.32] differs from the direct computation:

$$\mathcal{CI}^{\text{direct}}(x) = \sum_{i=1}^{n} x_i \mathcal{CI}_i \tag{3.34}$$

REMARK 3.5.– *If we assume that $\mathcal{FP}_i = 100\%$, ω_i is constant for a market-cap portfolio because $x_i \propto \mathcal{MC}_i^{-1}$. Therefore, the exact value of the carbon intensity reduces to:*

$$\mathcal{CI}^{\text{exact}}(x) = \frac{\sum_{i=1}^{n} \mathcal{CE}_i}{\sum_{i=1}^{n} Y_i} \tag{3.35}$$

Let us illustrate the discrepancy between the exact and direct computation. In Table 3.5, we report the $\mathcal{CE}(x; W)$ of three S&P 500 index portfolios. The first one corresponds to the total value of the S&P 500 index as if the investor buys the 500 companies that make up the index. Therefore, the carbon emissions of the S&P 500 index are, respectively, equal to 1.66 GtCO$_2$e, 2.01 GtCO$_2$e and 3.75 GtCO$_2$e for the three scopes. If we consider an investment of \$1 bn, these figures become 10.5 ktCO$_2$e, 48.9 ktCO$_2$e and 91.4 ktCO$_2$e. For an S&P 500 index portfolio whose value is \$5 bn,

20 We note that the variable W vanishes.

we multiply these numbers by a factor of five. If we compute the carbon intensity, we note that it does not depend on the portfolio notional even if we use the exact approach. We also remark that the direct computation of the carbon intensity is less than the exact computation. For instance, we obtain 161.7 versus 129.8 $tCO_2e/\$$ mn for scope $1+2$. As explained before, there is no ordering between these two measures. For instance, in Figure 3.8, we simulate portfolios of size n by randomly choosing n stocks from the S&P 500 index and drawing their weights from the uniform distribution. For each simulated portfolio x, we report the exact and direct values of $\mathcal{CI}(x)$. We verify that the direct value can be less than or greater than the exact value. However, when we fix n to 500, the direct value is systematically lower than the exact value.

Scope	$\mathcal{CE}(x;W)$			$\mathcal{CI}(x)$	
	S&P 500	\$1 bn	\$5 bn	Exact	Direct
1	1.66×10^9	40.5×10^3	202.5×10^3	133.8	99.2
$1+2$	2.01×10^9	48.9×10^3	244.7×10^3	161.7	129.8
$1+2+3$	3.75×10^9	91.4×10^3	457.2×10^3	302.1	245.2

Source: Trucost reporting year 2020 and Authors' calculations.

Table 3.5. *Carbon emission (in tCO_2e) and intensity ($tCO_2e/\$$ mn) of S&P 500 index portfolios*

For the last year, we have observed a trend from some investors preferring to decarbonize their portfolios using carbon emissions and not carbon intensities. This is equivalent to defining the carbon constraint as $\mathcal{CE}(x;W)\leq\mathcal{CE}^{+}$ (or $\mathcal{CE}(x;W)\leq(1-\mathcal{R})\,\mathcal{CE}(b;W)$). We have:

$$\begin{aligned}\mathcal{CE}(x;W)\leq\mathcal{CE}^{+} &\Leftrightarrow \sum_{i=1}^{n}\frac{W_i\mathcal{FP}_i}{\mathcal{MC}_i}\mathcal{CE}_i\leq\mathcal{CE}^{+}\\ &=\sum_{i=1}^{n}x_i\mathcal{CI}_i^{\mathcal{MV}}\leq\frac{\mathcal{CE}^{+}}{W}\end{aligned}\qquad[3.36]$$

where $\mathcal{CI}_i^{\mathcal{MV}}$ is a carbon intensity measure normalized by the market value $\mathcal{MV}_i$ of the issuer i:

$$\mathcal{CI}_i^{\mathcal{MV}}=\frac{\mathcal{FP}_i}{\mathcal{MC}_i}\mathcal{CE}_i=\frac{\mathcal{CE}_i}{\mathcal{MV}_i}\qquad[3.37]$$

and $\mathcal{MV}_i=\mathcal{MC}_i/\mathcal{FP}_i$. Therefore, problem [3.21] becomes:

$$x^{\star}(\mathcal{R})=\arg\min\frac{1}{2}(x-b)^{\top}\Sigma(x-b)\qquad[3.38]$$

$$\text{s.t.}\begin{cases}\mathbf{1}_n^{\top}x=1\\ x\geq\mathbf{0}_n\\ \sum_{i=1}^{n}x_i\mathcal{CI}_i^{\mathcal{MV}}\leq(1-\mathcal{R})\sum_{i=1}^{n}b_i\mathcal{CI}_i^{\mathcal{MV}}\end{cases}$$

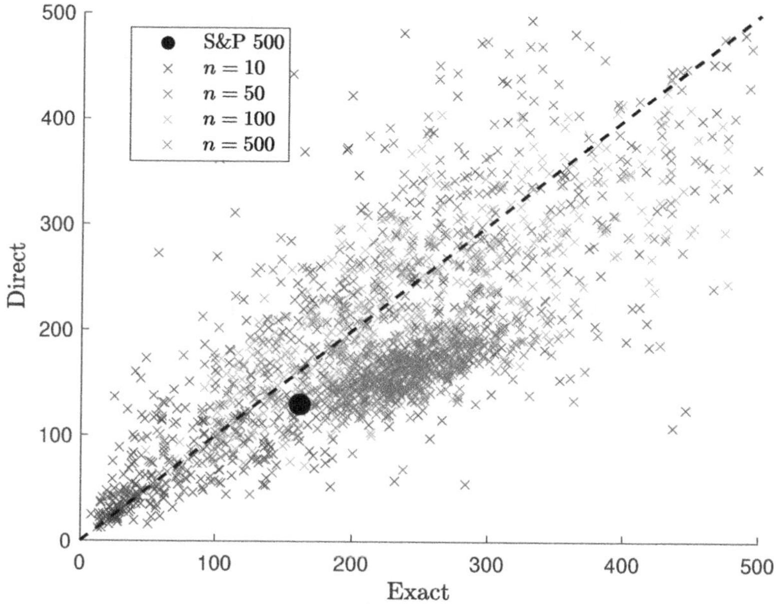

Source: Trucost reporting year 2020 and authors' calculations.

Figure 3.8. *Exact versus direct computation of scope $1 + 2$ carbon intensity (S&P 500 index, October 2021)*

Again, we obtain a quadratic program, which is appealing because it is very efficient to solve from a computational point of view (Perrin and Roncalli 2020).

Figure 3.9 shows the results of this portfolio decarbonization program. The first top-left panel compares the tracking errors when targeting carbon emission or carbon intensity reduction. We obtain higher tracking errors in the last case. Indeed, since absolute emissions are concentrated on few issuers, it is possible to reduce carbon emissions by reducing the allocation on these issuers, whereas the distribution of carbon intensities is less skewed, implying that the constraint potentially impacts the weights of more issuers. The overlap of the two optimized portfolios is given in the second top-right panel. We note that we obtain similar holdings, but the overlap decreases dramatically when the reduction is greater than 80%. In the bottom-left panel, we compute the emission and intensity reduction levels $\hat{\mathcal{R}}(\mathcal{CE}) = 1 - \mathcal{CE}(x^\star)/\mathcal{CE}(b)$ and $\hat{\mathcal{R}}(\mathcal{CI}) = 1 - \mathcal{CI}(x^\star)/\mathcal{CI}(b)$ using the direct and exact formulas. Finally, the fourth bottom-right panel shows that the carbon intensity is underestimated when we consider the direct formula.

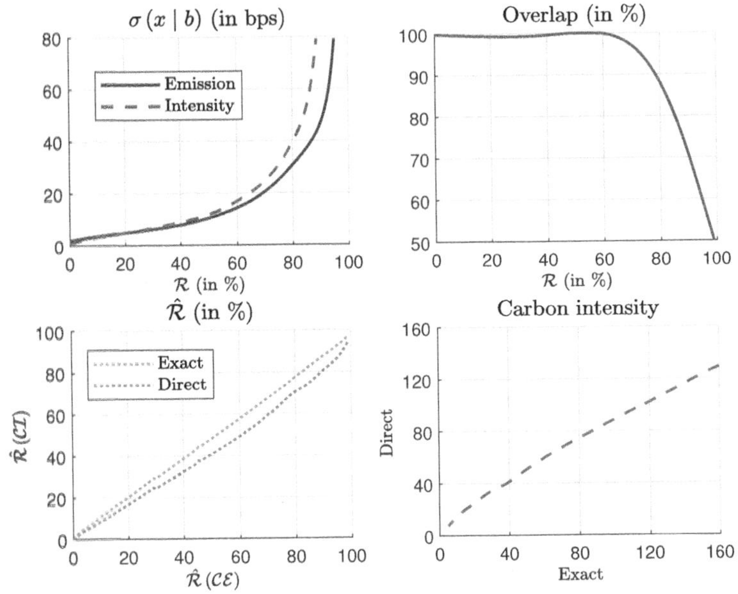

Source: Trucost reporting year 2020 and authors' calculations.

Figure 3.9. *Portfolio decarbonization with carbon emissions (S&P 500 index, October 2021, scope* $1+2$*)*

3.3.3. *Portfolio alignment*

While portfolio decarbonization is a static problem, portfolio alignment implies a dynamic strategy in order to comply with a given climate policy. Therefore, the dynamic problem is trickier since it involves several rebalancing decisions and depends on the future behavior of corporate issuers. Two main climate policies are generally used: the Paris-aligned benchmark (PAB) approach and the net zero emission (NZE) scenario for 2050.

3.3.3.1. *Paris-aligned benchmarks*

The goal of the Paris agreement on climate change is to limit global warming to well below 2°C. In this context, the EU Technical Expert Group on sustainable finance (hereafter TEG) has proposed to create two climate benchmark labels[21]: climate

21 According to the TEG (2019a), "*a climate benchmark is defined as an investment benchmark that incorporates – next to financial investment objectives – specific objectives related to greenhouse gas (GHG) emission reductions and the transition to a low-carbon economy through the selection and weighting of underlying benchmark constituents*".

transition benchmark (CTB) and Paris-aligned benchmark (PAB). These labels are structured along the following common principles:

1) a year-on-year self-decarbonization of 7% on average per annum, based on scopes 1, 2 and 3 emissions;

2) a minimum carbon intensity reduction $\mathcal{R}^-$ compared to the investable universe;

3) a minimum exposure to sectors highly exposed to climate change.

For the CTB label, the minimum reduction $\mathcal{R}^-$ is set to 30%, whereas it is equal to 50% for the PAB label. Other constraints are also imposed such as issuer exclusions (controversial weapons and societal norm violators), a minimum green share revenue or some activity exclusions. Nevertheless, in what follows, we focus on the three principles that correspond to the main constraints.

3.3.3.1.1. Decarbonization pathway

Let t_0 be the base date of the climate benchmark. The initial carbon intensity reduction must be equal to $\mathcal{R}^-$ (30% for CTB and 50% for PAB). Then, the carbon intensity must decrease by 7% every year. The minimum reduction $\mathcal{R}(t_0, t)$ of the carbon intensity between the current date t and the base date t_0 is then equal to:

$$\mathcal{R}(t_0, t) = 1 - (1 - 7\%)^{t-t_0} \cdot \left(1 - \mathcal{R}^-\right) \quad [3.39]$$

Figure 3.10 shows this decarbonization pathway for CTB and PAB indices when the base year is 2021. We observe that the difference between the two trajectories mainly concerns the start of the period, because the two reductions converge when t is sufficiently large ($t > 2040$).

At date t, the CTB and PAB labels impose the following inequality constraint for the portfolio $x(t)$:

$$\mathcal{CI}(x(t)) \leq (1 - \mathcal{R}(t_0, t)) \cdot \mathcal{CI}(b(t_0)) \quad [3.40]$$

We note the importance of the base year t_0 since it defines the reference level of the carbon intensity. Indeed, the reference level is equal to $\mathcal{CI}(b(t_0))$ and not $\mathcal{CI}(b(t))$. If we collectively and significantly reduce the carbon emissions and the corporations make an effort to fight climate change, we can assume that $\mathcal{CI}(b(t)) \leq \mathcal{CI}(b(t_0))$, meaning that the carbon intensity of the traditional benchmark will be reduced. Let T be a reference date for the climate scenario (T may be equal to 2030 or 2050), we must observe that $\mathcal{CI}(b(T)) \ll \mathcal{CI}(b(t_0))$ and even $\mathcal{CI}(x(T)) \approx \mathcal{CI}(b(T))$. Otherwise, this implies that we will be unsuccessful in fighting climate change, reducing GHG emissions and respecting the Paris agreement.

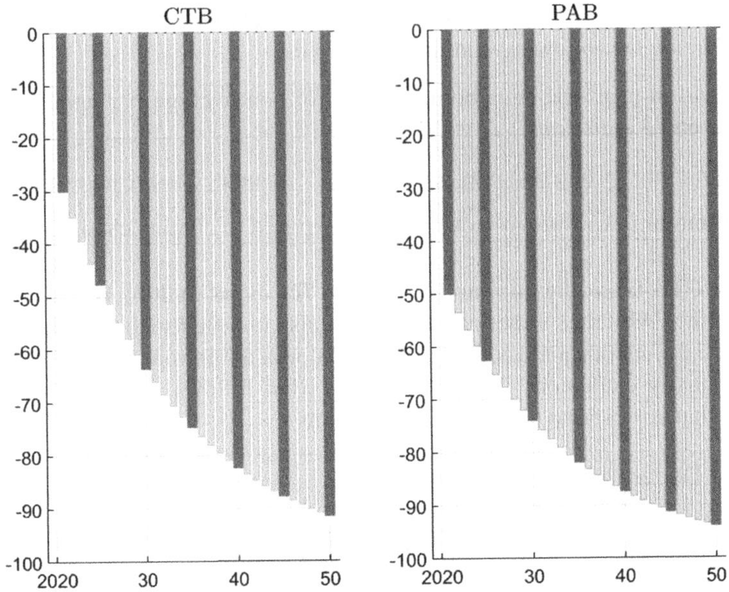

Source: authors' calculations.

Figure 3.10. *Decarbonization pathway of CTB and PAB labels*

3.3.3.1.2. Climate impact sector (CIS)

The CTB and PAB labels require that the exposure to sectors highly exposed to climate change is at least equal to the exposure in the investment universe. TEG (2019a) distinguishes two types of sectors: (1) high climate impact sectors (HCIS or $\mathcal{CIS}_{\mathcal{H}igh}$) and (2) low climate impact sectors (LCIS or $\mathcal{CIS}_{\mathcal{L}ow}$). The first category is made up of sectors that are key to the low-carbon transition. They correspond to the following NACE classes[22]: A. Agriculture, Forestry and Fishing; B. Mining and Quarrying; C. Manufacturing; D. Electricity, Gas, Steam and Air Conditioning Supply; E. Water Supply; Sewerage, Waste Management and Remediation Activities; F. Construction; G. Wholesale and Retail Trade; Repair of Motor Vehicles and Motorcycles; H. Transportation and Storage; L. Real Estate Activities. Let $\mathcal{CIS}_{\mathcal{H}igh}(x) = \sum_{i \in \mathcal{CIS}_{\mathcal{H}igh}} x_i$ be the HCIS weight of portfolio x. At each rebalancing date t, we must verify that:

$$\mathcal{CIS}_{\mathcal{H}igh}(x(t)) \geq \varphi_{\mathcal{CIS}} \cdot \mathcal{CIS}_{\mathcal{H}igh}(b(t)) \qquad [3.41]$$

where $\varphi_{\mathcal{CIS}} = 1$.

22 NACE is the European Union's classification of economic activities. It is made up of 21 classes.

TEG (2019b, Appendix B, pages 26–170) has published a mapping between the NACE classes and several sector classification structures: BICS (Bloomberg), GICS (MSCI and S&P), ICB (FTSE) and TRBC (Refinitiv). In the case of the GICS, there are 129 sub-industries out of a total of 185 that are classified as high climate impact sectors. This represents 69.73% of the sub-industries. This classification has been criticized because this figure is very high[23]. This would mean that almost all activities are critical for building a low-carbon economy. Therefore, only two sectors are classified in low climate impact sectors (Communication Services and Financials), but more than half of the Health Care and Information Technology sub-industries are viewed as high climate impact sectors. The original idea of the CIS constraint was to continue financing the sectors that are essential for reaching a low-carbon economy (e.g. energy and utilities) and at the same time promoting investments in green issuers instead of brown issuers in these sectors. Nevertheless, the constraint [3.41] is not very restrictive with this broad HCIS measure. Moreover, this constraint encourages substitutions between sectors or industries and not substitutions between issuers within a same sector. Therefore, the trade-off is not necessarily between green electricity and brown electricity, but, for example, between electricity generation and health care equipment.

Besides the broad HCIS measure, we propose a narrow measure, which is more restrictive and focuses on a small number of sub-industries. The mapping between the NACE classes and the GICS sectors is given in Table 3.10. We note that the narrow HCIS measure concerns four main sectors: energy, industrials, utilities and real estate. This scope is complemented by one industry group (Transportation) and four industries (Food Products, Metals and Mining, Construction Materials, Food and Staples Retailing). To compare the narrow and broad measures, we report in Table 3.6 the weights and the carbon intensity (scope $1 + 2 + 3$) of high climate impact sectors when the investment universe is the S&P 500 index. Since the carbon intensity of the S&P 500 is equal to 245 tCO_2e/\$ mn, it is equal to 380 tCO_2e/\$ mn for the sectors of the broad HCIS classification. We verify that the broad HCIS universe has a higher carbon intensity than the reference universe. Nevertheless, the broad HCIS universe represents 54.59% of the S&P 500 index, which is a very high figure. This means that more than half of the S&P 500 index has a high climate impact. If we consider the narrow HCIS universe, the weighted average carbon intensity is equal to 681 tCO_2e/\$ mn, which is 80% larger than the carbon intensity of the broad HCIS universe. Moreover, the narrow HCIS universe represents less than 20% of the S&P 500 index. These figures are more consistent than those obtained for the broad measures. Indeed, it is difficult to justify that 69% of the Health Care sector, 48% of the Information Technology, but 0% of the Financial sector are classified as high climate impact sectors when we consider the broad HCIS measure.

23 Certainly too high to be credible.

Sector	S&P 500		Narrow HCIS		Broad HCIS	
	b_s	$\mathcal{CI}_s$	b_s	$\mathcal{CI}_s$	b_s	$\mathcal{CI}_s$
Communication services	10.89%	80				
Consumer discretionary	13.57%	190			10.22%	185
Consumer staples	6.10%	355	2.73%	348	6.10%	355
Energy	2.81%	790	2.81%	790	2.81%	790
Financials	11.13%	67				
Health care	12.74%	126			8.56%	152
Industrials	7.97%	330	7.97%	330	6.32%	368
Information technology	27.50%	99			13.30%	139
Materials	2.45%	966	0.44%	850	2.45%	966
Real estate	2.55%	198	2.55%	198	2.55%	198
Utilities	2.30%	2 669	2.30%	2 669	2.30%	2 669
Total	100.00%	245	18.79%	681	54.59%	380

Table 3.6. *Weights and carbon intensity of high climate impact sectors*

3.3.3.1.3. One-period optimization problem

The optimization program is an extension of problem [3.21] where we introduce the two inequality constraints [3.40] and [3.41]:

$$x^{\star}(t) = \arg\min_{x(t)} \frac{1}{2}\sigma^2(x(t) \mid b(t)) + \lambda\tau(x(t) \mid x^{\star}(t-1)) \qquad [3.42]$$

$$\text{s.t.} \begin{cases} \mathbf{1}_n^\top x(t) = 1 \\ x(t) \geq \mathbf{0}_n \\ \mathcal{CI}(x(t)) \leq (1 - \mathcal{R}(t_0, t)) \cdot \mathcal{CI}(b(t_0)) \\ \mathcal{CIS}_{\mathcal{H}igh}(x(t)) \geq \varphi_{\mathcal{CIS}} \cdot \mathcal{CIS}_{\mathcal{H}igh}(b(t)) \\ \left| \sum_{i \in \mathcal{S}ector_j} x_i(t) - \sum_{i \in \mathcal{S}ector_j} b_i(t) \right| \leq \delta_j \end{cases}$$

where $\lambda \geq 0$. The objective function depends on the tracking error risk:

$$\sigma(x(t) \mid b(t)) = \sqrt{(x(t) - b(t))^\top \Sigma(t)(x(t) - b(t))} \qquad [3.43]$$

Since we face a dynamic process, it is useful to control the one-way turnover of the portfolio between $t-1$ and t defined as[24]:

$$\tau(x(t) \mid x^{\star}(t-1)) = \frac{1}{2}\|x(t) - x^{\star}(t-1)\|_1 = \frac{1}{2}\sum_{i=1}^{n}|x_i(t) - x_i^{\star}(t-1)| \qquad [3.44]$$

24 It is equal to 0 if the portfolio does not change between $t-1$ and t and 100% if there is no common exposure between $t-1$ and t.

We can impose a maximum turnover constraint $\tau(x(t) \mid x^\star(t-1)) \leq \tau^+$. Nevertheless, the convergence towards an optimal solution is not guaranteed. Therefore, we prefer to introduce the turnover in the objective function and control the turnover aversion with the coefficient λ. For instance, we minimize the tracking error risk if λ is set to 0 and the turnover risk if λ is set to $+\infty$. Otherwise, λ measures the trade-off between tracking error and turnover risks. We can also impose some limits concerning sector deviation. This is the purpose of the fifth constraint.

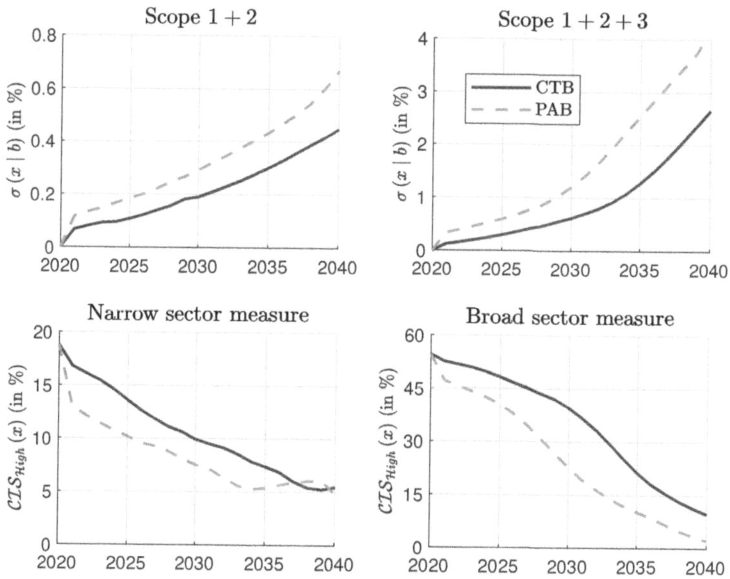

Source: Trucost reporting year 2020 and authors' calculations.

Figure 3.11. *The impact of scope 3 on CTB and PAB labels*

We note that the computation of $x^\star(t)$ at time t requires us to know the covariance matrix $\Sigma(t)$, the carbon intensities $\mathcal{CI}_i(t)$, the investable universe $b(t)$ and the previously optimized portfolio $x^\star(t-1)$. At the current year t_1, these observations are only available for the dates between t_0 and t_1. Therefore, it is impossible to compute $x^\star(t_1+1)$, $x^\star(t_1+2)$, etc. Nevertheless, we can perform an exercise by assuming that the world does not change. In this case, the covariance matrix of asset returns, the carbon intensities of issuers and the investable universe remain constant, implying that $\Sigma(t) = \Sigma(t_0)$, $\mathcal{CI}_i(t) = \mathcal{CI}_i(t_0)$ and $b(t) = b(t_0)$. The results of the sequential optimization process starting from the base year $t_0 = 2021$ and ending at the target date $T = 2040$ are given in Figures 3.11 and 3.12 and Table 3.7 when the turnover aversion coefficient λ is set to zero. We can make a number of observations. First, we have optimized the portfolios by only considering the decarbonization pathway of CTB and PAB labels in Figure 3.11. The difference between the top-left and top-right panels is the choice of the scope. The top-left panel presents the tracking error risk $\sigma(x(t) \mid b(t))$ of optimized portfolios when we consider scope $1+2$. When we include scope 3, we obtain the top-right panel. We observe that the magnitude is very different. Indeed, the tracking error risk is less than 80 bps for $t \leq 2040$ when we consider scope $1+2$, whereas it can reach 4% for the PAB label when we include scope 3. In the following, we only consider this last case since CTB and PAB labels require all CO_2 emissions to be covered, and in particular upstream and downstream emissions. In this context, we have computed the HCIS weight $\mathcal{CIS}_{\mathcal{H}igh}(x^{\star}(t))$ of scope 1+2+3 optimized portfolios. We verify that the exposure to high climate impact sectors decreases over time when we impose a higher carbon emission reduction. For example, $\mathcal{CIS}_{\mathcal{H}igh}(x^{\star}(2040))$ is equal to 5.49% (narrow) and 9.73% (broad) for the CTB label[25], whereas it is equal to 18.79% (narrow) and 54.59% (broad) for the current investable universe (see Table 3.6).

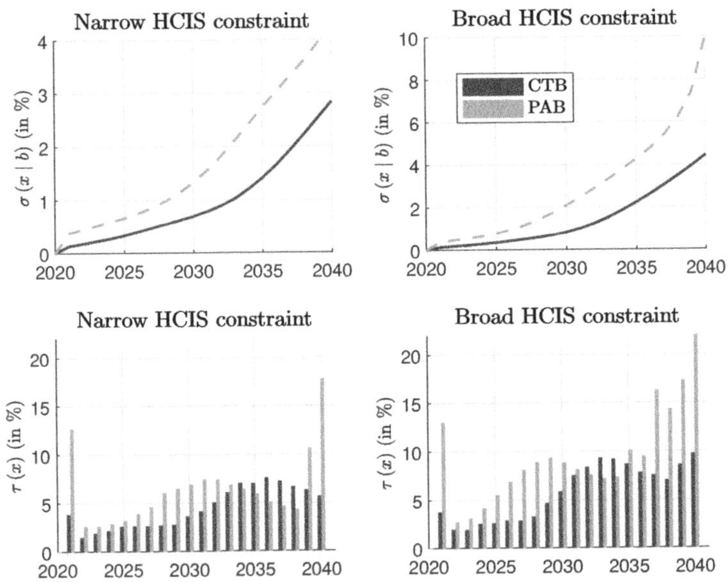

Source: Trucost reporting year 2020 and authors' calculations.

Figure 3.12. *The impact of the HCIS constraint on CTB and PAB labels*

Therefore, we introduce the HCIS constraint and obtain the results given in Figure 3.12. We observe that the constraint using the broad HCIS universe is tighter than the constraint using the narrow HCIS universe. Indeed, the tracking error risks

25 For the PAB label, $\mathcal{CIS}_{\mathcal{H}igh}(x^{\star}(2040))$ is equal to 4.86% (narrow) and 2.23% (broad).

are similar with or without the narrow HCIS universe[26]. This is not the case when we compare the tracking error risks obtained with or without the broad HCIS universe[27]. For the CTB label, $\sigma(x^\star(2040) \mid b(2040))$ is, respectively, equal to 2.66% when we only consider the carbon emission reduction, 2.83% when we add the narrow HCIS constraint and 4.43% when we consider the broad HCIS constraint (see Table 3.7). The figures become 4.19%, 4.43% and 9.97% for the PAB label. Therefore, the introduction of the HCIS constraint is not neutral. This is particularly true for the broad universe in the medium term ($t > 2025$) and the narrow measure in the long term ($t > 2040$). In Figure 3.12, we have also reported the one-way turnover $\tau(x^\star(t) \mid x^\star(t-1))$. We observe a high turnover at the beginning of the period[28] and at the end of the period. In the last case, the carbon reductions are very high, and we obtain concentrated portfolios that are far from diversified. In order to illustrate this lack of diversification, we compute the effective number of bets which is the inverse of the Herfindahl index $\mathcal{H}(x) = \sum_{i=1}^{n} x_i^2$. It is currently equal to 70.56 for the S&P 500 index, meaning the S&P 500 index portfolio is equivalent to an equally weighted portfolio of 70.56 stocks (see Table 3.8). Until 2030, the effective number of bets of optimized portfolios is relatively close to this of the S&P 500 index and greater than 65. Nevertheless, we observe a drop in terms of diversification in 2040. For instance, it is equal to 5.48 for the PAB label with the broad HCIS constraint.

Year	CTB			PAB		
	Scope 3	+ Narrow CIS	+ Broad CIS	Scope 3	+ Narrow CIS	+ Broad CIS
2021	0.12%	0.13%	0.13%	0.33%	0.37%	0.39%
2022	0.16%	0.18%	0.19%	0.39%	0.44%	0.47%
2023	0.20%	0.23%	0.23%	0.46%	0.51%	0.56%
2024	0.24%	0.27%	0.28%	0.52%	0.59%	0.65%
2025	0.29%	0.33%	0.35%	0.59%	0.66%	0.78%
2030	0.62%	0.69%	0.83%	1.20%	1.33%	2.09%
2035	1.28%	1.40%	2.23%	2.55%	2.72%	4.25%
2040	2.66%	2.83%	4.43%	4.19%	4.43%	9.97%

Table 3.7. *Tracking error risk of CTB and PAB labels*

REMARK 3.6.– *In our implementation, the CTB and PAB labels only differ because of the minimum reduction $\mathcal{R}^-$. In this case, we can show that the CTB approach is lagging behind the PAB approach by 4.64 years. This explains the difference between the two labels.*

26 We have to compare the top-right panel in Figure 3.11 and the top-left panel in Figure 3.12.

27 We have to compare the top-right panel in Figure 3.11 and the top-right panel in Figure 3.12.

28 Because of the implementation of the minimum reduction $\mathcal{R}^-$.

Year	CTB			PAB		
	Scope 3	+ Narrow CIS	+ Broad CIS	Scope 3	+ Narrow CIS	+ Broad CIS
2020	70.56	70.56	70.56	70.56	70.56	70.56
2021	69.95	70.29	70.15	68.37	69.30	68.78
2022	69.77	70.24	70.53	68.06	68.93	68.30
2023	69.48	70.22	69.88	67.59	68.43	67.95
2024	69.07	69.73	69.68	67.02	68.12	67.11
2025	68.66	69.52	69.08	66.58	67.46	66.75
2030	66.26	67.24	66.42	66.64	68.82	66.85
2035	67.36	69.61	66.15	76.35	76.42	44.72
2040	76.26	75.67	41.45	49.97	42.61	5.48

Table 3.8. *Effective number of bets*

REMARK 3.7.– *In practice, the ex post tracking error risk will be lower than those computed in this section, in particular when* t *is greater than* 2025. *The reason is that we assume that the future investable universe* $b(t)$ *is equal to the current investable universe* $b(t_0)$. *In fact, we expect that the market-cap benchmark will follow a decarbonization pathway at a level, which will certainly be lower than the CTB and PAB labels (e.g.* $2 - 4\%$ *instead of the* 7% *objective).*

REMARK 3.8.– *Our empirical results are very different from those obtained by Bolton et al. (2022). Indeed, these authors found that the tracking error of optimized portfolios is less than* 2% *by 2050. In fact, their calculations use "scope* $1 + 2$, *and only scope 3 upstream first-tier data, which only covers the direct supply chain", whereas our computation is based on the total sum of carbon scopes and includes full indirect upstream emissions. On the other hand, "first-tier indirect emissions are defined as GHG Protocol scope 2 emissions, plus the company's first-tier upstream supply chain – their direct suppliers". Therefore, they omitted a significant part of the upstream emissions, in particular as we get close to the final product. This makes a big difference. By only considering scope 3 direct upstream emissions, the tracking error risk is approximately equal to the figure obtained with scope* $1 + 2$. *In this case, we confirm that the tracking error is less than* 2% *by 2050 for the CTB label. Nevertheless, the EU regulation considers the entire scope 3, and not only the direct upstream. Moreover, we also observe that adding the broad HCIS constraint is a supplementary source of tracking error risk. These results show that the cost for investors may be higher when we consider both scope 3 and the HCIS constraint. Nevertheless, as argued before, we can assume that the investable universe in 10, 20 or 30 years' time will be very different and more aligned than the current one.*

So far, we have illustrated the sequential optimization process when the turnover aversion risk λ is set to 0. In fact, the optimization process is static since we do not need to know the optimized portfolio $x^\star(t-1)$ to compute $x^\star(t)$. If $\lambda > 0$, the computation of $x^\star(t)$ depends on the past values $x^\star(t-1), x^\star(t-2), \ldots, x^\star(t_0)$ and the optimization process becomes sequential. In this case, it is a special case of the multi-period optimization problem.

3.3.3.2. *Net zero objective*

In section 3.3.3.1, we have performed a portfolio alignment by considering a global decarbonization path for the portfolio but without taking into account the decarbonization path of the issuers. In this section, we go further by using net zero metrics. The simplest solution is to consider the carbon intensity trend for each issuer. Nevertheless, we can use other more sophisticated measures (Le Guenedal et al. 2022).

3.3.3.2.1. A simple net zero metric

Using the historical Trucost data between 2013 and 2019, we project the trend of total absolute emissions (scope $1+2+3$) between 2020 and 2050 – for some assets, we have the historical data for the year 2020 and the trend is then projected between 2021 and 2050. Among the 500 constituents of the S&P 500 index in October 2021, 450 issuers have full reporting, allowing us to build a consistent trajectory. Filling in upstream/downstream values around reported values allows us to improve the coverage up to 493 issuers which corresponds to 98.6% of the index with a synthetic trend trajectory[29]. The trend is normalized at the initial year t_0 such that we have:

$$\widehat{\mathcal{CI}}_i^{\text{trend}}(t) = m_i^{\text{trend}}(t) \cdot \mathcal{CI}_i(t_0) \qquad [3.45]$$

where the multiplicative factor $m_i^{\text{trend}}(t)$ is linear:

$$m_i^{\text{trend}}(t) = 1 + \beta_i^{\text{trend}} \cdot (t - t_0) \qquad [3.46]$$

Therefore, we have $m_i^{\text{trend}}(t_0) = 1$ for all the assets. Since we have the following relationship between the multiplicative factor and the reduction:

$$m_i^{\text{trend}}(t) = 1 - \mathcal{R}_i(t_0, t) \qquad [3.47]$$

the reduction $\mathcal{R}_i(t_0, t)$ between t_0 and t of the asset i is then the opposite of the projected trend:

$$\mathcal{R}_i(t_0, t) = -\beta_i^{\text{trend}} \cdot (t - t_0) \qquad [3.48]$$

In Figure 3.13, we have reported the multiplicative factor $m_i^{\text{trend}}(t)$ for the 500 constituents of the S&P 500 index. 70% of trends are upward, meaning that only 30% of the issuers have begun to substantially reduce their carbon emissions.

29 It is important to note that the pivot year, from which we shift from historical emissions to projected emissions, is an important parameter that plays a key role, in particular in the context of a dynamic use. This problem is illustrated in Le Guenedal et al. (2022). Nevertheless, we consider the 2019 pivot year because it corresponds to the last reporting year with a high coverage.

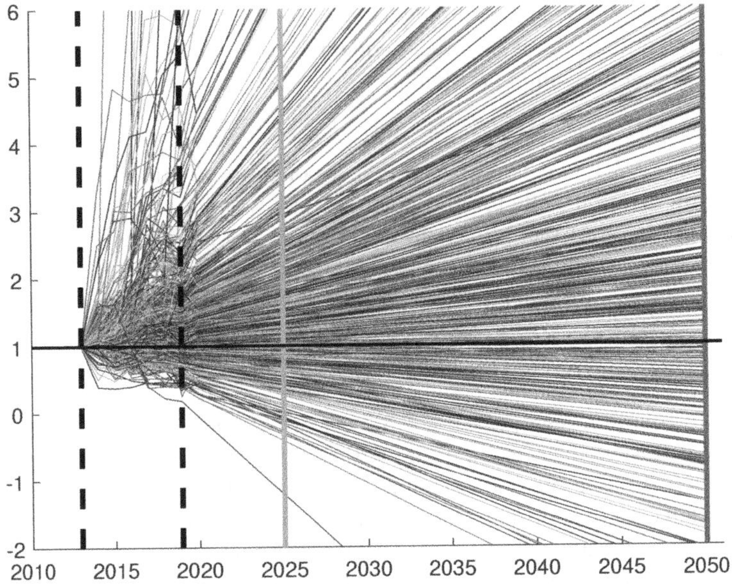

Source: Trucost reporting year 2019/2020 and authors' calculations.

Figure 3.13. *Carbon emission trends $m_i^{\text{trend}}(t)$ of the S&P index constituents*

3.3.3.2.2. NZE optimization problem

The optimization problem is the same as the previous one except that we explicitly introduce the NZE trajectories for the individual carbon intensity trajectories. For instance, we can formulate the NZE problem as follows:

$$x^{\text{NZE}}(t) = \arg\min_x \frac{1}{2}(x-b)^\top \Sigma (x-b) \qquad [3.49]$$

$$\text{s.t.} \begin{cases} \mathbf{1}_n^\top x = 1 \\ x \geq \mathbf{0}_n \\ \mathcal{CI}^{\text{NZE}}(x) \leq (1-\mathcal{R}(t_0,t)) \cdot \mathcal{CI}^{\text{NZE}}(b(t_0)) \end{cases}$$

where $\mathcal{CI}^{\text{NZE}}(x)$ uses the projected trends:

$$\mathcal{CI}^{\text{NZE}}(x) = \sum_{i=1}^{n} x_i \cdot \widehat{\mathcal{CI}}_i^{\text{trend}}(t) \qquad [3.50]$$

We can compare the solution $x^{\text{NZE}}(t)$ with the one $x^{\text{DCN}}(t)$ obtained by considering the current carbon intensities $\mathcal{CI}_i(t_0)$ instead of the estimated values $\widehat{\mathcal{CI}}_i^{\text{trend}}(t)$. The latter can be viewed as a pure decarbonization portfolio. In order to measure the discrepancy between the NZE portfolio $x^{\mathrm{NZE}}(t)$ and the decarbonization portfolio $x^{\mathrm{DCN}}(t)$, we compute the active share between the weights of the two portfolios:

$$\mathcal{AS}\left(x^{\mathrm{NZE}}(t), x^{\mathrm{DCN}}(t)\right) = \frac{1}{2}\left\|x^{\mathrm{NZE}}(t) - x^{\mathrm{DCN}}(t)\right\|_1 \qquad [3.51]$$

The results are given in Table 3.9. We observe that the divergence between the NZE portfolio and the decarbonization portfolio increases with the target date and the reduction level $\mathcal{R}(t_0, t)$.

Year	$\mathcal{R}(t_0,t)$							
	10%	20%	30%	40%	50%	60%	70%	80%
2025	1.0%	1.3%	2.0%	3.2%	4.4%	7.8%	18.4%	48.9%
2030	1.1%	1.4%	2.7%	4.8%	9.2%	15.3%	28.5%	58.3%
2035	1.1%	1.8%	3.3%	5.9%	11.0%	17.3%	30.6%	60.0%
2040	1.1%	2.0%	3.6%	6.3%	11.5%	18.0%	31.3%	60.6%
2045	1.2%	2.1%	3.8%	6.5%	11.8%	18.3%	31.6%	60.8%
2050	1.2%	2.1%	3.8%	6.7%	11.9%	18.5%	31.7%	60.9%

Table 3.9. *Active share between the NZE portfolio and the decarbonization portfolio*

REMARK 3.9.– *This divergence will increase in the future when we have more exhaustive data and robust NZE measures. For instance, Le Guenedal et al. (2022) use issuer targets and trend slopes to perform NZE portfolios and show that they are highly sensitive to the ambition of carbon issuers.*

3.4. Conclusion

In this survey dedicated to portfolio construction with climate risk measures, we first present several metrics and then show how climate risk measures can be integrated into traditional portfolio allocation problems. Some years ago, carbon emissions and intensity were the two main climate risk measures. In this case, portfolio optimization reduced to portfolio decarbonization. Since 2018, we have observed major development of new climate risk metrics. For example, one of the main challenges is to define carbon transition pathways. This means defining a carbon reduction scenario at the global level and the carbon trajectory of corporate issuers. Even if this research axis is not mature, it completely changes the practice of portfolio alignment, whose earlier approaches can be viewed as an enhanced method of portfolio decarbonization. Alongside the usefulness of this survey for both academics and professionals, this chapter also shows two important results. First, decarbonization is more difficult when we consider scope 3 carbon emissions. This is a major issue because asset owners and managers are encouraged by regulators and the society to go beyond scope 1 + 2. Second, portfolio alignment requires new metrics that are more difficult to estimate, implying more uncertainties about the portfolio solution. As a result, these two combined factors (scope 3 and portfolio alignment) will have a significant impact on investors in the coming years. Implementing a net zero carbon portfolio alignment with scope 3 emissions and carbon trajectories is therefore a big challenge for investors. In particular, we can predict that carbon investing will have a major effect on asset allocation with respect to traditional investing.

3.5. Appendices

3.5.1. *Appendix 1: Scope 3 emissions*

The scope 3 emissions of Trucost we have used in this analysis are defined as follows: other indirect greenhouse gas emissions, such as from the extraction and production of purchased materials and fuels, transport-related activities in vehicles not owned or controlled by the reporting entity, electricity-related activities (e.g. T&D losses) not covered in scope 2, outsourced activities, waste disposal, etc. (in line with GHG protocol standards). It seems that this definition does not contain the use of the sold product, end of life treatment of the sold product, downstream transportation and distribution, etc.

3.5.2. *Appendix 2: Data*

<table>
<tr><th colspan="2">NACE</th><th colspan="2">GICS</th></tr>
<tr><th>Code</th><th>Sector</th><th>Code</th><th>Sector</th></tr>
<tr><td>A</td><td>Agriculture, forestry and fishing</td><td>302020</td><td>Food products</td></tr>
<tr><td rowspan="2">B</td><td rowspan="2">Mining and quarrying</td><td>10</td><td>Energy</td></tr>
<tr><td>151040</td><td>Metals and mining</td></tr>
<tr><td>C</td><td>Manufacturing</td><td>20</td><td>Industrials</td></tr>
<tr><td>D</td><td>Electricity, gas, steam
and air conditioning supply</td><td rowspan="2">55</td><td rowspan="2">Utilities</td></tr>
<tr><td>E</td><td>Water supply
Sewerage, waste management
and remediation activities</td></tr>
<tr><td>F</td><td>Construction</td><td>151020</td><td>Construction materials</td></tr>
<tr><td>G</td><td>Wholesale and retail trade
Repair of motor vehicles and motorcycles</td><td>301010</td><td>Food and staples retailing</td></tr>
<tr><td>H</td><td>Transportation and storage</td><td>2030</td><td>Transportation</td></tr>
<tr><td>L</td><td>Real estate activities</td><td>60</td><td>Real estate</td></tr>
</table>

Source: authors' calculations.

Table 3.10. *The narrow measure of high climate impact sectors*

3.6. References

Andersson, M., Bolton, P., Samama, F. (2016). Hedging climate risk. *Financial Analysts Journal*, 72(3), 13–32.

Apel, M., Betzer, A., Scherer, B. (2021). Real-time transition risk. *SSRN*, 3911346.

Ardia, D., Bluteau, K., Boudt, K., Inghelbrecht, K. (2021). Climate change concerns and the performance of green versus brown stocks. Working paper, National Bank of Belgium, 395.

Bennani, L., Le Guenedal, T., Lepetit, F., Ly, L., Mortier, V., Roncalli, T., Sekine, T. (2018). How ESG investing has impacted the asset pricing in the equity market. *SSRN*, 3316862.

Bolton, P., Kacperczyk, M.T., Samama, F. (2022). Net-zero carbon portfolio alignment. *SSRN*, 3922686.

Bouchet, V. and Le Guenedal, T. (2020). Credit risk sensitivity to carbon price. *SSRN*, 3574486.

Derwent, R.G. (2020). Global warming potential (GWP) for methane: Monte Carlo analysis of the uncertainties in global tropospheric model predictions. *Atmosphere*, 11(5), 486.

Drei, A., Le Guenedal, T., Lepetit, F., Mortier, V., Roncalli, T., Sekine, T. (2019). ESG investing in recent years: New insights from old challenges. *SSRN*, 3683469.

Engle, R.F., Giglio, S., Kelly, B., Lee, H., Stroebel, J. (2020). Hedging climate change news. *Review of Financial Studies*, 33(3), 1184–1216.

EU Technical Expert Group on Sustainable Finance (2019a). TEG final report on climate benchmarks and benchmarks' ESG disclosures. September 2019.

EU Technical Expert Group on Sustainable Finance (2019b). Handbook of Climate Transition Benchmarks, Paris-aligned Benchmarks and Benchmarks' ESG Disclosures. 20 December 2019.

Financial Stability Board (2020). The implications of climate change for financial stability. Report, FSB, 1–32.

Görgen, M., Jacob, A., Nerlinger, M., Riordan, R., Rohleder, M., Wilkens, M. (2020). Carbon risk. *SSRN*, 2930897.

Harris, J. (2015). The carbon risk factor. *SSRN*, 2666757.

Intergovernmental Panel on Climate Change (2021). Climate change 2021: The physical science basis. 6th assesment report, 1–3949.

International Energy Agency (2021). Net zero by 2050: A roadmap for the global energy sector. Report, July, 1–223.

Le Guenedal, T., Girault, J., Jouanneau, M., Lepetit, F., Sekine, T. (2020). Trajectory monitoring in portfolio management and issuer intentionality scoring. *SSRN*, 3630302.

Le Guenedal, T., Drobinski, P., Tankov, P. (2021). Measuring and pricing cyclone-related physical risk under changing climate. *SSRN*, 3850673.

Le Guenedal, T., Lombard, F., Roncalli, T., Sekine, T. (2022). Net zero carbon metrics. *SSRN*, 4033686.

Pedersen, L.H., Fitzgibbons, S., Pomorski, L. (2021). Responsible investing: The ESG-efficient frontier. *Journal of Financial Economics*, 142(2), 572–597.

Perrin, S. and Roncalli, T. (2020). Machine learning optimization algorithms & portfolio allocation. In *Machine Learning for Asset Management: New Developments and Financial Applications*, Jurczenko, E. (ed.). ISTE, London, and John Wiley & Sons, Hoboken, NJ.

Raynaud, J., Voisin, S., Tankov, P., Hilke, A., Pauthier, A. (2021a). The alignment cookbook: A technical review of methodologies assessing a portfolio's alignment with low-carbon trajectory or temperature goal. Report, Institut Louis Bachelier.

Raynaud, J., Voisin, S., Tankov, P. (2021b). Portfolio alignment to a 2°C trajectory: Science or art? *SSRN*, 3644171.

Roncalli, T. (2013). *Introduction to Risk Parity and Budgeting*. Chapman & Hall/CRC Financial Mathematics Series, Boca Raton, FL.

Roncalli, T., Le Guenedal, T., Lepetit, F., Roncalli, T., Sekine, T. (2020). Measuring and managing carbon risk in investment portfolios. *SSRN*, 3681266.

Roncalli, T., Le Guenedal, T., Lepetit, F., Roncalli, T., Sekine, T. (2021). The market measure of carbon risk and its impact on the minimum variance portfolio. *Journal of Portfolio Management*, 47(9), 54–68.

4

Hedging Climate Risks: A Cross-asset Approach

Climate change risk is now at the forefront for all investors. Recent research has suggested hedging climate risk factors through mimicking portfolio methodologies. In this chapter, we extend the climate risk hedging literature by investigating the benefits of correcting for errors-in-variables bias inherent in standard beta estimates in a cross-asset setup. The empirical application shows that correcting for these errors is important for metrics such as statistical errors or turnover but does not always guarantee a better fit out-of-sample.

4.1. Introduction

Climate change has become one of the major concerns for investors. In this chapter, we investigate the issue of hedging climate change risks from an investment perspective.

A first approach to hedge climate risks consists of taking long positions in "green" assets that are assumed ex ante to benefit from the increase of climate change risks, and short positions in "brown assets" that are assumed to lose from the realization of climate risks. While simple and sensible, this approach remains arbitrary as the choice of asset classes aiming to replicate/hedge the climate risk factors is not motivated by a statistical efficiency criterion.

Chapter written by Emmanuel JURCZENKO and Jérôme TEILETCHE.

For a color version of all figures in this chapter, see www.iste.co.uk/jurczenko/climate.zip.

This book chapter solely reflects the views of the authors, not necessarily those of their respective employers. We remain solely responsible for any mistake or error.

A more general solution to hedge climate risks is to construct climate factor-mimicking portfolios (FMPs henceforth) which are well-diversified tradable portfolios of so-called base assets that isolate the risk exposure of the non-traded climate risk factor to be hedged while keeping all other risks constant[1]. Engle et al. (2020) have been the first to apply the FMP methodology for hedging climate change risk by extracting climate risk factors from textual analysis of news sources. Using characteristic-based equity portfolios, they show the ability of the mimicking portfolio approach of Lamont (2001) in hedging innovations in climate news both in-sample and out-of-sample. In a related work, Faccini et al. (2022) follow a similar textual methodology but introduce further granularity in climate risk measures by introducing four climate textual risk factors related to natural disasters, global warming, international summits and US climate policy, respectively. They show that only the climate-policy textual factor is priced in the US stock market, especially post-2012. Ardia et al. (2022) show that the unexpected change in their Media Climate Change Concerns index explains return differences between "green" and "brown" S&P500 firms – with greenness measured by firms' carbon emissions intensity. Alekseev et al. (2022) introduce a new methodology based on changes in investor holdings that outperform traditional climate hedging methods.

Besides climate risk hedging applications, the academic asset pricing literature has focused on ways to improve the estimation of FMPs when factors are measured imprecisely, and as a result, OLS estimates are biased due to the errors-in-variables (EIV) problem. This feature is well known for macroeconomic variables but is also true for climate risk factors, especially when they are constructed from textual analysis[2]. Giglio and Xiu (2021) have recently proposed to mitigate the EIV bias by an instrumental variable (IV) estimation procedure that is using the first principal components (PCs) of the base asset returns as instruments for the observed risk factors. They show that using the PC-fitted risk factors instead of the original ones helps to resolve the errors-in-variables bias inherent in non-tradable risk factors.

In this chapter, we compare the hedging performance of climate FMPs that rely on standard OLS betas against the hedging performance of FMPs that rely on PC IV beta estimates and a new instrumental variable estimation methodology that we introduce. In particular, we rely on recent developments in machine-learning methodologies and notably supervised learning where the individual components of the PCAs are selected by a LASSO regression before applying PC regression approach to denoise the climate risk targets. This target PCA approach (Bai and Ng 2008) allows us to consider the predictive power of assets for each targeted climate

1 Other approaches have been devised, such as Andersson et al. (2016), but FMP remains the dominant one, so that we focus on this in this chapter.

2 One could argue that physical measures of climate risk are more precisely measured.

factor[3]. Another contribution we bring is to adopt a cross-asset perspective on climate hedge portfolios, while most of the existing research has mainly focused on individual stocks.

In our empirical analysis, we compare the ability of various FMP methodologies to replicate the characteristics of eight climate textual risk factors. The analysis is done both in-sample and out-of-sample, using a set of investible assets that are representative of cross-asset portfolios and can be invested through liquid and cost-efficient vehicles – such as futures/swaps derivatives or ETFs. Overall, the results prove the superiority of regularized versions of the factor-mimicking portfolio approach.

The remainder of this chapter is organized as follows. In section 4.2, we present the general FMP framework, including the discussion on FMP estimation in the presence of measurement errors. In section 4.3, we investigate the empirical abilities of the various FMP methods in hedging climate risk factors. We conclude in section 4.4.

4.2. Factor-mimicking portfolios methodology

Building climate hedging portfolios is usually done through the use of factor-mimicking portfolios (FMPs). In this section, we first review the various FMP methodologies that have been designed in the literature and present them in a unified framework. We then discuss how correction technics can help improve their estimation.

4.2.1. *General FMP approach*

Factor-mimicking portfolios have been the subject of a vast academic literature (Balduzzi and Robotti 2008). In general terms, they consist of forming portfolios of investable assets (so-called "base assets") that replicate ("mimic") the behavior of one or several (potentially non-investable) factors. In asset pricing literature, two main FMP approaches have been proposed: two-pass cross-sectional regressions (CSR) and maximum correlation portfolios (MCP).

Both approaches have in common a reliance, for a set of N assets, on factor models of the type:

$$\boldsymbol{R}_t = \boldsymbol{\mu} + \boldsymbol{B}\boldsymbol{F}_t + \boldsymbol{\varepsilon}_t \quad [4.1]$$

3 In a recent study, Luo et al. (2022) investigate the benefits of regularization methodologies for FMP based on norms of portfolio weights, including an application to Engle et al. (2020)'s climate factors.

where $\boldsymbol{R}_t$ and $\boldsymbol{\mu}$ are the $(N \times 1)$ vectors of base asset realized and expected returns, respectively; $\boldsymbol{F}_t$ is the $(K \times 1)$ vector of innovations ("surprises") in the K factors that we want to replicate; $\boldsymbol{B}$ is the $(N \times K)$ matrix of asset betas with respect to the factors and $\boldsymbol{\varepsilon}_t$ is the $(N \times 1)$ vector of mean-zero disturbances.

In the two-pass CSR approach, pioneered by Fama and MacBeth (1973), the betas of the factor model [4.1] are estimated from time-series regressions of base asset returns onto the factors. In the second step, the FMPs are obtained jointly by regressing cross-sectionally the returns of the base assets onto the estimated betas. The estimated slope coefficients are the returns of the FMPs. While in the original Fama-MacBeth approach, the second-pass cross-sectional regressions are usually performed through ordinary least squares (OLS), meaning that the base asset returns are equally weighted, several authors have also recommended alternative weighting schemes to deal with the heteroskedasticity and cross-correlations in [4.1], such as weighted least squares (WLS; Litzenberger and Ramaswany 1979) or generalized least squares (GLS; Lehman and Modest 1988). In the MCP approach, pioneered by Breeden et al. (1989) and Lamont (2001), the FMPs are determined separately by regressing the base asset returns on each factor – meaning that the individual betas of the factor model [4.1] are estimated for each factor through univariate regressions – or equivalently by projecting each factor on the base asset returns. The FMP weights are given by the coefficients estimated from these univariate regressions.

Following Huberman et al. (1987), the vector of weight of the k-th climate factor-mimicking portfolio $\boldsymbol{w}_k$ is the solution to the variance-minimization program:

$$\text{Min} \frac{1}{2} \boldsymbol{w}_k^T \boldsymbol{\Omega} \, \boldsymbol{w}_k \qquad [4.2]$$

$$s.t. \boldsymbol{B}^T \boldsymbol{w}_k = \boldsymbol{\beta}_k$$

where $\boldsymbol{\Omega}$ is the $(N \times N)$ variance–covariance matrix of the base assets and $\boldsymbol{\beta}_k$ is a $(K \times 1)$ vector of target climate betas. By convention, we assume that each FMP has a unit beta with respect to the climate risk factor it is tracking and pre-specified betas with respect to the other climate risk factors, meaning that the vector $\boldsymbol{\beta}_k$ has 1 in k-th entry and β_{kl} for $l \neq k$. As there are generally more investable assets than climate risk factors $(N \gg K)$, there is an infinite number of portfolios that can satisfy the target beta constraints. Among all those portfolios, program [4.2] states that the FMP portfolio must be the one that maximizes the explanatory power of the factor model [4.1]. Indeed, constraining the betas of the portfolio and minimizing its variance simultaneously is equivalent to minimizing the portfolio's idiosyncratic risk[4]. As a result, the k-th FMP is maximally correlated with the k-th climate risk factor. In general terms, this means that FMPs are well-diversified portfolios that target some specific betas with respect to a set of climate risk factors of interest.

Solving program [4.2], the portfolio weights for the full set of the FMPs are given by the column vectors of the $(N \times K)$ weighting matrix (see section 4.5.1):

$$\boldsymbol{W}_K^* = \Omega^{-1}\boldsymbol{B}\left(\boldsymbol{B}^T\Omega^{-1}\boldsymbol{B}\right)^{-1}\boldsymbol{B}_K, \quad [4.3]$$

where $\boldsymbol{B}_K$ is the $(K \times K)$ matrix collecting the target betas for the various FMPs, with $\boldsymbol{B}_K = (\boldsymbol{\beta}_1\boldsymbol{\beta}_2 \dots \boldsymbol{\beta}_K)$. The associated return of the k-th FMP is obtained as $FMP_{k,t} = \boldsymbol{w}_k^{*T}\boldsymbol{R}_t$ [5].

In Table 4.1, we show how it is possible to recover all the traditional factor-mimicking portfolio solutions from the general formula [4.3]. In particular, we can see that all FMP construction methods can be distinguished by three types of specifications:

i) the covariance matrix Ω that can be equal to the full sample matrix (GLS-CSR, Target-beta MCP and Unit-beta MCP), the original sample matrix fully shrunk towards a diagonal matrix (WLS-CSR) or a scalar multiple of the identity matrix (OLS-CSR);

ii) the factor loading matrix $\boldsymbol{B}$ that can be obtained by running either single-factor (univariate) time-series regressions (all MCPs) or multifactor (multivariate) time-series regressions (all CSRs);

iii) the target exposure matrix $\boldsymbol{B}_K$ that can be equal to either an identity matrix (all CSRs) or a general matrix (all MCPs).

4.2.2. *Errors-in-variable estimates*

In theory, expression [4.3] provides the best linear unbiased estimator of the FMP weights. In practice, however, the individual betas are estimated imprecisely due to the presence of measurement errors in the climate risk factors. Since the

---

4 Substituting the residual risk matrix $\boldsymbol{\Omega}_\varepsilon$ for the total risk matrix $\boldsymbol{\Omega}$ in the objective function [4.2] leads to the same FMP solution. See Melas et al. (2010).

5 Note that while the FMP betas are by construction equal to the targeted ones, since $\boldsymbol{B}^T\boldsymbol{W}_K^* = \boldsymbol{B}_K$, their covariances are generally different from the true factor covariances, since $Cov(\boldsymbol{W}_K^{*T}\boldsymbol{R}_t) = \boldsymbol{B}_K{}^T\left(\boldsymbol{B}^T\boldsymbol{\Omega}^{-1}\boldsymbol{B}\right)^{-1}\boldsymbol{B}_K \neq Cov(\boldsymbol{F}_{t+h})$.

estimated betas are used in the construction of the FMP, this introduces an error-in-variables (EIV) bias in the OLS weight estimates.

To correct for this bias, we follow the instrumental variables (IV) estimation approach of Connor and Korajczyk (1991) and Giglio and Xiu (2020) that uses the first principal components (PCs) of the base asset returns as instruments for the risk factors to estimate the betas in [4.1][6]. This approach can be viewed as an application of the two-stage least squares (2SLS) regression method. In the first step, we project the climate risk factors ($\boldsymbol{F}_{t+h}$) against the L first PCs of the asset returns ($\boldsymbol{F}^{PC}_{t+h}$), with $L \leq N$, to obtain the fitted values of the climate risk factors ($\widehat{\boldsymbol{F}}^{PC}_{t+h}$). In the second step, we regress the base asset returns against the values of the climate risk factors estimated from PCA to obtain the EIV-corrected betas that are substituted in [4.3] to estimate the FMP weights (see section 4.5.2).

A sufficient condition for this PCA procedure to remove the measurement error in the climate risk factors is that the first PCs of the base asset returns span the same space as the true risk factors[7]. In practice, however, this might not be necessarily the case if some of the latent factors are weak, i.e. affect only a small subset of base assets or affect all base assets only weakly. To mitigate this issue, we modify the IV estimation approach by replacing the traditional unsupervised PCA by an alternative supervised statistical approach, the target PCA that uses the predictive power of the base assets for each climate risk factor when extracting the statistical factors (Bair et al. 2006; Bai and Ng 2008). Following Bai and Ng (2008), we use a "soft" thresholding LASSO penalty rule to perform for each climate risk factor the subset selection of the base asset returns. We then perform a traditional PCA on the selected base asset returns to extract the first L PCs onto which the climate risk factor of interest is projected. We finally regress the individual base asset returns against the values of the climate risk factors estimated through target PCA.

6 The improvement of FMP is the subject of a growing literature. See, for example, Luo et al. (2022).

7 Formally, consider the classical measurement error model: $\boldsymbol{F}_{t+h} = \boldsymbol{F}^{*}_{t+h} + \boldsymbol{\eta}_{t+h}$, with $\boldsymbol{F}^{*}_{t+h}$ the true climate risk factors and $\boldsymbol{\eta}_{t+h}$ the measurement errors such that $Cov(\boldsymbol{\eta}_{t+h}, \boldsymbol{F}^{*}_{t+h}) = \boldsymbol{0}$. The regression of the observed climate risk factors onto the first PCs of the asset returns yields the coefficients $\widehat{\boldsymbol{b}}_K = Cov(\boldsymbol{F}^{PC}_{t+h}, \boldsymbol{F}_{t+h})^{T}\, Var(\boldsymbol{F}^{PC}_{t+h})^{-1}$. If the first PCs can span the same space as the true climate risk factors, this means that there exists an invertible $(L \times K)$ matrix $\boldsymbol{L}$, such that $\boldsymbol{F}^{PC}_{t+h} = \boldsymbol{L}\boldsymbol{F}^{*}_{t+h}$. We can then write $\widehat{\boldsymbol{b}}_K = Var(\boldsymbol{F}^{*}_{t+h})\boldsymbol{L}^{T}\,[\boldsymbol{L}\, Var(\boldsymbol{F}^{*}_{t+h})\, \boldsymbol{L}^{T}]^{-1} = \boldsymbol{L}^{-1}$. The fitted values of the climate risk factors are then equal to $\widehat{\boldsymbol{F}}^{PC}_{t+h} = \widehat{\boldsymbol{b}}_K \boldsymbol{F}^{PC}_{t+h} = \boldsymbol{L}^{-1}\boldsymbol{F}^{PC}_{t+h} = \boldsymbol{L}^{-1}\boldsymbol{L}\, \boldsymbol{F}^{*}_{t+h} = \boldsymbol{F}^{*}_{t+h}$.

Methodology	Properties	Specifications	Portfolio weight matrix W_K
Two-pass cross-sectional regression approach (CSR)			
GLS-CSR (Lehman and Modest 1988)	Same	$\boldsymbol{\Omega}$: Unconstrained $\boldsymbol{B}$: Multivariate $\boldsymbol{B}_K = \boldsymbol{I}_K$	$\boldsymbol{\Omega}^{-1}\boldsymbol{B}(\boldsymbol{B}^T\boldsymbol{\Omega}^{-1}\boldsymbol{B})^{-1}$
WLS-CSR (Litzenberger and Ramaswany 1979)	Same	$\boldsymbol{\Omega} = Diag(\boldsymbol{\sigma}^2)$: Uncorrelated assets $\boldsymbol{B}$: Multivariate $\boldsymbol{B}_K = \boldsymbol{I}_K$	$Diag(\boldsymbol{\sigma}^{-2})\boldsymbol{B}(\boldsymbol{B}^T Diag(\boldsymbol{\sigma}^{-2})\boldsymbol{B})^{-1}$
OLS-CSR (Fama and MacBeth 1973)	Minimum variance portfolio with unit beta to the k-th factor and zero betas with respect to all other factors	$\boldsymbol{\Omega} = \sigma^2\boldsymbol{I}_N$: Uncorrelated assets with constant variance $\boldsymbol{B}$: Multivariate $\boldsymbol{B}_K = \boldsymbol{I}_K$	$\boldsymbol{B}(\boldsymbol{B}^T\boldsymbol{B})^{-1}$
Maximum correlation portfolio approach (MCP)			
Unit-beta MCP (Grinold and Kahn 2000)	Minimum variance portfolio with unit beta to the k-th factor and non-null associated betas with respect to all other factors	$\boldsymbol{\Omega}$: Unconstrained $\boldsymbol{B}$: Univariate only $\boldsymbol{B}_K = (\boldsymbol{B}^T\boldsymbol{\Omega}^{-1}\boldsymbol{B})Diag(\boldsymbol{B}^T\boldsymbol{\Omega}^{-1}\boldsymbol{B})^{-1}$	$\boldsymbol{\Omega}^{-1}\boldsymbol{B}\, Diag(\boldsymbol{B}^T\boldsymbol{\Omega}^{-1}\boldsymbol{B})^{-1}$
Target-beta MCP (Huberman et al. 1987; Breeden et al. 1989)	Minimum variance portfolio with pre-specified beta to the k-th factor and non-null associated betas with respect to all other factors	$\boldsymbol{\Omega}$: Unconstrained $\boldsymbol{B}$: Univariate only $\boldsymbol{B}_K = \boldsymbol{B}^T\boldsymbol{\Omega}^{-1}\mathbf{B}$	$\boldsymbol{\Omega}^{-1}\boldsymbol{B}$
MCP (Lamont 2001)	Minimum variance portfolio with beta to the k-th factor equals to the R^2 in the regression of the k-th factor on the base asset returns and non-null associated betas with respect to all other factors	$\boldsymbol{\Omega}$: Unconstrained $\boldsymbol{B}$: Univariate only $\boldsymbol{B}_K = (\boldsymbol{B}^T\boldsymbol{\Omega}^{-1}\boldsymbol{B})Diag(\boldsymbol{\Omega}_K)$	$\boldsymbol{\Omega}^{-1}\boldsymbol{B}\, Diag(\boldsymbol{\Omega}_K)$

NOTE.– The table summarizes the various specifications of the FMPs. $\boldsymbol{\Omega}$ is the $(N \times N)$ covariance matrix of the base asset returns, $\boldsymbol{\Omega}_K$ is the $(K \times K)$ covariance matrix of the factors, $\boldsymbol{B}$ is the $(N \times K)$ factor loading matrix, $\boldsymbol{B}_K$ is the $(K \times K)$ matrix of target risk exposures and $\boldsymbol{I}_K$ is the $(K \times K)$ identity matrix.

Table 4.1. *Factor-mimicking portfolio construction methodologies*

4.3. Hedging climate risk factors

In this section, we implement the Fama-MacBeth factor-mimicking approach to build climate hedge portfolios. We consider various climate textual risk factors and beta estimation methods. We then evaluate the in-sample and out-of-sample hedging performances of the different portfolios using a cross-asset universe.

4.3.1. *Setup*

To alleviate the contamination bias that arises when climate risk factors are correlated and measured with errors (Pukthuanthong et al. 2022), we consider the special case of $K=1$ so that the FMPs are constructed for each climate risk factor independently by running univariate regressions. To focus on the impact of the estimation errors in the betas, we fully shrunk the variance-covariance sample matrix towards the identity matrix. These choices imply that the FMP weight associated with the ith base asset to mimic the kth climate risk factor is obtained as:

$$w_{i,k} = \frac{b_{i,k}}{\sum_{i=1}^{N} b_{i,k}^2}, \qquad [4.4]$$

where $b_{i,k}$ is the estimated beta of the regression of the ith base asset onto the kth climate risk factor. In the following, we consider three different beta estimation methods:

i) **OLS**: regression of base assets on the original climate risk factor.

ii) **PCA**: regression of base assets on the denoised climate risk factor obtained by projecting the original climate risk factor on the L first PC components of the base asset returns. We select $L = [N/3]$, where N is the number of base assets and [.] is the rounding function to the closest integer.

iii) **tPCA**: regression of base assets on the denoised climate risk factor obtained by projecting the original climate risk factor on the L first PC components of the LASSO-selected base assets.

4.3.2. *Climate textual risk factors data*

We follow Alekseev et al. (2022) and consider a large set of measures constructed from textual analysis of climate news series, namely:

i) **WSJ**: The Wall Street Journal climate news index created by Engle et al. (2020);

ii) **CHN**: The Crimson Hexagon Negative climate news index used by Engle et al. (2020);

iii) **ClimPol**: The US Climate Policy textual factor of Faccini et al. (2022);

iv) **InterSum**: The International Summits textual factor of Faccini et al. (2022);

v) **GlobWarm**: The Global Warming textual factor of Faccini et al. (2022);

vi) **NatDisas**: The Natural Disasters textual factor of Faccini et al. (2022);

vii) **NarrClim**: The US Narrative Climate Policy textual factor of Faccini et al. (2022);

viii) **MCCC**: The Media Climate Change Concern index of Ardia et al. (2022).

All these climate risk factors are based on textual analysis of various media sources (journals or websites), using a differentiated set of climate-related topics. The database spans from 1990 to 2018. We selected this starting date as there is empirical evidence that climate risks were not priced before that date[8]. The ending date is determined by the availability of the climate textual risk factors.

The various climate textual risk factors are shown in Figure 4.1 after being normalized as z-scores (i.e. demeaned and standardized to a common volatility level; here 10% annualized) to facilitate their interpretation. In the hedging analysis, we follow the literature and use AR(1) innovations of climate risk factors as hedge targets.

While the textual risk factors differ across time, we observe some commonalities as well. In particular, we note that climate risk factors seem to pick in two main periods. A first period stems from 2007 to 2010 covering the 2007 United Nations (UN) Security Council talks on climate change, the Nobel Peace prize award given to Al Gore, the IPCC and the Copenhagen 2009 UN Climate Change Conference. A second period of elevated climate risks concern appears after 2016 that might be related to the decision of US President Donald Trump to withdraw from the Paris Agreement.

8 With notably the G7 Summit of July 1990 which backed the conclusions of the First Intergovernmental Panel on Climate Change (IPCC) on the impact of human emissions of carbon on global temperature (see Gregory 2021).

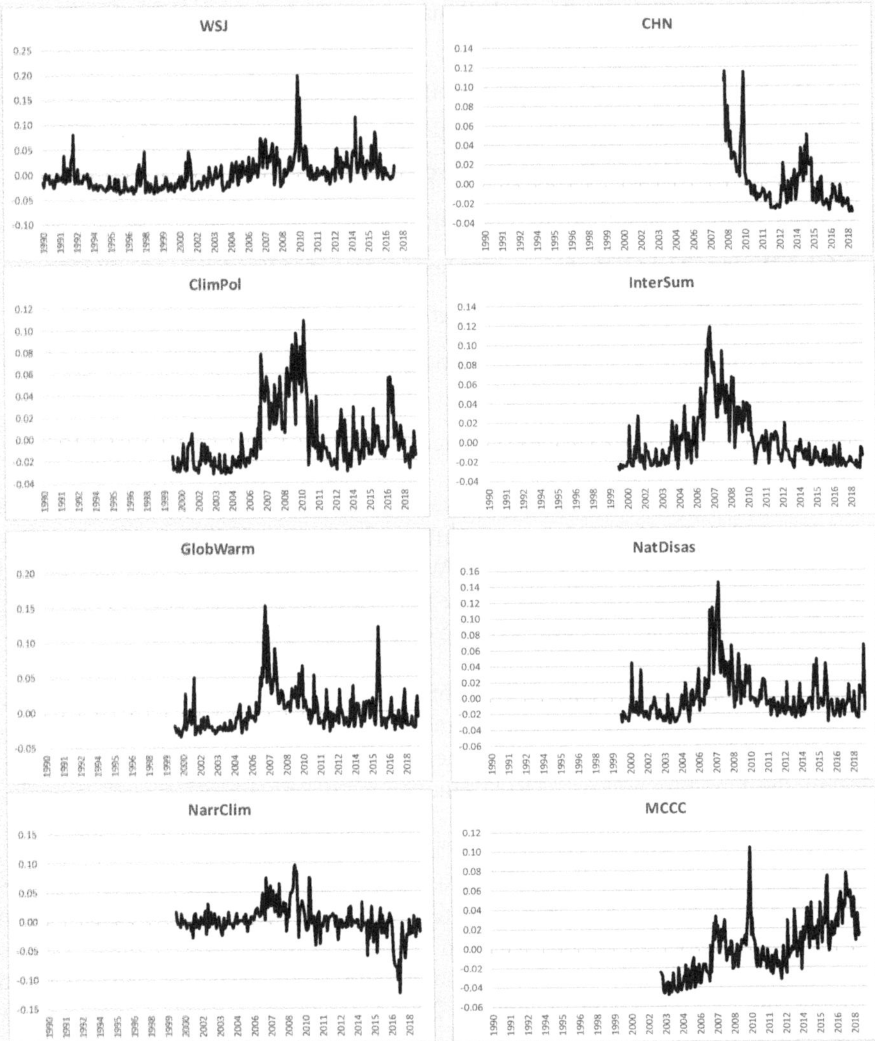

NOTE.– The figure represents the climate risk factors used in the empirical analysis. WSJ and CHN are from Engle et al. (2020); ClimPol, InterSum, GlobWarm, NatDisas and NarrClim are from Faccini et al. (2022). MCCC is from Ardia et al. (2022). All the climate risk factors are centered and standardized to a 10% volatility.

Figure 4.1. *Climate textual risk factors*

To investigate the similarities between climate risk factors in more detail, we display their correlations in Table 4.2. Most correlation numbers are positive but small, meaning that the climate textual risk factors are consistent but different. The climate risk factor displaying the highest correlation with other factors is the MCCC of Ardia et al. (2022), which could act as the reference climate textual risk factor.

	CHN	ClimPol	InterSum	GlobWarm	NatDisas	NarrClim	MCCC
WSJ	0.49	-0.01	0.06	0.30	0.24	-0.02	0.56
CHN		0.18	0.29	0.37	0.19	0.12	0.60
ClimPol			0.14	0.21	0.08	-0.06	0.17
InterSum				0.24	0.01	0.30	0.16
GlobWarm					0.43	0.16	0.50
NatDisas						0.14	0.26
NarrClim							-0.07

Table 4.2. *Correlation of AR(1) innovations of climate textual risk factors. Sources: Engle et al. (2020), Faccini et al. (2022), Ardia et al. (2022), author's calculations. The calculations are done over the common period, July 2008–June 2017*

4.3.3. *Base assets data*

As for any factor-mimicking portfolio, building climate FMPs requires a set of base assets. While most of the literature has relied on equity portfolios/factors, we adopt in this chapter a cross-asset perspective. Our cross-asset universe is equivalent to the one that has been used in the recent literature on time-series momentum (Moskowitz et al. 2012; Huang et al. 2020).

From Bloomberg. we collect future prices for nine developed country equity indices, 13 developed government bonds, 24 commodities and nine currencies against the US dollar. Consistently with the literature, whenever future prices are not available, we rely on total return versions of similar indices (such as MSCI indices for equity indices). All returns are defined as excess returns over cash, which renders our FMP implementations more realistic and alleviates issues related to leverage funding or shorting.

In Table 4.3, we display the descriptive statistics of the 46 base assets. The mean and volatility are annualized. Significant variations exist in mean return and volatility across different futures contracts.

Asset class	Contract	Mean	Volatility
Equities	ASX	3.39%	13.16%
	CAC	4.10%	18.52%
	DAX	5.62%	20.60%
	IBEX	5.73%	21.05%
	MIB	2.94%	22.21%
	Topix	-0.25%	19.65%
	AEX	6.41%	18.64%
	FT100	3.51%	14.07%
	SPX	6.58%	14.27%
Rates	AU-10	3.75%	6.99%
	CA-10	3.64%	5.92%
	JP-10	3.08%	4.19%
	UK-10	3.35%	6.79%
	US-10	3.71%	5.81%
Commodities	ALUM	2.18%	18.79%
	BRENT	11.33%	33.89%
	CATTLE	2.46%	13.87%
	COCOA	-1.21%	29.28%
	COFFEE	-3.32%	36.48%
	COPPER	5.95%	23.56%
	CORN	-5.31%	25.23%
	COTTON	-2.09%	26.76%
	CRUDE	6.49%	34.63%
	GASOIL	10.84%	33.69%
	GOLD	2.20%	15.44%
	HEATOIL	10.73%	32.72%
	HOGS	4.08%	28.65%
	NICKEL	6.42%	33.09%
	PLATIN	3.64%	20.34%

Commodities (cont'd)	SILVER	4.06%	27.76%
	SOYBEANS	4.74%	23.24%
	SOYMEAL	11.20%	26.93%
	SOYOIL	-1.68%	23.42%
	SUGAR	5.55%	31.93%
	UNLEAD	11.87%	38.85%
	WHEAT	-6.79%	28.34%
	ZINC	5.15%	23.95%
Foreign exchange	AUD	2.36%	11.30%
	EUR	0.06%	9.98%
	CAD	0.23%	7.88%
	JPY	-0.72%	10.77%
	NOK	1.01%	10.82%
	NZD	3.75%	11.53%
	SEK	0.42%	11.24%
	CHF	0.83%	10.87%
	GBP	0.87%	9.19%

Table 4.3. *Base asset returns Source: Bloomberg, author's calculations*

4.3.4. *In-sample hedging results*

We begin with an in-sample analysis of the hedging properties of FMPs over the full sample for each climate textual risk factor.

In Figure 4.2, we provide the composition of the various FMPs. We observe that **OLS** FMPs are characterized by much larger base asset exposures in general. Most FMPs tend to be long equities, precious metals and agricultural commodities, while they tend to be short industrial metals and short USD. For rates and energy, the sign of the exposures differs across methodologies and/or climate risk measures considered.

Associated statistics are displayed in Table 4.4. Through leverage (measured as the sum of absolute positions), we observe that the EIV-corrected FMPs (**PCA** and **tPCA**) are much less extreme for all climate hedge targets. As leverage is often associated with risk or limited for some investors, this is a key advantage. Conversely, **OLS** leads to highly leveraged portfolios.

We then display a set of statistics aiming at evaluating the goodness of fit of the FMPs relatively to climate risk factors over the full sample, and more precisely the correlation of climate FMPs with their underlying climate textual risk factors, RMSE (Root Mean Square Errors) and MAE (Mean Absolute Error). In-sample correlation metrics do point to the superiority of **OLS** – which is an automatic result of in-sample regressions. For EIV-corrected FMPs, **PCA** seems to be superior from a correlation perspective. Measures of average errors consistently point to the superiority of **PCA** followed by **tPCA** approach, while **OLS** is frequently offering the worst fit.

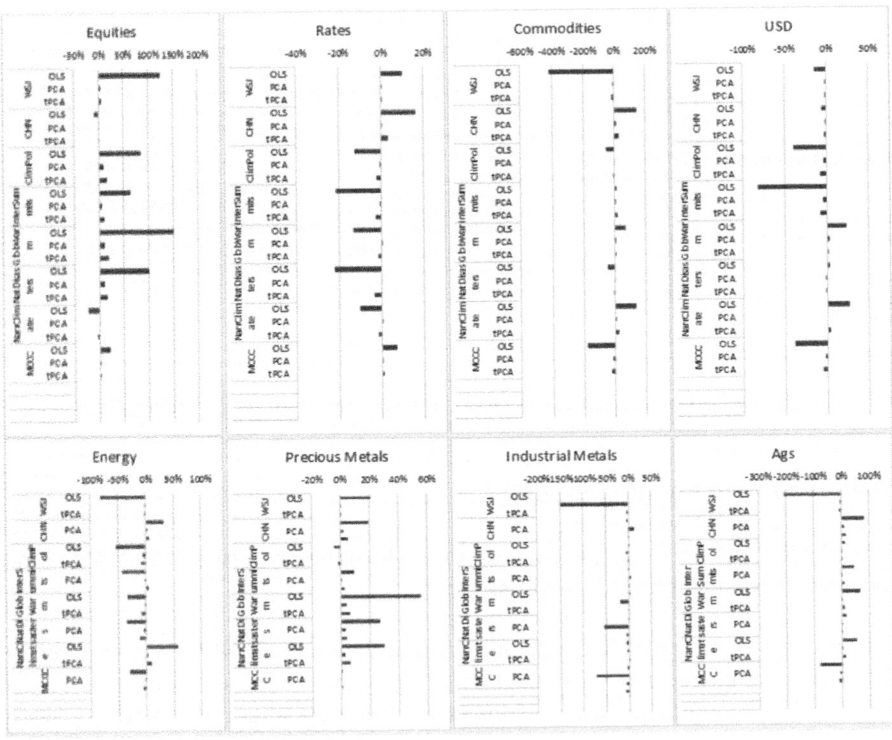

Figure 4.2. *Climate FMP composition. Source: authors' calculations*

	WSJ			CHN			ClimPol			InterSummits		
	OLS	**PCA**	**tPCA**	**OLS**	**PCA**	**tPCA**	**OLS**	**PCA**	**tPCA**	**OLS**	**PCA**	**tPCA**
Leverage	804%	18%	25%	249%	21%	47%	328%	33%	54%	398%	23%	47%
Correlation with original factor	0.12	0.09	0.10	0.21	0.18	0.20	0.27	0.24	0.24	0.24	0.17	0.09
RMSE	20.03%	2.41%	2.44%	8.01%	1.80%	2.13%	6.75%	1.84%	1.95%	6.50%	1.62%	1.94%
MAE	15.27%	1.69%	1.72%	6.16%	1.37%	1.71%	5.15%	1.34%	1.44%	4.90%	1.19%	1.47%

	GlobWarm			NatDisasters			NarrClimate			MCCC		
	OLS	**PCA**	**tPCA**	**OLS**	**PCA**	**tPCA**	**OLS**	**PCA**	**tPCA**	**OLS**	**PCA**	**tPCA**
Leverage	484%	27%	65%	405%	37%	63%	491%	22%	81%	265%	22%	36%
Correlation with original factor	0.22	0.17	0.17	0.30	0.26	0.23	0.19	0.15	0.17	0.19	0.18	0.17
RMSE	9.95%	2.18%	2.41%	6.54%	1.99%	2.17%	11.44%	2.26%	2.72%	9.40%	1.90%	2.10%
MAE	7.60%	1.57%	1.85%	5.15%	1.46%	1.61%	8.97%	1.68%	2.13%	6.94%	1.49%	1.64%

Table 4.4. *Climate FMP in-sample hedging properties Source: authors' calculations*

4.3.5. *Out-of-sample hedging results*

We next perform an out-of-sample exercise of hedging abilities. Climate FMPs are re-estimated every month based on available samples so far, and we build a new time-series based on the realized performance of these FMP compositions over the next month. Following the literature (Alekseev et al. 2022), we evaluate the hedging performances over the period post-2010 as investors were paying little attention to climate risk factors before this date.

In Table 4.5, we report various metrics associated with these out-of-sample analyses: the out-of-sample correlation, the mean absolute error, the portfolio monthly turnover. For each metric/period/climate risk factor, we highlight in bold which estimation method seems the best.

	WSJ	CHN	ClimPol	InterSum	GlobWarm	NatDisas	NarrClim	MCCC
				Correlation				
OLS	**0.111**	**0.049**	-0.107	-0.143	0.031	0.054	0.093	0.215
PCA	0.110	-0.005	-0.095	-0.168	-0.009	0.063	0.101	0.174
tPCA	0.111	-0.004	**-0.072**	**-0.141**	**0.041**	**0.086**	**0.108**	**0.226**
				Mean absolute errors				
OLS	0.099	0.039	0.049	0.045	0.063	0.075	0.071	0.053
PCA	**0.020**	**0.012**	**0.016**	**0.010**	**0.016**	**0.012**	**0.019**	**0.015**
tPCA	0.021	0.018	0.017	0.013	0.017	0.019	0.022	0.016
				Turnover				
OLS	60	19	42	44	41	65	34	24
PCA	**7**	**5**	**6**	**8**	**7**	**8**	**4**	**3**
tPCA	18	16	12	17	16	28	13	10

Table 4.5. *Climate FMP out-of-sample hedging properties. Source: author's calculations*

For almost all methodologies and all climate textual risk factors, we observe, as expected, a deterioration out-of-sample in all the goodness-of-fit metrics. Indeed, some correlations even turn negative, while MAEs are larger. From the comparison of the different estimation methodologies, there is a clear hierarchy for mean absolute errors and turnover where **PCA** is the most effective followed by **tPCA**, while **OLS** seems quite off-the-mark, most probably due to higher sensitivity to estimation errors in the climate betas. This is consistent with the in-sample results.

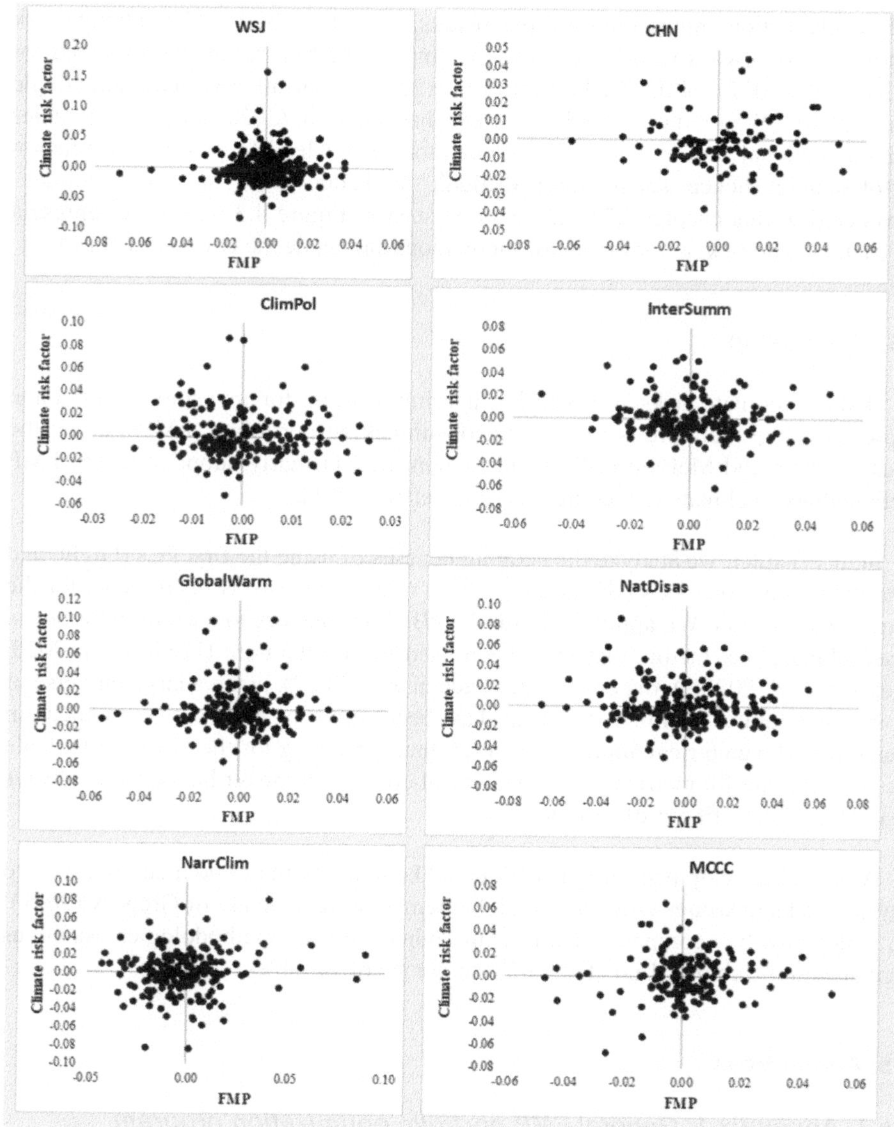

Figure 4.3. *Out-of-sample tPCA FMP climate hedge performance Sources: Ardia et al. (2020), Engle et al. (2020) and Faccini et al. (2020), author's calculations*

In terms of correlation, there is no absolute hierarchy, but **tPCA** frequently outperforms **OLS** while **PCA** is the best performer most of the time. This indicates that regularization might improve the results. We also observe that climate risk factors are not always well-hedged out-of-sample. The best results are obtained for Ardia et al. (2022)'s MCCC, Faccini et al. (2022)'s Narrative Climate and Engle et al. (2020)'s WSJ indices. On the contrary, Faccini et al. (2022)'s Climate Policies and International Summits textual risk factors do not lead to positive correlation out-of-sample, hence seem more difficult to hedge with the methodologies considered in this chapter. This can also be seen in Figure 4.3 where we represent for each climate risk factor the factor innovations alongside the **tPCA**-based FMP.

4.4. Conclusion

Hedging climate risks is becoming a primary issue for investors. A common approach in the literature is to form factor-mimicking portfolios (FMPs) along the lines of Fama and McBeth (1976) and Lamont (2001). Early adoption of the FMP methodology in climate risk context is Engle et al. (2021).

In this chapter, we analyze the hedging benefits of using the first PCs (Giglio and Xiu 2020) and target PCs (Bai and Ng 2008) of asset returns as instruments for the climate risk factors. We apply the Fama-MacBeth framework to a set of eight news-based climate risk factors that have been used in the literature (Engle et al. 2020; Alekseev et al. 2022; Ardia et al. 2022; Faccini et al. 2022), using a large universe of cross-asset futures (similar to what has been used in time-series momentum literature). The empirical application shows that correcting for the errors-in-variable bias is important for metrics such as statistical errors or turnover but does not always guarantee a better fit out-of-sample.

A natural development of our work could be to apply the same methodologies to physical risks measures (such as the global temperature anomaly of Gregory (2021)) or, alternatively, compare them with other robust methodologies such as regularization technics (Luo et al. 2022) to our set of variables.

4.5. Appendices

4.5.1. *Appendix 1: General FMP portfolio optimization program*

For each macro factor k, the factor-mimicking portfolio (FMP) can be obtained as the solution of the following variance minimization problem:

$$\min_{w_k} \frac{1}{2}\, \boldsymbol{w}_k^T \boldsymbol{\Omega}\, \boldsymbol{w}_k \qquad [4.A1.1]$$

$$s.t.\, \boldsymbol{B}^T \boldsymbol{w}_k = \boldsymbol{\beta}_k$$

where $\boldsymbol{\Omega}$ is the positive definite $(N \times N)$ covariance matrix, $\boldsymbol{B}$ is the $(N \times K)$ factor loading matrix (univariate or multivariate) and $\boldsymbol{\beta}_k$ is an $(K \times 1)$ vector of portfolio factor exposures, with the k-th entry equal to 1 and all others equal to β_{kl}. The Lagrangian of the system [4.A1.1] is given by

$$L(\boldsymbol{w}_k; \boldsymbol{\lambda}_k) = \frac{1}{2}\boldsymbol{w}_k^T\boldsymbol{\Omega}\,\boldsymbol{w}_k - \boldsymbol{\lambda}_k^T(\boldsymbol{B}^T\boldsymbol{w}_k - \boldsymbol{\beta}_k) \qquad [4.A1.2]$$

where $\boldsymbol{\lambda}_k$ is the $(K \times 1)$ vector of the Lagrangian multipliers. The first-order condition of [4.A1.2] is given by:

$$\frac{\partial L(\boldsymbol{w}_k; \boldsymbol{\lambda}_k)}{\partial \boldsymbol{w}_k} = \boldsymbol{\Omega}\,\boldsymbol{w}_k - \boldsymbol{B}\boldsymbol{\lambda}_k = 0 \qquad [4.A1.3]$$

Pre-multiplying by the inverse matrix $\boldsymbol{\Omega}^{-1}$ and solving for the optimal weights, we have:

$$\boldsymbol{w}_k = \boldsymbol{\Omega}^{-1}\boldsymbol{B}\boldsymbol{\lambda}_k \qquad [4.A1.4]$$

Premultiplying [4.A1.4] by $\boldsymbol{B}^T$ leads to:

$$\boldsymbol{B}^T\boldsymbol{w}_k = \boldsymbol{B}^T\Omega^{-1}\boldsymbol{B}\lambda_k = \boldsymbol{\beta}_k \qquad [4.A1.5]$$

We infer:

$$\lambda_k = \left(\boldsymbol{B}^T\Omega^{-1}\boldsymbol{B}\right)^{-1}\boldsymbol{\beta}_k \qquad [4.A1.6]$$

Equation [4A.1.4] can then be rewritten as:

$$\boldsymbol{w}_k^* = \Omega^{-1}\boldsymbol{B}\left(\boldsymbol{B}^T\Omega^{-1}\boldsymbol{B}\right)^{-1}\boldsymbol{\beta}_k \qquad [4.A1.7]$$

Taken together, the factor-mimicking portfolio weight matrix is:

$$\boldsymbol{W}_K^* = \Omega^{-1}\boldsymbol{B}\left(\boldsymbol{B}^T\Omega^{-1}\boldsymbol{B}\right)^{-1}\boldsymbol{B}_K \qquad [4.A1.8]$$

where $\boldsymbol{B}_K$ is the $(K \times K)$ target factor beta matrix, with $\boldsymbol{B}_K = (\boldsymbol{\beta}_1\ \boldsymbol{\beta}_2 \ldots\ \boldsymbol{\beta}_K)$.

4.5.2. *Appendix 2: Principal components instrumental variables FMP estimator*

We assume that the individual asset returns have a macro factor structure that can be represented as:

$$\boldsymbol{R}_t = \boldsymbol{\mu} + \boldsymbol{B}\boldsymbol{F}_t + \boldsymbol{\varepsilon}_t \qquad [4.A2.1]$$

where $\boldsymbol{R}_t$ and $\boldsymbol{\mu}$ are the $(N \times 1)$ vector of the base asset returns and associated expected returns, respectively; $\boldsymbol{B}$ is the $(N \times K)$ matrix of asset loadings to factors; $\boldsymbol{F}_t$ is the $(K \times 1)$ vector of zero-mean observable factors and $\boldsymbol{\varepsilon}_t$ is the $(N \times 1)$ vector of zero-mean disturbances.

Let $\overline{\boldsymbol{R}}$ correspond to the $(N \times T)$ matrix of asset centered returns, $\boldsymbol{F}$ denote the $(K \times T)$ matrix of observed factors and $\boldsymbol{\varepsilon}$ be the $(N \times T)$ matrix of disturbances. The factor model can then be represented in matrix form as:

$$\overline{\boldsymbol{R}} = \boldsymbol{B}\boldsymbol{F} + \boldsymbol{\varepsilon} \qquad [4.A2.2]$$

Given the base asset returns and the factors of interest, the principal components instrumental variables (IV) FMP weight estimator proceeds as follows:

1) **PC extraction step:** Extract the first L principal components of base asset returns by conducting the PCA of the covariance matrix $\widehat{\boldsymbol{\Omega}} = (T)^{-1}\overline{\boldsymbol{R}}\overline{\boldsymbol{R}}^{\boldsymbol{T}}$, so that:

$$\boldsymbol{F}^{PC} = \boldsymbol{E}_L^{\boldsymbol{T}}\,\overline{\boldsymbol{R}} \qquad [4.A2.3]$$

where $\boldsymbol{F}^{PC}$ is the $(L \times T)$ matrix of PCs and $\boldsymbol{E}_L$ corresponds to the $(N \times L)$ eigenvector matrix associated with the largest L eigenvalues of $\widehat{\boldsymbol{\Omega}}$, with $L \leq N$.

2) **Fitted factor estimation step:** Regress the K observed economic factors $\boldsymbol{F}_t$ onto the L statistical factors extracted from the PCA to obtain the fitted values of the observed factors and their multivariate factor exposures, i.e.

$$\widehat{\boldsymbol{F}}^{PC} = \widehat{\boldsymbol{b}}_K\boldsymbol{F}^{PC} \text{ and } \widehat{\boldsymbol{b}}_K = \boldsymbol{F}\boldsymbol{F}^{PC^{\boldsymbol{T}}}\left(\boldsymbol{F}^{PC}\boldsymbol{F}^{PC^{\boldsymbol{T}}}\right)^{-1} \qquad [4.A2.4]$$

The observable fitted factors and their exposures can be written equivalently more compactly as:

$$\widehat{\boldsymbol{F}}^{PC} = \boldsymbol{F}\boldsymbol{P} \qquad [4.A2.5]$$

where $\boldsymbol{P}$ is a $(T \times T)$ projection matrix, with $\boldsymbol{P} = \boldsymbol{F}^{PC^{\boldsymbol{T}}}\left(\boldsymbol{F}^{PC}\boldsymbol{F}^{PC^{\boldsymbol{T}}}\right)^{-1}\boldsymbol{F}^{PC}$.

3) **Times-series regression step:** Regress the excess asset returns $\overline{\boldsymbol{R}}$ onto the K fitted values of the observable factors $\widehat{\boldsymbol{F}}^{PC}$ to obtain the instrumental variables (IV) macro beta estimates, i.e.

$$\widehat{\boldsymbol{B}}^{PCIV} = \overline{\boldsymbol{R}}\,\widehat{\boldsymbol{F}}^{PC^{\boldsymbol{T}}}\left(\widehat{\boldsymbol{F}}^{PC}\widehat{\boldsymbol{F}}^{PC^{\boldsymbol{T}}}\right)^{-1} \qquad [4.A2.6]$$

Substituting in [4.A2.6] $\widehat{\boldsymbol{F}}^{PC}$ by its expression [4.A2.5] and rearranging lead to:

$$\widehat{\boldsymbol{B}}^{PCIV} = \overline{\boldsymbol{R}}\,\boldsymbol{P}^{\boldsymbol{T}}\boldsymbol{F}^{\boldsymbol{T}}(\boldsymbol{F}\boldsymbol{P}\boldsymbol{P}^{\boldsymbol{T}}\boldsymbol{F}^{\boldsymbol{T}})^{-\boldsymbol{1}} \quad [4.A2.7]$$

where $\boldsymbol{P}$ is the $(T \times T)$ projection matrix defined in step 2.

The instrumental variables estimator of the FMP weights is then obtained by substituting $\widehat{\boldsymbol{B}}$ by $\widehat{\boldsymbol{B}}^{PCIV}$, i.e.:

$$\boldsymbol{W}_K^{*PCIV} = \widehat{\Omega}^{-\boldsymbol{1}}\widehat{\boldsymbol{B}}^{PCIV}\left(\widehat{\boldsymbol{B}}^{PCIV^T}\widehat{\Omega}^{-\boldsymbol{1}}\widehat{\boldsymbol{B}}^{PCIV}\right)^{-\boldsymbol{1}}\boldsymbol{B}_K \quad [4.A2.8]$$

where $\boldsymbol{B}_K$ is the $(K \times K)$ target beta matrix. We can remark that the IV estimator [4.A2.8] provides an unbiased estimator of the FMP weights when the first PCs of the base asset returns can recover the same space as the true latent factors (Bai and Ng 2002).

4.6. References

Alekseev, G., Giglio, S., Maingi, Q., Selgrad, J., Stroebel, J. (2022). A quantity-based approach to constructing climate risk hedge portfolios. *SSRN*, Working paper.

Andersson, M., Bolton, P., Samama, F. (2016). Hedging climate risk. *Financial Analysts Journal*, 72, 1–20.

Ardia, D., Bluteau, K., Boudt, K., Inghelbrecht, K. (2022). Climate change concerns and the performance of green versus brown stocks. National Bank of Belgium, Working Paper Research no. 395.

Bai, J. and Ng, S. (2008). Forecasting economic time series using targeted predictors. *Journal of Econometrics*, 146, 304–317.

Bair, E., Hastie, T., Paul, D., Tibshirani, R. (2006). Prediction by supervised principal components. *Journal of the American Statistical Association*, 101, 119–137.

Balduzzi, P. and Robotti, C. (2008). Mimicking portfolios, economic risk premia, and tests of multi-beta models. *Journal of Business & Economic Statistics*, 26, 354–368.

Breeden, D., Gibbons, M., Litzenberger, R. (1989). Empirical tests of the consumption-oriented CAPM. *Journal of Finance*, 44, 231–262.

Connor, G. and. Korajczyk, R. (1988). Risk and return in an equilibrium APT: Application of a new test methodology. *Journal of Financial Economics*, 21, 255–289.

Connor, G. and Korajczyk, R. (1991). The attributes, behavior, and performance of U.S. mutual funds. *Review of Quantitative Finance and Accounting*, 1, 5–26.

Engle, R., Giglio, S., Kelly, B., Lee H., Stroebel, J. (2020). Hedging climate change news. *The Review of Financial Studies*, 33, 1184–1216.

Faccini, R., Matin, R., Skiadopoulos, G. (2022). Are climate change risks priced in the US stock market? *SSRN*, Working paper.

Fama, E. and MacBeth, J. (1973). Risk, return, and equilibrium: Empirical tests. *Journal of Political Economy*, 81, 607–636.

Giglio, S. and Xiu, D. (2021). Asset pricing with omitted factors. *Journal of Political Economy*, 129, 1947–1990.

Gregory, R. (2021). The pricing of global temperature shocks in the cost of equity capital. *Journal of International Financial Markets, Institutions & Money*, 72, 1–12.

Grinold, R. and Kahn, R. (2000). *Active Portfolio Management: Quantitative Theory and Applications*, 2nd edition. McGraw Hill, New York.

Huang, D., Li, J., Wang, L., Zhou, G. (2020). Time-series momentum: Is it there? *Journal of Financial Economics*, 135, 774–794.

Huberman, G., Kandel, S., Stambaugh, R. (1987). Mimicking portfolios and exact arbitrage pricing. *Journal of Finance*, 42, 1–9.

Lamont, O. (2001). Economic tracking portfolios. *Journal of Econometrics*, 105, 161–184.

Lehmann, B. and Modest, D. (1988). The empirical foundations of the arbitrage pricing theory. *Journal of Financial Economics*, 21, 213–254.

Litzenberger, R. and Ramaswany, K. (1979). The effects of dividends on common stock prices: Theory and empirical evidence. *Journal of Financial Economics*, 7, 163–195.

Luo, D., Rytchkov, O., Zhong, X. (2022). Regularized mimicking portfolios. *SSRN*, Working paper.

Melas, D., Suryanarayanan, R., Cavaglia, S. (2010). Efficient replication of factor returns: Theory and applications. *Journal of Portfolio Management*, 36, 39–51.

Moskowitz, T., Ooi, Y., Pedersen, L. (2012). Time series momentum. *Journal of Financial Economics*, 104, 228–250.

Pukthuanthong, K., Roll, R., Wang, J., Zhang, T. (2022). A new method for factor-mimicking portfolio construction. *SSRN*, Working paper.

5

A Framework for Achieving Net-Zero-Carbon Alpha Portfolios

5.1. Introduction

The world is on a trajectory towards anthropogenic climate change. Over the last several decades, a growing body of research has been pointing to the direct linkage between climate change and the amount of carbon put into the atmosphere through human activities (see, for example, Manabe and Wetherald (1967) and Keeling et al. (1976), etc.). The latest such research is the UN-backed IPCC Working Group I report, "Climate Change 2021: The Physical Science Basis"[1], which was approved by its 195 member countries and released in August 2021. In it, the 234 scientists involved from 66 countries did not mince words; climate change is real, and human activity is the primary source of global warming. Furthermore, according to the report, the Earth's temperature is likely to breach the 1.5°C (2.7°F) pre-industrial threshold deemed environmentally catastrophic within 20 years, thus requiring drastic action to reduce greenhouse gas emissions. Fortunately, the scientists also concluded that there is still time to limit the most severe damages associated with climate change if immediate action is taken.

Chapter written by Sebastian LANCETTI.

For a color version of all figures in this chapter, see www.iste.co.uk/jurczenko/climate.zip.

1 See Working Group I, "Climate Change 2021: The Physical Science Basis", IPCC Sixth Assessment Report, 2021.

Climate change has potentially large and negative effects on global ecology, society and economy. Because of these effects, the high-order complexity and interaction involved in estimating future climate paths, and adaptive measures that will need to be taken, its exact impact is yet to be determined. However, a range of likely scenarios can be estimated, and most of the outcomes are grim[2]. Specifically, a 2°C level of global mean temperature rise will negatively affect the global GDP by -11% versus the zero-climate change (0°C) scenario (see Swiss Re Institute 2021)[3].

Governments and organizations around the world are beginning to consider and try to mitigate the impact of climate change.

According to Stroebel and Wurgler (2021), investors are most concerned about these effects on firms and their investment portfolios. The two authors highlighted that over the next five years, regulatory risk is viewed as the top climate-related risk as governments adapt rules to limit and mitigate climate damage (e.g. by limiting carbon emission, enacting carbon tax, etc.). Moreover, over a longer horizon of 30 years, physical risk associated with climate change is considered the top risk[4].

Climate change is not a risk solely for investors. The transition for the world's energy supplies, industrial processes, etc. to a lower-carbon emission world requires massive investment and innovation. A Bloomberg survey revealed that mitigating the worst effects of climate change would require an investment of at least 92 trillion dollars by 2050[5]. Innovation in the green space outpaces those in other technologies. For investors, if we can identify promising green technology and invest in it early, healthy financial returns may result.

Quantitative equity investors can position their portfolios for the coming low-carbon transition. We believe the advantages of such portfolios are threefold. Low-carbon portfolios are less likely to hold companies that will be affected by any potential regulatory risk that could lead to a decrease in the portfolio value. They are also more likely to hold companies that are positioned to benefit as the world transitions to a low-carbon economy, either through providing innovative products and services that may help to mitigate climate change, have low-carbon emission, or both. Also important, low-carbon portfolios are likely positioned for client demand as societal awareness of climate change continues to increase.

2 For negative consequences associated with climate change, see "Climate Science Special Report: Fourth National Climate Assessment, Volume I.", 2017.

3 See "World economy set to lose up to 18% GDP from climate change if no action taken, reveals Swiss Re Institute's stress-test analysis", 2021.

4 Physical risks are direct risks of climate change impacting assets and means of production.

5 See "Greening Energy to Fight Climate Threat May Cost $92 Trillion", 2021.

5.2. Carbon emission in the capital market

The developed world has been gradually decarbonizing over the last 10 years. Even if carbon footprint is still increasing in emerging markets, it is anticipated that this index's carbon footprint will also follow the developed world indices' carbon footprint and start falling, as various EM countries such as China hit peak carbon emission before 2030[6].

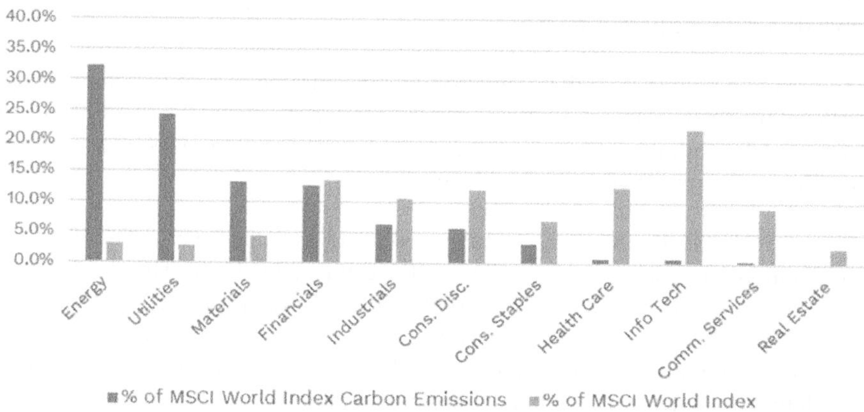

Figure 5.1. *MSCI World Index carbon emissions and weight by GICS sector (December 2020) Source: PanAgora, MSCI. For Illustrative purposes only. Certain information ©2021 MSCI ESG Research LLC.Reproduced by permission; no further distribution*

The emission distribution of global companies has a heavy tail, with only 20 companies producing more than one-third of global emissions and the top 100 companies producing 70%[7]. Measured levels of carbon emissions for individual companies vary broadly across industries, as they are generally concentrated in a few specific sectors; most notably, utilities, materials and energy. Within the MSCI World Index, nearly 70% of the index's carbon emissions come from these three sectors. As such, it is rather intuitive that a passive carbon reduction approach would underweight these sectors.

5.3. Passive approach to zero-carbon portfolios

There are two potential passive approaches to explore for lowering portfolio carbon levels: complete sector divestment and a passive optimization that seeks to

6 For more details, see the definition of carbon footprint and Figure 5.4 in the Appendix.

7 For additional details, see "CDP Carbon Majors Report 2017".

minimize active risk (i.e. tracking error). Portfolio carbon levels are defined as the total amount of scope 1 and scope 2 emissions across companies within the portfolio. Scope 1 emissions are direct greenhouse gas (GHG) emissions from company-owned or company-controlled sources (e.g. emissions from a company's factory). Scope 2 emissions are indirect GHG emissions related to the electricity purchased and used by a company. Please note that we do not consider scope 3 emissions (scope 3 emissions are the result of activities from assets not owned or controlled by the reporting organization) due to the poor coverage and lack of reliable data over the period studied.

As seen in Figure 5.2, carbon emission is concentrated in a few sectors. Simply not allocating to the largest-emitting sectors (materials, energy and utilities) and proportionately scaling up the remaining companies' position sizes is a simplistic approach to reducing portfolio carbon levels.

Figure 5.3 presents the active results, risks and average levels of scope 1 and scope 2 emission reduction of the Sector Divestment backtest. This backtest has delivered positive excess results since 2009, with an annualized active risk of 1.49% while averaging 77% less carbon emissions than the index. While this is a relatively passive implementation of carbon reduction, the resultant active risk level is more in line with that of an active quantitative long-only strategy.

A second approach to reducing carbon portfolio levels is through an optimization process. Here, we simulated several global, long-only portfolios that were optimized to achieve increasing levels of GHG reduction (0–90% for scope 1 only and scopes 1 and 2 combined in 10% increments) while minimizing the tracking error to the benchmark.

Annualized Active Net Result	1.47%
Annualized Active Risk	1.49%
Average Scope 1 Emission Reduction	83%
Average Scope 2 Emission Reduction	44%
Average Scopes 1 + 2 Emission Reduction	77%

Figure 5.2. *Simulated results sector divestment (ex-materials, energy and utilities) portfolio versus MSCI World Index Source: PanAgora and Trucost. Copyright © 2021, S&P Global Market Intelligence (and its affiliates, as applicable)*

The simulations were conducted using the MSCI World Index constituents for the period of January 2009 to December 2020 and assume monthly rebalancing and no leverage.

The following constraints were imposed: active security weight limit +/- 2%, country active weight limit +/-3%, region active weight limit +/- 1%, sector active weight limit +/- 1%, portfolio beta constraints 0.9975/1.0025, average daily volume holding limit 200% and average daily volume trading limit 30%. Active factor risk was also limited to +/-20% using the MSCI Barra GEMLT model.

Also note that the results presented are net of estimated transaction costs and gross of management fees.

Table 5.1 shows the results of these simulations. Our backtests indicate that it may have been possible to achieve a 90% reduction in scope 1 emissions while maintaining an active risk of 38 basis points. The impact on results in this scenario was minimal, with net results (i.e. after assumed transaction costs) essentially flat over the full period.

Within the backtests seeking to reduce the combined scope 1 and scope 2 emissions, a reduction of up to 80% was possible before active risk began to exceed 1%. As we might expect, going out on the carbon reduction scale progressively increased the realized active risk. The impact on net results was relatively consistent for a reduction between 0 and 60%, averaging -9 basis points. Interestingly, as scopes 1 and 2 reduction levels exceeded 70%, active results turned positive, up 27 basis points in the 80% reduction backtest. We believe this is driven by the optimized portfolio's greater underweight to the high-emission sectors, which generally underperformed during the backtest period. Also, it is to be noted that while up to -90% carbon reduction is possible, the optimization will fail if we try to reduce carbon beyond the -90% reduction level. This is because all companies in the investment universe currently have a net-positive carbon emission. Without additional tools, a long-only equity portfolio cannot achieve net-zero carbon at the current time. We investigate an additional tool that may allow us to achieve net-zero on the portfolio level in a later section.

	Greenhouse Gas Reduction Percentage									
	0%	-10%	-20%	-30%	-40%	-50%	-60%	-70%	-80%	-90%
Annualized Active Risk										
Scope 1	0.12%	0.12%	0.13%	0.13%	0.13%	0.15%	0.15%	0.18%	0.23%	0.38%
Scope 1 + 2	0.13%	0.12%	0.13%	0.14%	0.16%	0.21%	0.31%	0.44%	0.65%	1.05%
Annualized Active Net Result										
Scope 1	-0.10%	-0.10%	-0.07%	-0.05%	-0.03%	-0.01%	0.01%	0.02%	-0.02%	-0.01%
Scope 1 + 2	-0.10%	-0.12%	-0.09%	-0.09%	-0.09%	-0.08%	-0.07%	0.05%	0.27%	0.47%

Table 5.1. *Simulated results for carbon-reduction portfolio versus the MSCI World Index Source: PanAgora, Trucost. Copyright © 2021, S&P Global Market Intelligence (and its affiliates, as applicable)*

The optimization backtests generally held underweight positions to the heavy-emission sectors, with the higher percentage reduction backtests frequently optimizing the portfolio to the largest sector underweights allowable by the optimization constraints (+/-1%). As with the sector divestment approach, the persistent underweights to sectors (driven by non-alpha-seeking factors) may introduce unintended risks or periods of performance variability, particularly when the high-emission sectors' returns vary widely from the broader index.

While a complete divestment from the heaviest-emitting sectors and an optimized carbon-reduction portfolio both succeed in reducing overall portfolio carbon levels, we believe that they represent suboptimal investment approaches. We believe that complete divestment from sectors is an approach that is too blunt and limits the investment opportunity set, particularly to firms within the energy, utility and material sector, which may currently be, or will become, an integral part of solving climate change (see Cohen et al. 2021). Furthermore, the divestment approach introduces a tracking-error level that is akin to more active approaches. Finally, divestment strategies only consider climate change from the risk side, not from the alpha side. The optimized portfolios, while able to achieve significant levels of carbon reduction, leave much to be desired in terms of alpha. An active approach designed to generate alpha through stock selection may be a superior use of an investor's active-risk budget.

5.4. Active approach to zero-carbon portfolios

An active approach to zero-carbon portfolios seeks to both limit the risk associated with owning carbon emitting companies and actively select and own companies that provide what the world needs as it transitions to a low-carbon economy. Since the companies selected under the active approach provide a solution to a critical global problem, an active approach, if done correctly, should be able to drive a positive portfolio return (i.e. alpha) as well as limit the risks associated with climate transition[8]. In the same vein as investing purely for alpha, an active approach to zero-carbon investing requires the art and science of stock selection, except in this case, we select companies that achieve both better-than-anticipated climate and alpha performance.

There are many investment strategies[9] possible under the active framework; here, we present a few examples of sensibility for such strategies. We consider their efficacy from both environmental and alpha perspectives. They are delineated by the time perspective of the data used.

5.4.1. *Backward-looking data: carbon efficiency*

Most climate-related data is backward looking with a reporting lag. Despite these limitations, the data can be used to generate alpha and drive a portfolio towards lower-carbon companies. One example of such data is scopes 1 and 2 carbon emissions, exactly those used in the passive approach in the previous section.

The sensibility of strategy based on carbon emission data is that carbon is a necessary output for all companies to deliver their products and services since there are currently no listed net-zero or net-negative companies. Thinking about this in a different way, carbon emission is a necessary input for all companies to deliver their products and services since, without it, these products and services cannot be produced[10]. With carbon emission as an input, comparable companies can be graded on the efficiency of how they use these carbon emission inputs, like all other production inputs, such as materials, labor, capital, etc. Companies that are more carbon-efficient have superior carbon emission profiles and should generate higher long-horizon financial returns, as they are more operationally efficient[11].

---

8 For more on active quantitative sustainable investing, see Chen and Mussalli (2020) and Sorensen et al. (2021).

9 These strategies are called "factors" or "signals" in quantitative investment parlance. We will use the term "factors" going forward.

10 There is also a physical interpretation of carbon emission as a production input. Carbon emission occurs as a by-product of combustion. That is, fossil fuel must be burnt somewhere in the production process to generate carbon emission, and these fossil fuels constitute a resource input.

11 As with all ESG factors, their alpha efficacy is driven by materiality considerations and may differ depending on the regions of the world.

5.4.2. *Present-time data: "nowcasting" of environmental news*

Most environmental, social and governance (ESG) data leave much to be desired, and one major criticism is that they tend to be backward looking and exhibit significant lag[12], as evidenced above. However, with some basic intuition and a reasonable command of modern statistical techniques, we may make use of the many big-data sources that are now available to complement backward-looking ESG data. One such development is ESG news, which is an area of active recent research[13]. The news comes from sources such as traditional news organizations or non-government organizations (NGOs), and the subset of environment-related news can provide a real-time view of how various companies perform on environmental matters.

We have found that corporate environmental news, especially negative news, tends to occur serially rather than incidentally. This serial nature of negative news can be because environmental mishaps and bad performance may either be the consequences of lax corporate regulation and control; or it may reflect a generally careless attitude towards environmental concerns. Negative news can also be indicative of a larger pattern of managerial or cultural deficiencies, which may have return implications. For example, before BP's well-publicized oil spill in the Gulf of Mexico, there were a series of warning signs that indicate deficient environmental and safety practices, which the astute investor could have picked upon via news[14].

5.4.3. *Forward-looking data: corporate climate alignment and adaptation plans*

Finally, to position for climate-change adaptations and profit from the oncoming low-carbon transition, it is important to understand corporate management's forward-looking plans. This type of information is not typically disclosed in more traditional, numerical-style ESG disclosures (such as carbon emission, gender ratio, etc.). However, management typically disseminates these forward plans in textual formats that are available in various alternative data sources, such as 10-K/Q reports, corporate sustainability reports (CSRs), etc. Through the use of natural language processing (NLP), we again have the ability to discover information on a large scale; information that is not easily available to fundamental investors or quants using

---

12 For more extensive discussions on ESG data and ratings, including those from commercial vendors, see "The Unreasonable Attractiveness of More ESG Data" (2021), "Four Things No One Will Tell You About ESG Data," 2019, and "Aggregate Confusion: The Divergence of ESG Ratings" (2019).

13 For example, see Serafeim and Yoon (2021).

14 For more details, see the National Academy of Engineering and National Research Council (2010).

more traditional statistical tools[15]. We have found that considering forward-looking plans may be particularly beneficial from an alpha perspective, in addition to potentially being able to predict forward commercial ESG ratings such as those from MSCI ESG[16].

5.4.4. *Case study: sustainable global equity strategy from PanAgora Asset Management*

With factors based on the sensibilities discussed above, we may build a quantitative alpha model that delivers both above-benchmark return through stock selection (i.e. an active strategy) and an attractive environmental profile[17].

Below, we present carbon footprint as proof of the environmental performance from a live portfolio at PanAgora Asset Management. Numerous other examples of active portfolios that produce similarly positive results can surely be found by interested asset owners and allocators.

The environmental performance of the Sustainable Global Equity portfolio is illustrated in Figure 5.3. In that figure, the carbon footprint is given for the benchmark, the portfolio (if the alpha model did not contain the environmental factors discussed in this section), the portfolio with the alpha model that includes the environmental factors discussed in this section, and the portfolio with the full alpha model, as well as carbon-constrained optimization discussed earlier. Portfolio carbon constraint is set at 50% relative to the benchmark, and carbon footprint is defined with the following formula:

$$Portfolio\ Carbon\ Footprint = \frac{\sum_i \left(\frac{Current\ Value\ of\ Investment_i}{Issuer's\ Market\ Capitalization_i} * Issuer's\ Scope\ 1/2\ GHG\ Emissions_i\right)}{Current\ Portfolio\ Value\ (\$M)}$$

As can be seen, the portfolio with alpha model without environmental factors has a 33% larger carbon footprint versus the benchmark. However, including environmental factors, the alpha model will drive down the carbon footprint of the resultant carbon portfolio significantly, to the tune of 48%, a level that is lower than the benchmark.

15 For another example of using NLP in the context of sustainable investment, see Sorensen et al. (2021).

16 For more discussion on predicting forward commercial ESG ratings, see Sorensen et al. (2021).

17 Interested readers can find the alpha performance of Sustainable Global Equity strategy in industry databases, such as eVestment.

Obviously, the level of carbon-footprint reduction depends on the weight that environmental factors are assigned in the alpha model, so the amount of carbon footprint reduction may be more or less than 48%. In the Sustainable Global Equity's alpha model, environmental factors are incorporated with a weight of around 15%. Finally, since the current environmental objective of Sustainable Global Equity is to realize a 50% carbon footprint reduction versus the benchmark, we use portfolio optimization with carbon constraint to reduce the portfolio's carbon footprint by a further 30%.

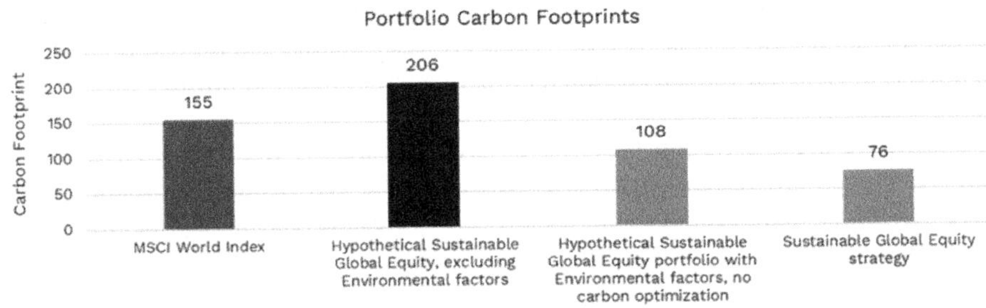

Figure 5.3. *Sustainable global equity environmental performance case study Source: PanAgora. Scope 1 and scope 2 emission data sourced from Trucost. Copyright © 2021, S&P Global Market Intelligence (and its affiliates, as applicable)*

5.5. Carbon offsets

Because most human activity still involves carbon emissions as of this writing, a public equities portfolio with a 100% carbon footprint reduction (i.e. a net-zero portfolio) cannot be achieved without additional tools. Carbon offset contracts may be of use in this context.

At a very high level, carbon offsets[18] are contracts linked to an action or activity that compensates for the emission of carbon dioxide or other greenhouse gases into the atmosphere. As this is still a developing and unregulated market, carbon offsets can represent a variety of underlying actions, and utmost care should be taken by the investor to ensure that the offsets they purchase are carbon-removal solutions rather than simply emission-reduction.

There are two major categories of carbon offsets, depending on the types of organizations that govern the rules under which carbon offsets are created. Institutional carbon offset schemes are issued by various governing bodies (such as the European Union) that set up an overall carbon emission level. The offsets are

18 For more details on carbon offsets, see University of Oxford (2020).

created in the institutional scheme mostly in an emission-reduction sense, as there is a limit on overall carbon emission levels. In contrast, voluntary carbon offset schemes are issued by independent third-party organizations that create offsets by financing a variety of projects, such as forestry development, solar fields and carbon capture development.

Due to the various ways by which carbon offsets may be created, there have been controversial tools in the climate change conversation. Some of the criticisms[19] include:

– How is carbon removal accounted for? Depending on the project, it is not always straightforward or even possible to accurately measure the amount of carbon removed.

– How is additionality accessed? Additionality is an important concept in carbon offsetting and means emission reduction or removal versus a baseline that is often counterfactual. The counterfactual nature of the baseline, since it cannot be observed directly, means that subjectivity is required.

– Is the reduction or removal permanent? Some carbon offset projects may not be permanent[20]. What happens to the offsets that may have already been spent should such a project suffer an adverse event?

– Is carbon offset an excuse to avoid the hard work of direct emission reduction? Quite simply, the most direct way to reduce emissions and achieve net-zero is to reduce emissions. Carbon offset may give these contract owners a false sense of security or satisfaction that they are doing enough to combat climate change and thereby avoiding the hard work that is necessary to truly achieve net-zero[21].

A full discussion on the merits and controversy associated with carbon offset contracts is outside the scope of this chapter. However, it can be argued that if sufficient caution and judgment is applied, carbon offsets can be a useful tool in an investor's toolbox to achieve a net-zero portfolio, especially in case of "hard-to-abate" emissions where mitigation options are either too costly or not technologically feasible[22].

19 For a more complete discussion, see Hecker and Dubourg (2020).

20 For example, carbon offsets issued on the backs of reforestation projects. However, forests are combustible and may burn down.

21 Much like the false idea that taking diet pills precludes the necessity of exercise and a balanced diet to achieve true and permanent weight loss.

22 For example, emissions associated with air travel. These may be eliminated to a large extent but unlikely to be fully eliminated in the foreseeable future.

Having said that, we can offer an example of how carbon offset may be used to achieve a net-zero portfolio. Suppose we have a \$100 million global equity portfolio that has roughly half of the index's carbon emissions, like the Sustainable Global Equity example. Since we only avoided 50% of the benchmark emission, from Figure 5.3, we would still need to offset approximately 5,000 metric tons annually. With each carbon offset typically representative of 1 metric ton of CO_2 and an average cost of between \$10 and \$20 per offset[23], offsetting the portfolio's carbon emissions would cost between \$50,000 and \$100,000, or five to 10 basis points of the portfolio, per annum.

We believe that a combination of stock selection and carbon offsets (depending on the type used and the price) could result in small – and possibly even minimal – performance drag versus the benchmark. An actively managed portfolio that produces consistent positive alpha would be able to cover the cost of carbon offset contracts. In a passive portfolio, their cost would increase the likelihood of underperforming the benchmark on a net basis. As such, we believe that an alpha-seeking strategy is the ideal approach when considering the use of carbon offsets.

5.6. Conclusion

As governments and private businesses grapple with addressing climate change, many investors are considering the implications that a net-zero carbon world might have on their portfolios and how to incorporate the risks and opportunities that exist as the world transitions to a lower-carbon future. We have demonstrated that portfolio carbon exposure may be reduced significantly through strict sector exclusions or by using a passive optimization. We consider both of these approaches to be suboptimal. Divestment from materials, energy and utilities may result in a portfolio that has an active risk of above 1.5%. Investors therefore may be better served using that sort of active-risk budget with an active, alpha-oriented strategy. While passive optimizations may achieve desired levels of carbon reduction, the lack of excess returns and persistent underweights to potential solution-oriented energy and material names may not fully align with investors' needs (i.e. with the dual objectives of portfolio sustainability and meeting return targets).

We believe that by using a well-calibrated active stock selection approach, it is possible to achieve a positive information ratio in addition to carbon-reduction goals. We maintain that going above and beyond the naïve approaches described above via an integration of a diverse set of carbon and environmental considerations within a broader quantitative stock selection model can both improve alpha

23 The dollar figures used here represent the price range of voluntary carbon offsets as of September 2021.

consistency and achieve a client's desired level of carbon reduction. Furthermore, this active approach can serve as a core building block, in combination with other carbon offset measures, for achieving a net-zero portfolio objective over the long term.

5.7. Appendix

We follow Task Force on Climate-Related Financial Disclosures (TCFD)'s definition for portfolio carbon footprint. For more details, please visit https://www.fsb-tcfd.org.

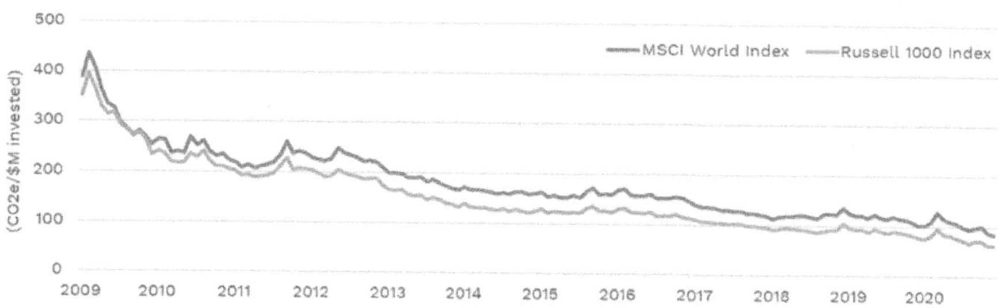

Figure 5.4. *MSCI World Index and Russell 1000 Index Carbon Footprint (January 2009–December 2020) Source: PanAgora, MSCI. For Illustrative purposes only. Certain information ©2021 MSCI ESG Research LLC. Reproduced by permission; no further distribution*

5.8. References

Berg, F., Kolbel, J., Rigobon, R. (2019). Aggregate confusion: The divergence of ESG ratings. *SSRN*, August 20.

Chen, M., and Mussalli, G. (2020). An integrated approach to quantitative ESG investing. *The Journal of Portfolio Management*, 46(3) 65–74. DOI: 10.3905/jpm.2020.46.3.065.

Chen, M., Mussalli, G., von Behren, R. (2021). The unreasonable attractiveness of more ESG data. *The Journal of Portfolio Management*, 48(1), 147–162. DOI: 10.3905/jpm.2021.1.281.

Chen, M., Mussalli, G., Amel-Zadeh, A., Weinberg, M. (2022). NLP for SDGs: Measuring corporate alignment with the sustainability development goals. *Journal of Impact and Sustainable Investing*, jesg.2021.1.035. DOI: 10.3905/jesg.2021.1.035.

Griffin, P. (2017). The carbon majors database CDP carbon majors report 2017. Report, CDP.

Hecker, J.X. and Dubourg, H. (2020). The long-view: The bullish case for carbon offsets and why they need to focus on carbon removal. Paper, JP Morgan ESG Wire.

Keeling, C.D., Bacastow, R.B., Bainbridge, A.E., Ekdahl Jr., C.A., Guenther, P.R., Waterman, L.S., Chin, J.F.S. (1976). Atmospheric carbon dioxide variations at mauna loa observatory, Hawaii. *Tellus A: Dynamic Meteorology and Oceanography*, 28(6).

Khan, M., Serafeim, G., Yoon, A. (2016). Corporate sustainability: First evidence on materiality. *The Accounting Review*, 91, 1697–1724.

Kotsantonis, S. and Serafeim, G. (2019). Four things no one will tell you about ESG data. *Journal of Applied Corporate Finance*, 31(2), 50–58.

Manabe, S. and Wetherald, R. (1967). Thermal equilibrium of the atmosphere with a given distribution of relative humidity. *Journal of the Atmospheric Sciences*, 24(3), 241–259.

Mathis, W. (2021). Greening energy to fight climate threat may cost $92 trillion. Bloomberg [Online]. Available at: https://www.bloomberg.com/news/articles/2021-07-21/greening-energy-to-fight-climate-threat-may-cost-92-trillion?leadSource=uverify%20wall.

National Academy of Engineering and National Research Council (2010). Interim report on causes of the deepwater horizon oil rig blowout and ways to prevent such events. The National Academies Press, Washington DC. DOI:10.17226/13047.

Serafeim, G. and Yoon, A. (2020). Stock price reaction to ESG news: The role of ESG ratings and disagreement. Harvard Business School Accounting and Management Unit Working Paper, no. 21–079 [Online]. Available at: https://ssrn.com/abstract=3765217.

Sorensen, E., Chen, M., Mussalli, G. (2021). The quantitative approach for sustainable investing. *The Journal of Portfolio Management*, 47(8).

Stroebel, J. and Wurgler, J. (2021). What do you think of climate finance? NBER Working Paper Series, Working paper, 29136.

Swiss Re Institute (2021). The economics of climate change: No action not an option [Online]. Available at: https://www.swissre.com/dam/jcr:e73ee7c3-7f83-4c17-a2b8-8ef23a8d3312/swiss-re-institute-expertise-publication-economics-of-climate-change.pdf.

University of Oxford (2020). The Oxford principles for net zero aligned carbon offsetting. Paper, September.

Working Group I (2021). Climate change 2021: The physical science basis. IPCC sixth assessment report.

Wuebbles, D.J., Fahey, D.W., Hibbard, K.A., Dokken, D.J., Stewart, B.C., Maycock, T.K. et al. (eds) (2017). Climate science special report: Fourth national climate assessment, Volume I. U.S. Global Change Research Program, Washington DC, 470, DOI: 10.7930/J0J964J6.

6

Active Paris-aligned Equity Investing

In this chapter, we develop a systematic equity strategy that can attain the climate goals of the Paris Agreement and incorporates climate-related alpha signals. We obtain immediate 50% relative and 7% year-on-year decarbonizations to achieve net zero by 2050 while maintaining minimum exposures to all sectors along with certain required exclusions. An important part of the portfolio is the incorporation of climate-related alpha signals: corporate target setting, carbon resource efficiency and green patents. Compared to a world equity benchmark, the strategy obtains an information ratio of 0.91 over the sample period from January 2013 to October 2021.

6.1. Introduction

Scientists predict “irreversible damage” to our planet from climate change if carbon emissions due to human activities are not significantly reduced[1]. But time is limited, with the best chances of stopping the devastation caused from climate change being an immediate significant reduction of global emissions and aiming for net zero by 2050 – the same level of carbon emissions as 1850–1900 before widescale industrialization. International treaties signed by governments across the globe collectively aim to achieve those goals. The most important of those treaties is the 2015 Paris Agreement, which is an international treaty on climate change with

---

Chapter written by Katharina SCHWAIGER, Jim SNOW, Viktoria-Sophie WENDT and Andrew ANG.

For a color version of all figures in this chapter, see www.iste.co.uk/jurczenko/climate.zip.

All the authors work at BlackRock, Inc. The authors thank Phil Hodges, James Kilburn, Katelyn Gallagher and He Ren. The views expressed here are those of the authors alone and not of BlackRock, Inc.

1 United Nations General Assembly, March 2019.

the objective of limiting global warming to well below 2°C, preferably to 1.5°C, by the end of the 21st century[2].

The Paris Agreement recognizes that not only do governments play an important role in these efforts, but achieving these climate objectives requires the investment community – both asset owners and asset managers – to allocate capital towards technologies, companies or countries that are fundamental for the energy transition. A Paris-aligned portfolio is one with an investment strategy that is consistent with achieving these goals. To help investors achieve Paris-aligned portfolios, the EU Technical Expert Group on Sustainable Finance (EU TEG) and the Institutional Investors Group on Climate Change (IIGCC) have developed minimum standards that portfolios must satisfy[3].

6.2. Standards of Paris-aligned benchmarks

The EU TEG's Paris-aligned guidelines are emerging as the global standard for what defines portfolio alignment with the goals of the Paris Agreement, which embeds the 1.5°C temperature target with explicit decarbonization goals. In particular, the TEG specifies: (1) certain exclusion criteria; (2) a maximum deviation of sector weights; (3) a decarbonization trajectory for the benchmarks; and (4) taking into account forward-looking guidance for corporate emissions. These guidelines are relevant for active portfolio managers as minimum guidelines against which to benchmark themselves.

We apply these guidelines to developed and emerging equity market exposures in the following ways.

First, we significantly decrease the overall greenhouse gas (GHG) emission intensity compared to the underlying investment universes. We reduce emission intensity by 50% versus the MSCI World, MSCI World Small Cap and MSCI Emerging Market indices, covering both scope 1 and 2 emissions (scope 3 emissions

2 Countries must set emission targets every year under the Paris Accord, but there are no consequences for countries failing to meet those targets. In contrast, the first major international treaty on climate change, the 1997 Kyoto Protocol, is legally binding.

3 See EU Technical Expert Group on Sustainable Finance, 2019, TEG Final Report on Climate Benchmarks and Benchmarks' ESG Disclosures, and IIGCC, 2021, Net Zero Investment Framework: Implementation Guide.

are gradually phased in over the next four years [4]). Emission intensity is measured as the total emissions to enterprise value using data from MSCI ESG Research[5].

Second, we remain sufficiently exposed to all sectors relevant to the fight against climate change. We do not divest from various high-risk industries or underweight high impact sectors to achieve the portfolio's goals[6] – we keep the exposure to these sectors at least equal to that of the benchmark. This is consistent with recognizing that all sectors, including energy, must move towards the goal of net zero. Being invested in all sectors is also consistent with capital markets reflecting all sectors of economic activity in society.

We further reduce GHG emission intensity on an annual basis. We commit to ensuring that our equity exposure will become less carbon-intensive over time by applying a 7% self-decarbonization reduction on average per annum, in accordance with the guidelines.

We exclude companies with exposure to controversial activities. In accordance with the TEG guidelines, we remove companies involved with controversial weapons, civilian firearms, nuclear weapons, thermal coal, tar/oil sands, tobacco and UN Global Compact violators[7]. We also exclude companies involved in environmental controversies; oil, coal and gas exploration/processing; and, high-intensity electricity generation. We summarize these screens in the Appendix (section 6.7).

Finally, we upweight companies based upon forward-looking alignment metrics. We increase exposure to companies using metrics such as industry-adjusted ESG score, emissions-to-sales, green patents-to-assets and corporate emission

---

4 While scope 3 plays an important role in reducing global emissions, estimating scope 3 is significantly harder than scope 1+2 measures. The EU TEG recommends a phase in period to resolve those estimation issues.

5 To adjust for company size, carbon intensity measures adjust carbon intensity by sales, revenue or enterprise value. The EU TEG requirements for the Paris Alignment benchmark specify using the denominator of enterprise value. This, however, does not equate to reducing absolute carbon dioxide or other GHG emitted (see Bolton et al. 2022).

6 High impact sectors are identified using NACE sector classifications. NACE is the industry and sector classification scheme used in the European Union and stands for *nomenclature statistique des activités économiques dans la Communauté européenne*. NACE is the sector classification scheme required under the EU Technical Expert Group (EU TEG) for Paris-aligned portfolios. See EU Technical Expert Group on Sustainable Finance (2019), Taxonomy Technical Report.

7 A list of divestment screens can be found at https://www.blackrock.com/uk/individual/literature/investment-guide/blackrock-baseline-screens-in-europe-middleeast-and-africa.pdf.

commitments from the science-based target initiative (SBTI). Incorporating this forward-looking information is an important form of climate-related alpha, as we now detail.

6.3. Climate-aware alpha drivers

Embedding a sustainable philosophy and transitioning to a net zero world potentially presents new investment opportunities to companies (see Chan et al. 2020). Companies that "do good" are defined as those making headway on the United Nation's Sustainable Development Goals (UN SDGs). In addition, those that also meet the ambitions as set out in the Paris Agreement may gain in firm value with greater investor demand in the short run, but perhaps the increase in prices over the short term may cause these companies to exhibit lower expected returns in the long run (see Avramov et al. (2021) and Pastor et al. (2021)). In addition, it is possible that more sustainable companies benefit from a higher demand of their products and services (see Edmans 2011; Eccles et al. 2014).

To capture these potential sustainable characteristics that may predict firm returns, we integrate climate-aware alphas into our equity portfolios. These are defined as "quality factor" insights and therefore stem from the same underlying economic rationale as classical factor premiums: rewards for bearing risk, structural or market impediments, and investors' behavioral biases (see Ang 2004). We present three examples of climate-aware alphas: carbon resource efficiency, green patents and corporate target setting.

6.3.1. *Carbon resource efficiency*

A main source of systematic risk under the Paris Agreement is a firm's carbon emissions (total and intensity). Our portfolio level constraint that aims to achieve a carbon emission reduction relative to the benchmark hedges the portfolio exposure to climate risk. However, we go a step further: Kazdin et al. (2021) demonstrate that firms with lower-carbon emission intensity exhibit higher excess returns. This is consistent with firms having lower-carbon emissions having higher productivity. Kazdin et al. (2021) also find that companies with low-carbon emission intensity exhibit other traditional quality factor characteristics based on earnings and earnings forecasts. To measure carbon emission intensity, we use the aforementioned MSCI dataset, which collates company-specific direct (scope 1) and indirect from power generation (scope 2) GHG emission data from companies' public documents and the Carbon Disclosure Project (CDP).

6.3.2. *Green patents*

Patents are an important form of creative output and companies focusing on innovation tend to enjoy a competitive advantage versus their peers (see, among others, Lee et al. (2019)). In particular, green patents play a crucial role in meeting the United Nations Sustainable Development Goals (UN SDGs) and include developments focused on water and waste treatment, lifesaving devices and clean energy. We collect data on global patents via Google Patents Data and then create sub-sets using International Patent Classification (IPC) codes relevant to a "greener" economy (Chan et al. 2020). Given the inherent uncertainty of Research and Development (R&D) in that not all investments result in commercial success, the innovative output captured in patents is risky. If we interpret patents as the capitalization of certain investments, with R&D devoted to patents being a form of investment following Cochrane (1991) and Zhang (2005), and others, then investors should earn a return premium for bearing these additional investment risks in new, unproven inventions. The special focus on green patents in a portfolio reflects not only that the SDGs are important to society, as recognized by the UN, but also that efforts to solve for these goals could result in profitable opportunities for firms.

6.3.3. *Corporate target setting*

As part of the transition to a low-carbon economy, an ever-increasing number of companies have publicly disclosed their temperature targets. One initiative that publicly collects company-level information is the science-based target initiative (SBTi) which aims to "mobilise companies to set science-based targets and boost their competitive advantage in the transition to the low-carbon economy"[8]. The SBTi is a collaboration between the CDP, the United Nations Global Compact, the World Resources Institute (WRI) and the World Wide Fund for Nature (WWF) and is one of the We Mean Business Coalition commitments. The initiative defines and promotes best practices in science-based target setting, offers resources and guidance to reduce barriers to adoption, and independently assesses and approves companies' targets.

As of October 2021, more than 2,000 companies have joined the SBTi and pledged to reduce their emissions in line with climate science. Companies can go one step further and not just commit to a reduction, but also set a target ranging from 1.5°C, well below 2°C, 1.5°C/well below 2°C, or 2°C. Of this group, more than half of the companies are from Europe, and approximately 20% each from Asia and North America. There are 656 companies from the European Union which have

8 See https://wwf.panda.org/discover/our_focus/climate_and_energy_practice/what_we_do/climate business/science_based_targets_initiative/.

signed, followed by the UK (350 companies), the US (327 companies) and Japan (167 companies). These represent companies from a broad mix of sectors including companies within the high-emitting sectors. With an increased focus on climate change over the past two years, coupled with governmental recovery packages linked to "green" policies, we assume the number of sign-ups will further accelerate in the near future.

We find that companies that commit to these targets tend to have higher gross profitability, which in itself is a sign of quality (see Novy-Marx 2013). Comparing the standardized gross profitability score of a global equity universe as represented by the MSCI World constituents and then de-constructing it into companies that have made the commitment, we find that committed companies almost entirely exhibit higher gross profitability (with a few outliers being the exception; see Figure 6.1) than companies without commitments. Committed companies also have lower-carbon emissions than their peers, and empirical evidence shows that they also have lower-carbon emissions one year and two years into the future, suggesting that making the commitment helps to achieve the goals.

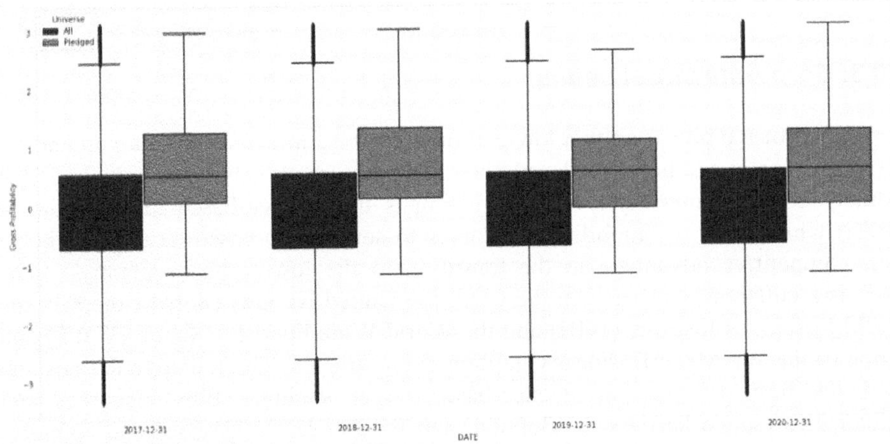

Figure 6.1. *Gross profitability z-scores with and without SBTi commitment. Source: MSCI, SBTI, Worldscope. Date range: 2017–2020. BlackRock calculations*

6.4. Empirical results

In this section, we explore the effects of incorporating the risk-mitigating and opportunity-oriented Paris Alignment guidelines, including our own climate aware-alpha insights, into our equity portfolios.

6.4.1. *Decarbonization pathway*

With the aim of becoming net zero by 2050 at a portfolio level, the aspect of decarbonization is the key component of any Paris-aligned strategy. As described in the earlier section, there are two types of decarbonization matter: relative versus a benchmark and self-decarbonization. The minimum requirements demand a 50% relative reduction alongside a 7% per annum decarbonization rate. An illustrative schematic of the resulting pathway is shown in Figure 6.2.

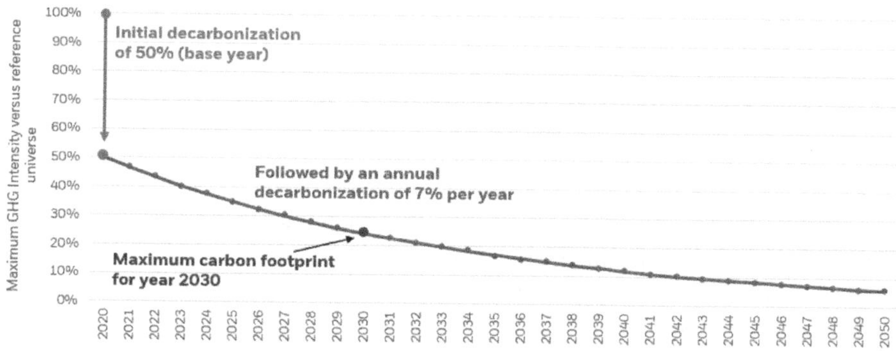

Figure 6.2. *Illustrative decarbonization path of a Paris-aligned portfolio. Source: EU TEG Guidelines. BlackRock calculations*

The base year, which is the year the portfolio starts its decarbonization trajectory, is 2020. The portfolio will gradually reduce its carbon emissions, as measured by carbon emission intensity, in each subsequent year. In year 0, the portfolio's carbon emission intensity is 50% lower than its benchmark, but in year 1, the portfolio needs to be an additional 7% lower, resulting in a reduction of 53.5% (7% of 50% plus the original 50%) versus the benchmark, and so on. Figure 6.2 shows the resulting pathway as a maximum upper bound versus the portfolio's benchmark.

6.4.2. *Climate-aware alpha*

Table 6.1 reports the simulated performance of climate-aware alpha portfolios – both as stand-alone signals and combined in a portfolio. We form optimal long-only portfolios (or characteristic portfolios, following the terminology of Grinold and Kahn (1999)) in Table 6.1 as follows. We take MSCI GEM3 as the risk model and neutralize with respect to industry and size exposures. Our alpha for each company is its industry-adjusted ESG Score (provided by MSCI ESG Research), its proprietary Carbon Resource Efficiency, Green Patents, or Corporate Target Setting z-score, or the final climate-aware alpha which combines the ESG score (60% weighting) with the three climate-aware alphas: 20% into carbon resource efficiency, 10% into green patents to assets and 10% into corporate target setting. We set the parameters of the mean-variance optimization to have an active risk of 1.0%. We construct the portfolios using mean-variance optimization:

$$Max\ \alpha_P - \frac{1}{2}\lambda\sigma_P^2$$

$$s.t.\ \ w'1 = 1$$

$$l_i \leq w'b_i \leq u_i$$

$$w \geq 0 \qquad [6.1]$$

where α_P is the portfolio's active climate-aware alpha; σ_P^2 is the portfolio's active risk; λ is a risk aversion parameter; w are the portfolio weights; 1 is a vector of ones; l_i and u_i are the lower and upper bounds, respectively, with respect to characteristics i: country, sector, high impact sector and industry; and b_i are the corresponding exposures of characteristic i. For any exclusionary screen, the portfolio weight is set to zero.

To construct Paris-aligned portfolios, the optimization in equation [6.1], we add constraints that specify that the carbon emissions (CE) must be improved by at least a certain minimum level, given by l_{CE}. This includes both the relative and the self-decarbonization rates:

$$w'CE \leq l_{CE} \qquad [6.2]$$

	MSCI World	**MSCI World Small Cap**	**MSCI Emerging Markets**
ESG score	0.85	0.63	1.00
Carbon resource efficiency	0.87	1.06	0.41
Green patents	0.92	0.19	0.20
Corporate target setting	0.74	0.82	0.22
Climate-aware alpha	1.00	0.60	0.89
Base Paris-aligned	1.00	0.39	0.15

Table 6.1. *Information ratios (IRs) of climate-aware alphas. Source: MSCI, SBTI, Google Patents. Date range: January 2013–December 2021. BlackRock calculations*

Table 6.1 shows that, over the sample period from January 2013 to October 2021, all signals outperform their respective underlying benchmark. The ESG score by itself produces relatively high IRs of 0.85, 0.63 and 1.00 in the world, world small cap and EM universes, respectively (see also Chan et al. 2020). Similarly to Chan et al. (2020) and Kazdin et al. (2021), we show that overweighting (underweighting) companies with a higher (lower) ratio of green patents to total assets or lower (higher)-carbon emission intensity can generate a positive alpha. The green patent signal produces the highest IR of 0.92 in the MSCI World universe, but still positive IRs of 0.19 and 0.20 in World Small Cap and EM universes, respectively. This is consistent with smaller companies and many companies in EM countries having fewer resources to invest in R&D leading to patents. Combining all four signals in the Climate-Aware Alpha line produces the highest IR of 1.00 in the largest MSCI universe.

Additionally, we look at a backtest that minimizes the tracking error to the benchmark while meeting the minimum Paris-aligned guidelines (labeled "Base Paris-aligned"). The tracking error is below 50 bps, but their IRs vary from 1.00 for the world universe, but are weaker for small cap and EM with 0.39 and 0.15, respectively. This is perhaps consistent with there being more attention on climate topics in developed markets and the relatively higher emissions in developed markets. In small caps and EM, adding climate-aware alphas produces significant increases in the IRs to 0.60 and 0.89, respectively. This speaks to the importance of climate-aware alphas which can generate excess returns.

6.4.3. *Incorporating climate-aware alphas and decarbonization*

Incorporating carbon reduction targets into diversified systematic equity portfolios does not mean changing their desired climate outcomes. In particular, Figure 6.3 confirms that the relative decarbonization and the annual decarbonization rate of the portfolio set out by the TEG guidelines can be achieved with minimal impact to the portfolio's climate-aware alpha exposures. The climate-aware alphas themselves also help to achieve the Paris targets while simultaneously generating outperformance.

Figure 6.3 graphs change in the exposures in climate-aware alphas α_P as a function of relative decarbonization and additional year-over-year absolute decarbonization rates in panels A and B, respectively. The base case is the alpha scores for each stock in the optimized portfolios in Table 6.1 based on the MSCI World universe, which is constructed without relative decarbonization or year-over-year decarbonization. As we impose constraints that the portfolio must achieve higher decarbonization targets, the alpha exposures change from the base case.

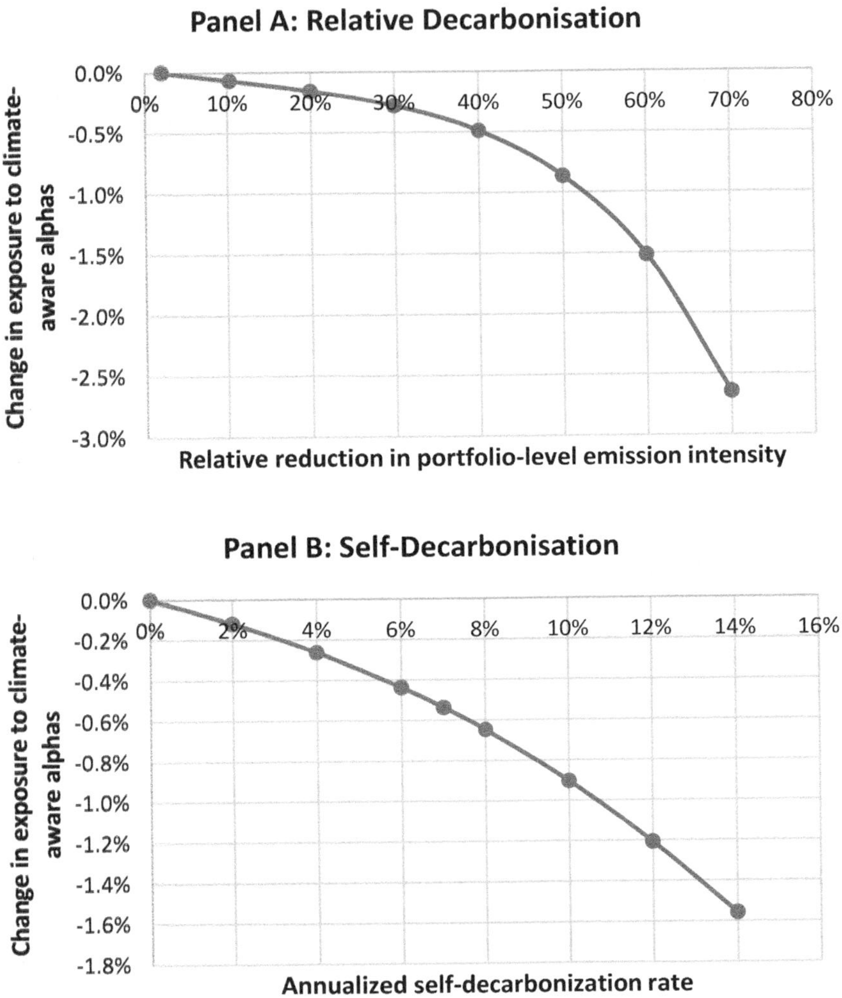

Figure 6.3. *Exposure change of climate-aware portfolios. Source: MSCI, SBTI, Google Patents. Date range: January 2013–December 2021. BlackRock calculations*

Panel A of Figure 6.3 shows that to achieve a 50% relative decarbonization, the alpha scores are less than 1% lower than the base case. Even to obtain a 70% portfolio-level decarbonization, the alpha exposure decrease is only 2.6%. Panel B shows that to achieve an additional 7% year-over-year absolute decarbonization rate, the change in the alpha exposures is 0.5%. It is important to note that the decarbonization in both panels A and B is not obtained by disinvesting from certain sectors; the Paris-aligned portfolios require a sector neutral construction. The intuition behind these results is that there are relatively few large gross emitters among all companies, and the optimization removes those companies while trading off alphas and risk. The relatively few large emitters allow the optimization to achieve very large reductions in GHG emissions to achieve the Paris goals.

Thus, the pathway to a 1.5°C scenario is achievable for core portfolios, while also maintaining a high degree of exposure to the "opportunity"-based climate-aware alphas (carbon resource efficiency, green patents to assets and corporate target setting) related to the climate transition.

6.4.4. *Systematic active Paris-aligned strategies*

Our next step is to combine all the minimum guidelines into an equity portfolio while maximizing climate-aware alphas alongside ESG scores. Incorporating social and governance aspects alongside environmental issues is in accordance with the spirit of the Paris Agreement in terms of preserving the planet for all global citizens. Failing good governance or creating social injustice would harm poorer countries more than affluent nations. By achieving a better aggregated ESG score, our portfolio minimizes this potential inequality.

We construct our systematic equity strategies using an optimization approach with a low ex ante tracking error of 1% on three universes: MSCI World, MSCI World Small Cap and MSCI Emerging Markets. The goal is to spend the tracking error budget on maximizing the portfolio level ESG score (60% weighting) alongside the three climate-aware alphas: 20% into carbon resource efficiency, 10% into green patents to assets and 10% into corporate target setting. The portfolios additionally incorporate exclusionary screens, the decarbonization rates (assuming 2013 as the base year for back-testing purposes) and a zero underweighting of high impact sectors, as described earlier. To avoid any unintended biases, the portfolios additionally constrain security, country and industry exposures versus the underlying benchmark.

Figure 6.4 shows the simulated performance of the Paris-aligned equity portfolios versus the underlying benchmarks. All three strategies outperform their respective benchmarks, driven by the addition of the climate-aware alphas with IRs of 0.91 for MSCI World, 0.52 for MSCI World Small Cap and 0.81 for MSCI Emerging Markets over the sample period from January 2013 to October 2021, while achieving a relative decarbonization by at least 50% and a 7% year-over-year self-decarbonization from 2014 onwards. Furthermore, by including appropriate security, country and industry exposure constraints, the Paris-aligned equity portfolios maintain similar diversified exposures as the underlying benchmarks. As a comparison, the minimum tracking error portfolio meeting the baseline Paris-aligned has an ex post tracking error below 50 bps and approximately zero excess returns.

Panel A: with climate-aware alphas

	Tracking error (ex ante)	Tracking error (ex post)	Excess return	IR
MSCI World	1%	1.23%	1.12%	0.91
MSCI World Small Cap	1%	1.33%	0.69%	0.52
MSCI EM	1%	1.36%	1.10%	0.81

Panel B: base Paris-aligned without climate-aware alphas

	Tracking error (ex ante)	Tracking error (ex post)	Excess return	IR
MSCI World	0.21%	0.35%	0.35%	1.00
MSCI World Small Cap	0.10%	0.23%	0.09%	0.39
MSCI EM	0.32%	0.45%	0.07%	0.15

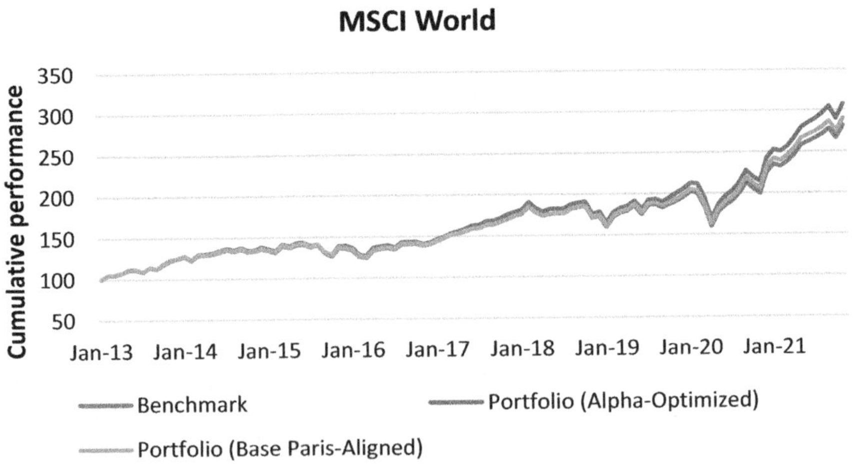

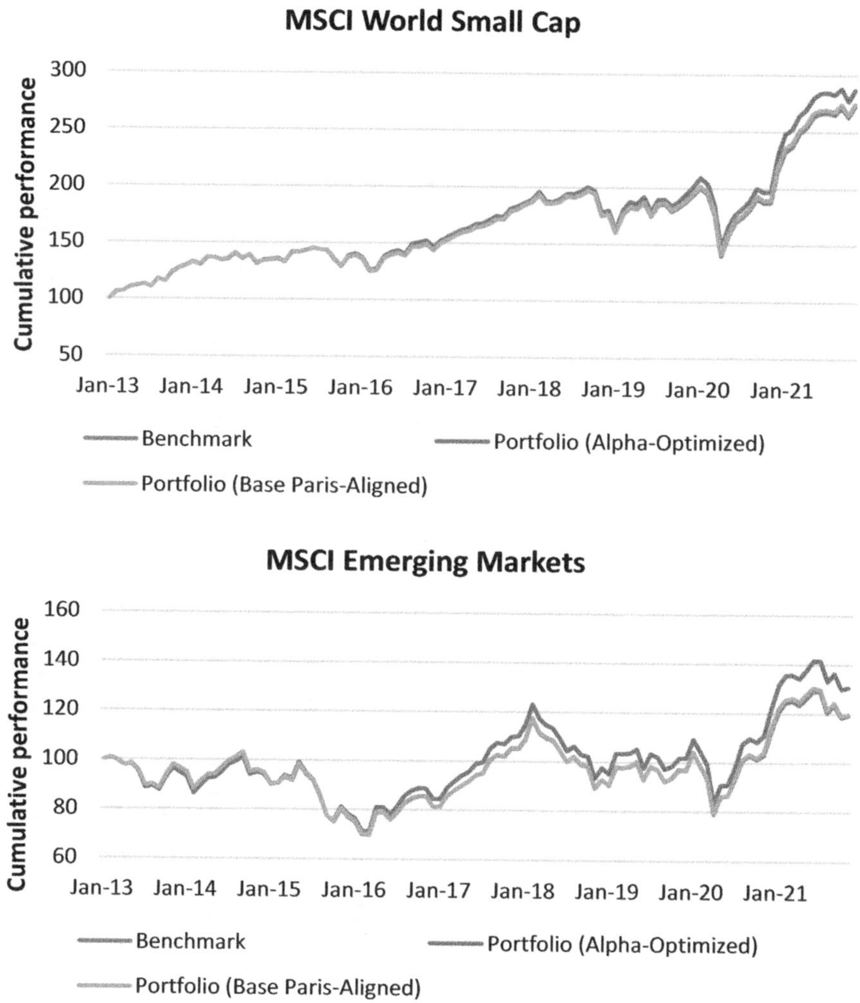

Figure 6.4. *Performance of Paris-aligned strategies. Source: MSCI, SBTI, Google Patents. Date range: January 2013 to December 2021. BlackRock calculations*

Paris-aligned strategies must demonstrate a clear pathway in reducing carbon emissions. While our base year of starting that pathway is 2020, we can look at a back-tested pathway to better understand the effects of exclusionary screens, decarbonization rates and climate-aware alphas in their contribution to reducing portfolio-level carbon emissions.

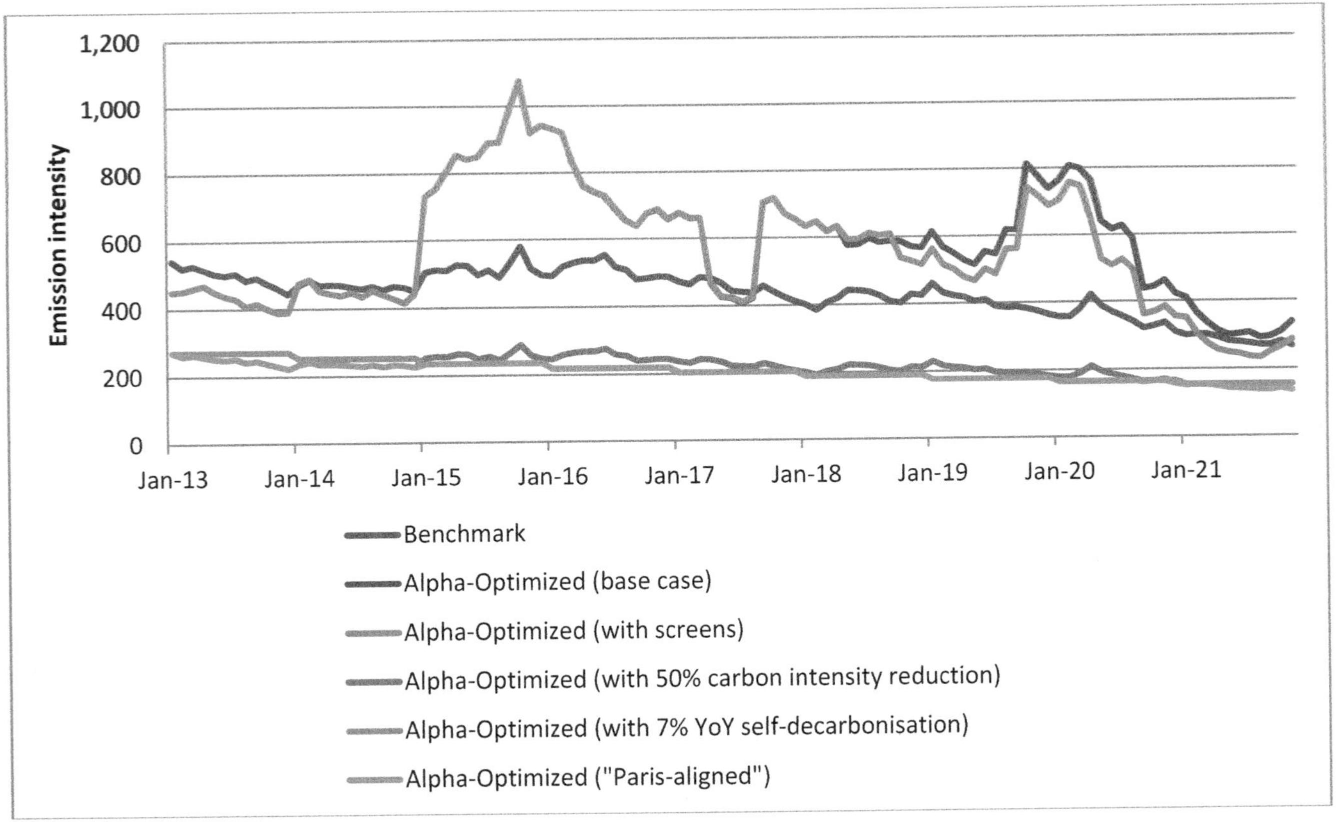

Figure 6.5. *Emission intensity of Paris-aligned strategies. Source: MSCI, SBTI, Google Patents. Date range: January 2013 to December 2021. BlackRock calculations*

In Figure 6.5, we plot the emission intensity of a global equity portfolio versus its benchmark based on MSCI World. There is no guarantee that the portfolio-level emission intensity will be lower than the benchmark just by optimizing on climate-aware alphas and actually, as we demonstrate, it is on average 34% higher over the sample period, as is demonstrated in the "Base Case" in Figure 6.5. Adding exclusionary screens ("with screens") does reduce the emission intensity, but again, this does not guarantee a reduction versus the benchmark. In this case, the starting reduction in carbon emissions is significant but there is no pathway reduction, which Figure 6.4 demonstrates with the "50% reduction" portfolio as compared to the year-on-year self-decarbonization ("7% self-decarbonization"). Each minimum standard contributes to achieving a Paris-aligned portfolio ("Paris-aligned") – but leaving out any of the components would result in the portfolio failing in its Paris-alignment goal.

6.5. Conclusion

The alignment with the Paris Agreement is critical not only to companies, countries and society but also to investors who are looking to build sustainable portfolios – to aim to limit global warming to 1.5°C by the end of the century by reducing net carbon emissions to zero by 2050. We have demonstrated how equity strategies can be future-proofed by aligning with the Paris Agreement, thus mitigating climate risk, and also benefitting from climate-aware opportunities that can generate excess returns.

6.6. Appendix: Paris-aligned equity strategy screens

Screen	Categories	Measure	Further details
Controversial weapons*	Controversial weapons	Zero tolerance	A legal requirement in various European countries, including Belgium and Luxembourg. The BlackRock Global Funds and BlackRock Strategic Funds ranges exclude companies with any tie to controversial weapons (cluster munitions, landmines, depleted uranium weapons, biological/chemical weapons, blinding lasers, non-detectable fragments and incendiary weapons)

Controversial weapons* (cont'd)	Nuclear weapons	Minimal tolerance	Chemical and biological, blinding laser, non-detectable fragments and incendiary weapons are viewed as equally controversial by a majority of clients and complete the category
Fossil fuels	Thermal coal	>5% revenue from production and power generation	Thermal coal is burned to generate electricity. It is more damaging to the environment than metallurgical coal and emits the highest proportion of carbon dioxide. Metallurgical coal is burnt at much higher temperatures in steel production
	Tar sands (oil sands)	>5% revenue from production	Otherwise known as oil sands, this extracting and refining oil process has been noted to generate 20% more carbon pollutants than conventional crude. More energy is needed for extraction and refinement
UN Global Compact*	Human Rights, Labor Standards, environment, anti-corruption	Violators	UNGC violation is a qualitative flag and recognized as a vital screen within the sustainable investing community. Human Rights violators are encapsulated within this screen. Companies identified by providers can differ, based on how the information is sourced and assessed, as well as how the companies are monitored over time
Civilian firearms	Civilian firearms producers and retailers	All producers >5% retailer	Demand has increased for this screen, especially in the US. Excluding civilian firearms appears to have de minimis impact on portfolio characteristics across regional exposures
Tobacco	Producer, aggregate revenues	All producers >5% aggregate revenues	Tobacco constraints are increasingly being demanded from both institutional and wealth clients

Activity exclusions			
Coal*	Exploration or processing activities	>1% revenues	
Oil*		>10% revenues	
Natural gas*		>50% revenues	
Environmental controversies*	(Very) severe environmental controversies	MSCI env. controversy score	
Electricity providers*	High GHG intensity electricity providers	>50% revenues	Companies who derive 50% or more of their revenues from electricity generation with a GHG intensity of lifecycle GHG emissions above 100 g CO_2e/kWh

Source: BlackRock Sustainable Investing, EU TEG Report, as of June 2021. The Paris-aligned benchmark guidelines explicitly require the exclusions denoted with an *. The remaining exclusions comprise BlackRock Sustainable Investing's baseline screens. For illustrative purposes only and subject to change without notice.

6.7. References

Ang, A. (2014). *Asset Management: A Systematic Approach to Factor Based Investing*. Oxford University Press, New York.

Avramov, D., Lioui, A., Liu, Y., Tarelli, A. (2021). Dynamic ESG equilibrium. Working paper.

Bolton, P., Kacperczyk, M., Samana, F. (2022). Net-zero carbon portoflio alignment. *SSRN*, September 16.

Chan, Y., Hogan, K., Schwaiger, K., Ang, A. (2020). ESG in factors. *Journal of Impact and ESG Investing*, 1, 26–45.

Cochrane, J.H. (1991). Production-based asset pricing and the link between stock returns and economic fluctuations. *Journal of Political Economy*, 104, 572–621.

Eccles, R.G., Ioannou, I., Serafeim, G. (2014). The impact of corporate sustainability on organizational processes and performance. *Management Science*, 60, 2835–2857.

Edmans, A. (2011). Does the stock market fully value intangibles? Employee satisfaction and equity prices. *Journal of Financial Economics*, 101, 621–640.

Grinold, R.C. and Kahn, R.N. (1999). *Active Portfolio Management: A Quantitative Approach for Producing Superior Returns and Controlling Risk*. McGraw-Hill, New York.

Kazdin, J., Schwaiger, K., Wendt, V., Ang, A. (2021). Climate alpha with predictors also improving company efficiency. *Journal of Impact and ESG Investing*, 2(2), 35–56.

Lee, C.M.C., Sun, S.T., Wang, R., Zhang, R. (2019). Technological links and predictable returns. *Journal of Financial Economics*, 132, 76–96.

Novy-Marx, R. (2013). The other side of value. *Journal of Financial Economics*, 108, 1–28.

Pastor, L., Stambaugh, R.F., Taylor, L.A. (2021). Sustainable investing in equilibrium. *Journal of Financial Economics*, 142, 550–571.

Zhang, L. (2005). The value premium. *Journal of Finance*, 60, 67–103.

7

Green Alpha

7.1. Introduction

In this chapter, we explore one specific approach of ESG integration in investment management. In particular, we introduce ESG – with a focus on environmental data – as stock-selection signals, alongside conventional value, momentum and profitability factors. We start from the MSCI ESG database, including the headline ESG score, three pillars (environmental, social and governance), and more importantly, nearly 300 key attributes that measure every aspect of a company's sustainability. We conduct an in-depth study of ESG rating's characteristics from a factor lens. Furthermore, we introduce a suite of novel stock-selection factors based on ESG data – from carbon emissions, clean technology, supply chain labor management, to corruptions and instability. Unlike conventional stock-selection factors whose performance has retreated considerably in recent years, sustainability metrics not only deliver highly uncorrelated excess returns, but also enjoy strong alpha since 2017. Therefore, "doing well by doing good" is certainly achievable.

7.2. Research methodology

First, we would like to lay out the foundations of this research. Given the unique features of ESG data, we need to carefully adjust country, sector and other style factors, before we can study the predictive power of sustainability factors.

Chapter written by Yin LUO.

For a color version of all figures in this chapter, see www.iste.co.uk/jurczenko/climate.zip.

7.2.1. *Region classification*

For the purpose of this chapter, we divide the global equity space into four geographic regions – US, Developed Europe (including the UK), Other Developed Markets (Japan, Canada, Australia and New Zealand, Hong Kong, and Singapore), and EM (LATAM, Emerging EMEA, Emerging Asia and China). Within each country, we focus on the all-cap universe. In the US, our research universe is the point-in-time Russell 3000 index. Similarly, we use the S&P/TSX Composite index for Canada, MSCI China A Onshore Index for China Onshore market, and S&P BMI for all other countries.

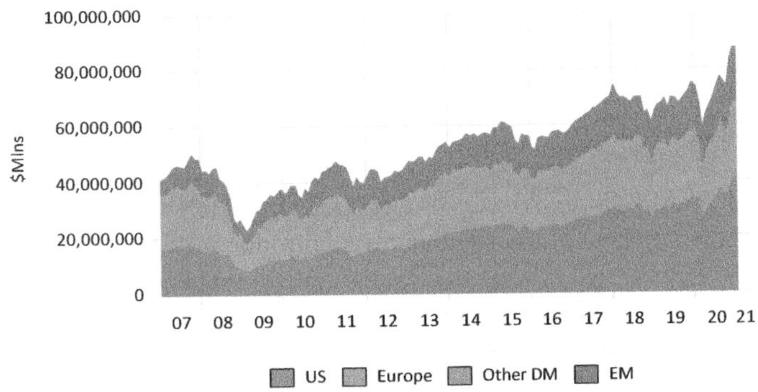

a) Market capitalization by region

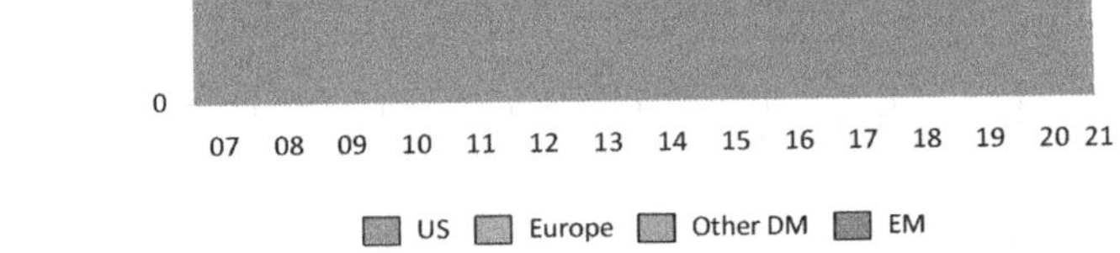

b) # of stocks by region

Figure 7.1. *QES research universe by region. Sources: Bloomberg Finance LLP, FTSE Russell, MSCI, Refinitiv, S&P Capital IQ, Wolfe Research Luo's QES*

As shown in Figure 7.1(A), by market capitalization, the US accounts for nearly half of the global equity market, while the other three regions are roughly of the same size. By number of stocks, however, the four regions are more balanced (see Figure 7.1(B)).

7.2.2. ESG-specific industry classification

The materiality of each key ESG issue varies greatly by industry, especially on the environmental and social pillars. For example, carbon emissions are critical for electric utilities, oil and gas exploration, and production industries, but are generally of little relevance to biotechnology firms. On the other hand, privacy and data security plays a dominant role for banks, but poses little risk to beverage companies. MSCI's ESG final ESG rating is adjusted within each of the roughly 60 industries defined by the vendor. These industries are based on the GICS (Global Industry Classification Standards) developed by MSCI and S&P Global.

For our modeling perspective, we need to strike a balance between sample size and sample homogeneity. There is no theory to suggest the "optimal" point. Therefore, it is mostly an empirical question. For this chapter, we divide our investment universe into 26 industries. Our 26 industries are mostly based on GICS Level 2 – industry groups. There are currently 24 industry groups. In the materials industry group, we separate three industries – chemicals and packaging, metals and mining, and paper and forest products.

As shown in Figure 7.2, in the US, the largest industries (by number of stocks) are BioPharma, banks, capital goods, and software and services. In Europe, the picture looks a bit different, where capital goods, real estate, BioPharma and diversified financials round up the top four. In other developed markets, capital goods, real estate, software and services, and chemicals and packaging are the four largest industries. Lastly, in EM, the top four are capital goods, technology hardware and equipment, chemicals and packaging, and BioPharma.

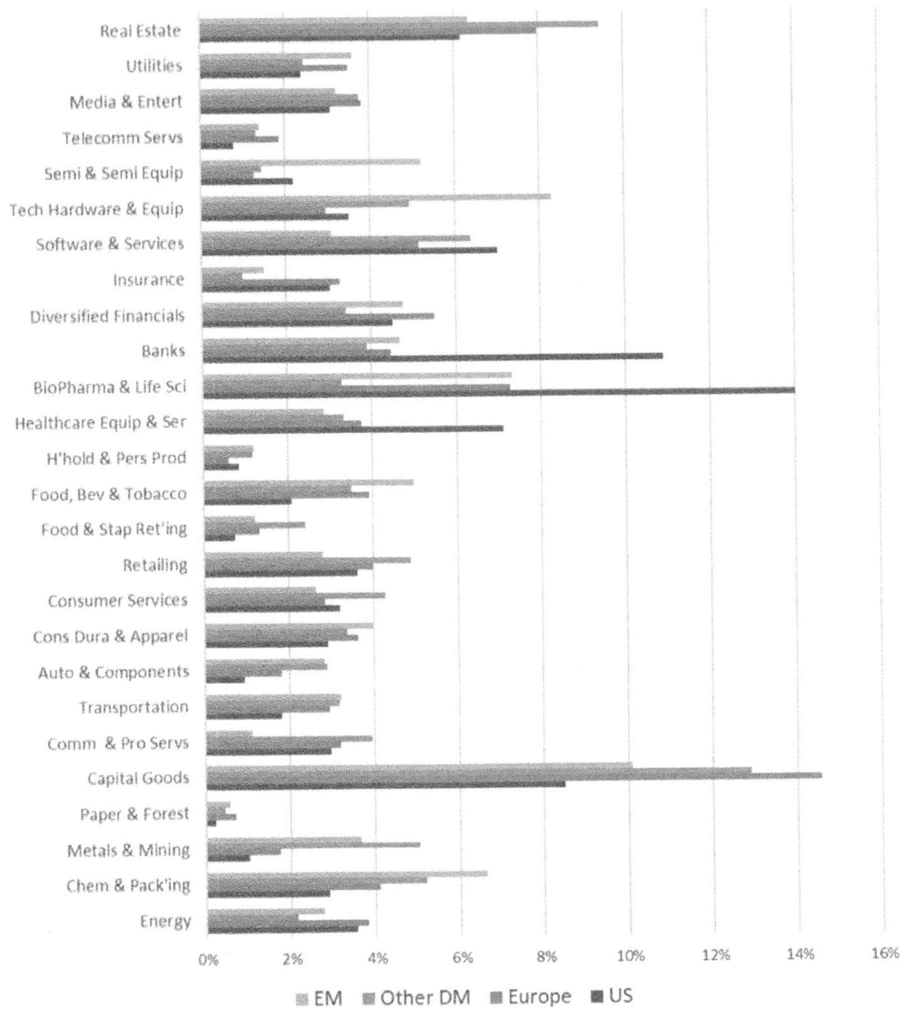

Figure 7.2. *QES research universe by ESG sectors (January 2021). Sources: Bloomberg Finance LLP, FTSE Russell, MSCI, Refinitiv, S&P Capital IQ, Wolfe Research Luo's QES*

7.2.3. *Common style factors*

In this chapter, we heavily use a small set of common equity style factors to assist us in understanding ESG investing. The list of 15 factors is shown in Figure 7.3. We choose these 15 factors, because they are widely used in the investment community for alpha generation, factor (or small beta) investing and risk management. They have also been thoroughly studied in academic research.

Style	Factor	Description	Direction
Value			
	Book-to-Market	Book value of equity/Market value of equity	High-Low
	Earnings Yield	Last reported trailing 4Q EPS/Share price	High-Low
Growth			
	FY2 Expected Earnings Growth	Consensus FY2 EPS versus Consensus FY1 EPS	High-Low
	Long-Term Expected Growth	Consensus long-term expected growth rate	High-Low
Price Momentum			
	1M Reversal	Last one-month total return	Low-High
	12M-1M Momentum	Last 12-month total return excluding the most recent month	High-Low
Analyst Revisions			
	3M EPS Revision	Consensus FY1 EPS revision, from three-month ago	High-Low
Quality			
	ROE	Trailing 4Q return on equity	High-Low
	Debt/Equity Ratio	Long-term debt/equity	Low-High
Risk			
	Size	Float-adjusted market capitalization	Low-High
	Illiquidity	Amihud [2002] illiquidity measure	High-Low
	Beta	CAPM beta	Low-High

Figure 7.3. *QES common factors.*
Sources: Wolfe Research Luo's QES

7.2.4. *Backtesting methodology*

In this research, we mostly follow a consistent set of backtesting methodology to assess the predictive power of various factors on future stock returns (see Luo et al. (2017) for details). The factor portfolios are constructed as long/short quintile portfolios[1]. All stocks in each of the four regions (US, Europe, other developed markets and EM) are sorted every month, based on the factor of interest. The top 20% stocks are selected for the long portfolio and the bottom 20% fall into the short

1 The portfolio construction process is relatively naïve by the standard of quantitative equity investing. However, it is easy to implement, fully transparent and consistent with the standard finance literature.

basket. Stocks in the long (and short) bucket are equally weighted. Factor portfolios are rebalanced monthly and returns are tracked daily without transaction cost. To capture pure factor performance, we adjust for industry effect in all regions, using our ESG industry classification standard defined in the previous section. For multi-country regions (i.e. Europe, other developed markets and EM), we further neutralize country effect. All factor returns are computed using local currency.

It is important to highlight that we do not adjust the expected direction of each factor. For example, a higher ESG rating is associated with a stronger ESG practice. Whether a higher rating will lead to higher or lower future stock returns is subject to debate. As we will elaborate in the rest of this chapter, the directions of most ESG-based signals are "intuitive", meaning green firms with better ESG footprints typically outperform brown companies that are lagging on sustainability. Consequently, our standard long/short portfolio may produce either a positive or negative average return, depending on whether higher scores are linked to better ESG practices or not.

7.3. MSCI ESG rating

Most asset owners and asset managers do not have enough in-house resources and expertise to assess each investee company's and investment instrument's ESG risks and opportunities. Therefore, third-party ESG rating agencies have flourished in recent years. Many academic studies on sustainable investment also rely heavily on ESG ratings (e.g. Liang and Renneboog 2017; Lins et al. 2017).

Different rating agencies use different frameworks, metrics and data sources. How ESG issues are being defined, measured, weighted and computed varies considerably across vendors (see Chatterji et al. 2016; Christensen, et al. 2019; Gibson et al. 2019; Berg, et al. 2020). Escrig-Olmgedo et al. (2019) find that ESG rating agencies have substantially improved their methodologies from 2008 to 2018, by incorporating new sustainability criteria, albeit more work is still needed. In this research, we choose one of the most widely used ESG rating data in the investment industry – MSCI ESG.

7.3.1. *MSCI ESG data*

MSCI ESG rating follows a rule-based methodology designed to measure a company's exposures to long-term, industry-material environmental, social and governance risks. MSCI ESG rating is primarily based on analysis from a large team of more than 200 analysts, supported by Artificial Intelligence (AI), machine learning and natural language processing. Currently, MSCI ESG data covers around

8,500 companies and more than 680,000 equity and fixed income securities globally. The list of 8,500 companies includes approximately 5,600 small-cap and private companies that are not in the MSCI ACWI index. ESG ratings for new additions to major MSCI indices (e.g. MSCI ACWI and US Investable Market Index) are available within one quarter from the effect date, while additions to other MSCI indices – typically small-cap – are rated within two quarters.

7.3.2. *Data coverage and average rating*

Over the past decade, MSCI has made a number of revisions, with a major coverage upgrade in 2012. As shown in Figure 7.4(A), data coverage improves considerably after 2012 in all regions of the world.

Figure 7.4(B) plots the distribution of the average ESG rating by the four regions over time. European firms clearly rank far ahead of companies in other regions, followed by stocks in other developed markets and the US. On average, EM companies struggle on the ESG front with a much lower average rating.

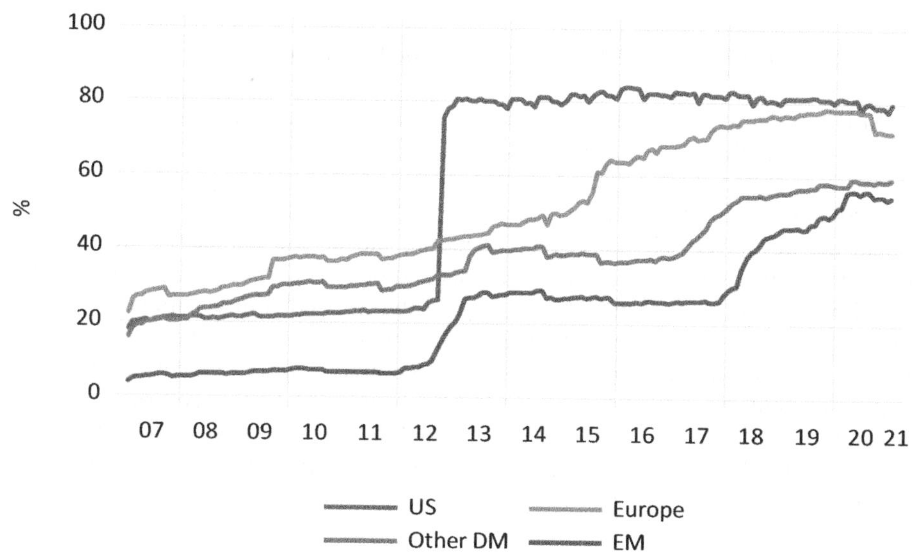

a) Data coverage

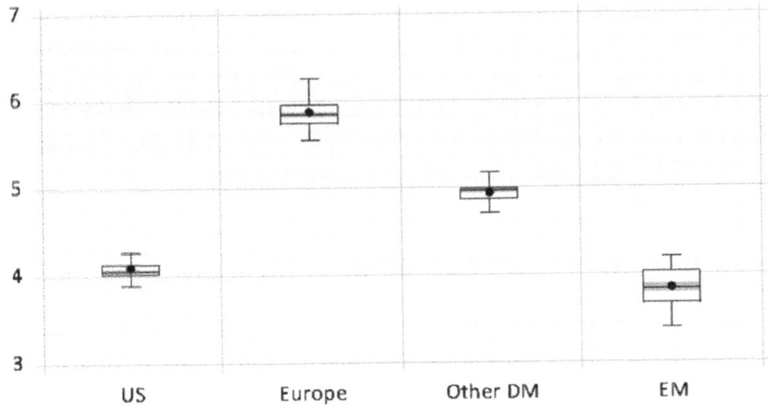

b) Average rating

Figure 7.4. *MSCI ESG rating coverage and average rating by geographic region. Sources: Bloomberg Finance LLP, FTSE Russell, MSCI, Refinitiv, S&P Capital IQ, Wolfe Research Luo's QES*

7.3.3. *An overview of MSCI ESG rating methodology*

MSCI collects data from a large range of data sources – company disclosures, regulatory and other specialized databases, and thousands of media sources.

MSCI essentially rates each company against its industry peers. The main reason is that the key ESG risks vary by sectors and industries. Companies in the same industry are likely to face similar ESG risks and opportunities. For example, carbon emission is a critical element for energy, materials and utility companies, but is much less relevant for firms in the financial sectors. On the other hand, privacy and data security are of paramount importance to communication services and financial companies, but generally do not apply to energy and materials firms. To help investors understand the importance of each ESG dimension for each sector and industry, MSCI provides an interactive ESG industry materiality map on its website[2].

The first two pillars – environmental and social – are GICS sub-industry-specific. MSCI identifies a few key issues under each theme in the environmental and social pillars for each sub-industry. Each key issue is assigned a weight between 5% and 30%, based on a combination of impact (high, medium and low) and time frame (short-, medium- and long-term). An issue with high impact in the short-term will receive a high weighting for that sub-industry.

2 See https://www.msci.com/our-solutions/esg-investing/esg-ratings/materiality-map.

Each company in a specific sub-industry is then evaluated along each key issue in three ways:

– **Risks:** a company is rated on a key issue based on a combination of its risk exposure and risk management. To score well on a key issue, the firm's management needs to focus on high-risk exposures with strong risk mitigation.

– **Opportunities:** ESG issues can be either risk or opportunity. For example, the development of clean energy technology may pose long-term risk for coal mining companies, but offer opportunity for renewable fuel producers. The ability (or inability) of management is amplified by the size of opportunity exposure.

– **Controversies:** a controversy case is defined by an event in which a company is alleged to have a significant and negative impact along a key issue. Examples include an oil spill, a mine safety accident, a regulatory investigation, anti-trust behavior, and so on. Each controversy is rated based on the magnitude of impact (from very severe to modest) and scope of impact (from extremely widespread to low).

The governance dimension is applied to companies in all sectors universally, with a floor of 33% weight. Each company starts from a perfect score of 10. Then a point is deduced if a company does not meet the requirement on one key issue.

Eventually, each rated company receives a numerical score between 0 and 10 and then the score is mapped to a categorical rating from CCC to AAA. It is critical to reiterate that MSCI's ESG rating is not absolute, but explicitly adjusted to be relative to a company's industry peers.

MSCI proactively reaches out to companies for verification and feedback. However, the MSCI ESG rating is not based on surveys, questionnaires or interviews. There is an official annual rating update and ongoing real-time monitoring/assessment.

7.3.4. *ESG pillars, themes and key issues*

MSCI identifies all the essential ESG-related topics and then divides them into three pillars, 10 themes and 35 key issues (see Figure 7.5). MSCI provides four scores for each key issue – an overall score, an exposure score, a management score and a weighting number. On the corporate governance dimension, MSCI rates each company alone ~100 binary events, for example, whether the lead director on the board is independent, whether a company's CEO pay is excessive, etc.

Pillar (3)	Themes (10)	Key Issues (35)
Environmental		
	Climate Change	
		Carbon Emissions
		Product Carbon Footprint
		Financing Environmental Impact
		Climate Change Vulnerability
	Natural Capital	
		Water Stress
		Biodiversity & Land Use
		Raw Material Sourcing
	Pollution & Waste	
		Toxic Emissions & Waste
		Packaging Material & Waste
		Electronic Waste
	Environmental Opportunities	
		Opportunities in Clean Tech
		Opportunities in Green Building
		Opportunities in Renewable Energy
Social		
	Human Capital	
		Labor Management
		Health & Safety
		Human Capital Development
		Supply Chain Labor Standards
	Product Liability	
		Product Safety & Quality
		Chemical Safety
		Financial Product Safety
		Privacy & Data Security
		Responsible Investment
		Health & Demographic Risk
	Stakeholder Opposition	
		Controversial Sourcing
		Community Relations

	Social Opportunities	
		Access to Communications
		Access to Finance
		Access to Health Care
		Opportunities in Nutrition & Health
Governance		
	Corporate Governance	
		Ownership & Control
		Board
		Pay
		Accounting
	Corporate Behavior	
		Business Ethics
		Tax Transparency

Figure 7.5. *MSCI ESG rating hierarchy. Sources: MSCI*

7.4. Characteristics of ESG – a factor perspective

First, let us gain some basic understanding of the MSCI ESG rating and how sustainability issues are related to common investment styles (e.g. value, growth, momentum and quality) and risk factors (such as size and liquidity). We also want to see whether there are any systematic country/regional and industry bias in the ESG data.

7.4.1. *The basics*

Since the MSCI ESG rating is mostly adjusted on an annual basis, we expect its signal autocorrelation (i.e. month-to-month serial correlation) to be relatively high. As shown in Figure 7.6(A), in the US, its monthly serial correlation is close to 99% most of the time, indicating the "sticky" nature of the ESG rating. Globally, we observe a similar pattern in that average monthly serial correlation is close to 99% in all regions (see Figure 7.6(B)). Even during MSCI's annual revision period, autocorrelation only drops to around 90%.

There is clear seasonality in the ESG rating data. As shown in Figure 7.6(C), serial correlation drops meaningfully in April and December, coinciding with MSCI's annual revisions. The revision date in other regions appears to be mostly around April (see Figure 7.6(D)).

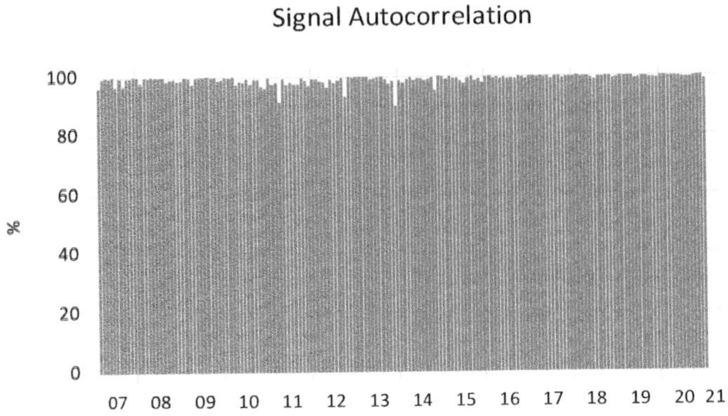

a) US

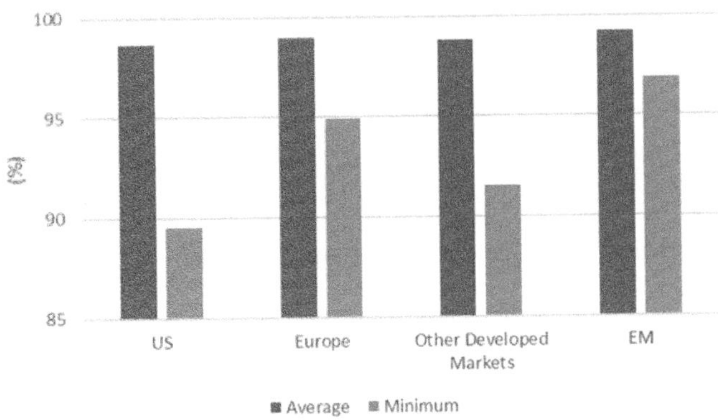

b) Annual revision

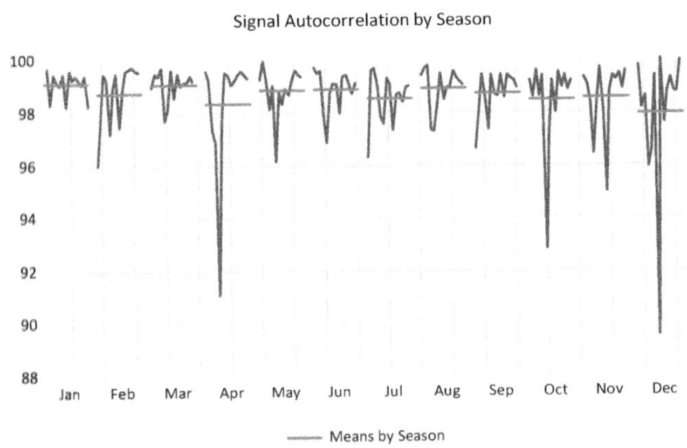

c) Seasonality, US

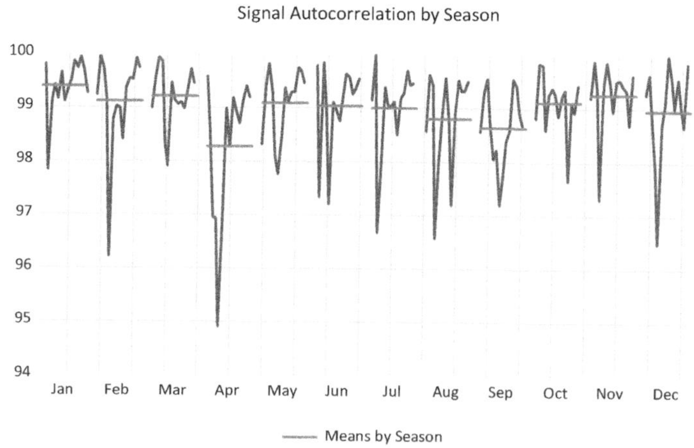

d) Seasonality, Europe

Figure 7.6. *MSCI ESG score, monthly serial correlation. Sources: Bloomberg Finance LLP, FTSE Russell, MSCI, Refinitiv, S&P Capital IQ, Wolfe Research Luo's QES*

7.4.2. *Difference across sectors*

As a reminder, the headline MSCI ESG rating is industry-adjusted. Therefore, to understand the differences of ESG characteristics across sectors, we use the non-industry-adjusted ESG score. We expect the sustainability profile to vary considerably across industries, both at the aggregated level and at the three pillars, especially along the environmental dimension.

7.4.2.1. *Non-industry-adjusted ESG score*

As shown in Figure 7.7(A), and as measured by the aggregate ESG rating, firms in the industrials, health care, information technology and utility sectors are ranked consistently higher than companies in the energy, materials and communication service industries in the US. In Europe, utility firms are clearly rated higher than companies in other sectors (see Figure 7.7(B)). In other developed markets, communication services and utility sectors are scored better than other industries (see Figure 7.7(C)). Lastly, in EM, information technology, communication services and utility companies are more likely to receive attractive ESG ratings (see Figure 7.7(D)).

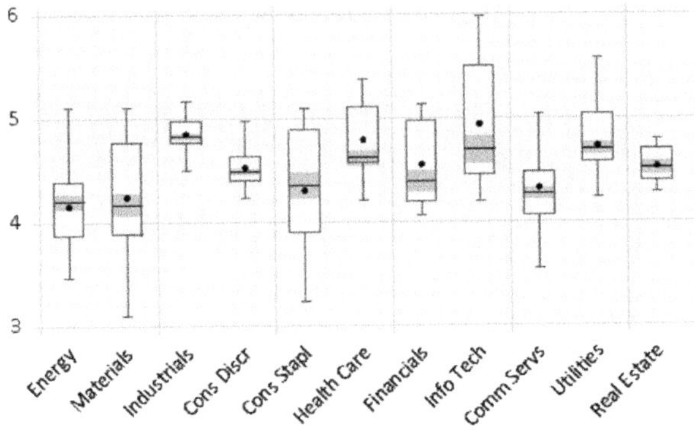

a) US

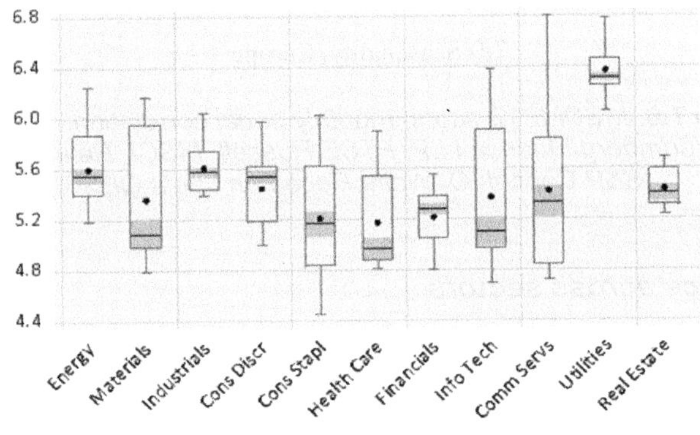

b) Europe

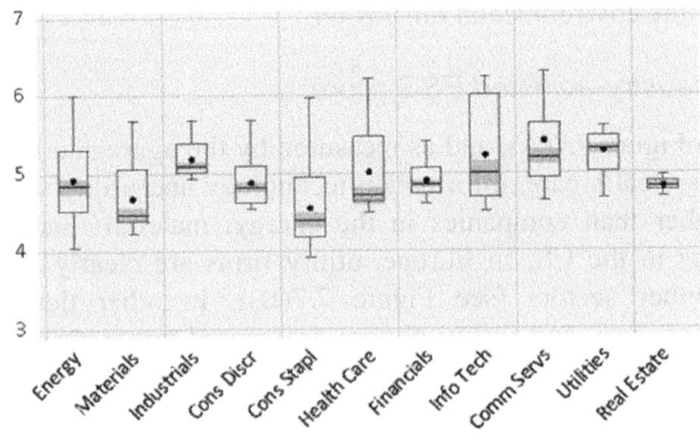

c) Other developed markets

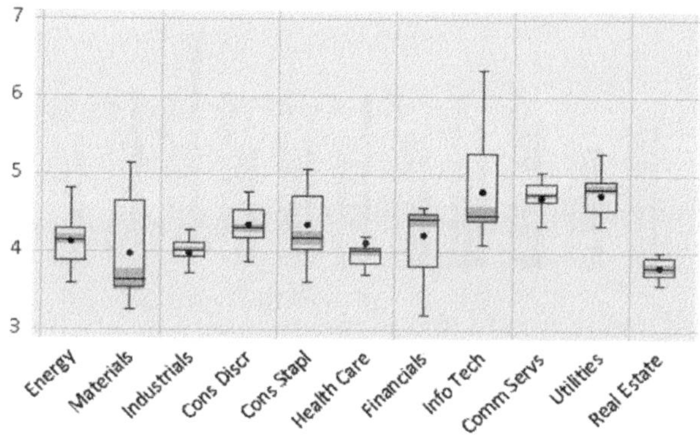

d) EM

Figure 7.7. *The distribution of MSCI ESG rating by sector. Sources: Bloomberg Finance LLP, FTSE Russell, MSCI, Refinitiv, S&P Capital IQ, Wolfe Research Luo's QES*

7.4.2.2. *Environmental, social and governance*

As shown in Figure 7.8, on the environmental front, firms in the communication services and health-care sectors are scored consistently higher than companies in the energy and consumer staple industries in all four regions. Conversely, although energy companies struggle to meet environmental expectations, they excel on the social aspect. On the governance pillar, the communication service sector in the US and Europe, along with the health-care industry in other developed markets and EM, appear to be the negative outliers.

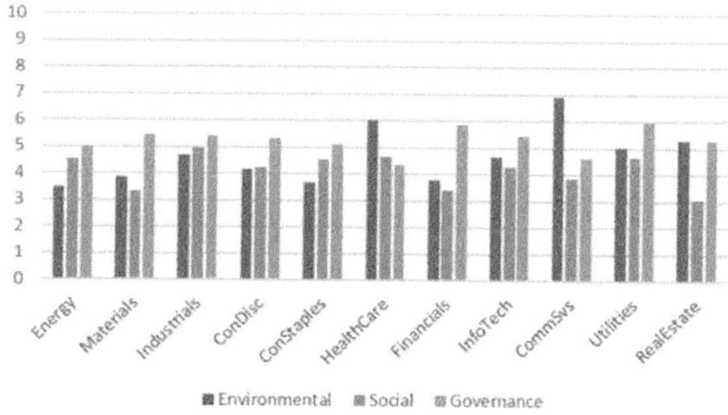

a) US

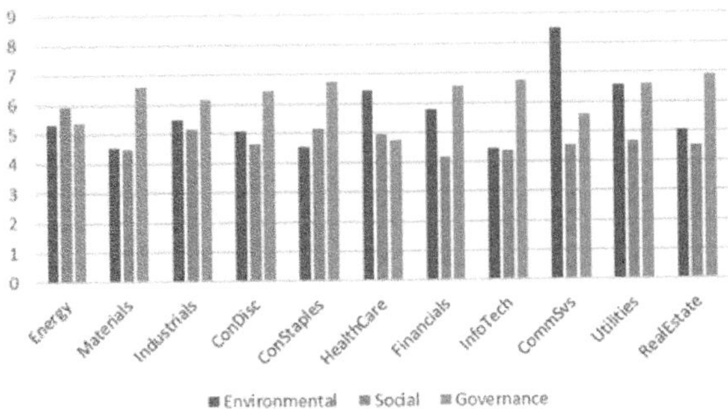

b) Europe

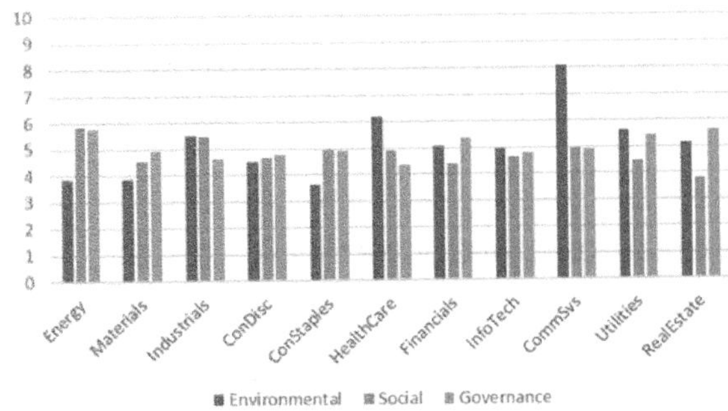

c) Other developed markets

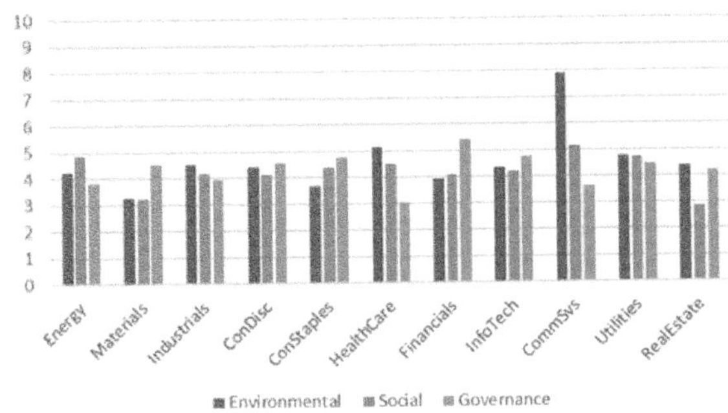

d) EM

Figure 7.8. *Average environmental, social, and governance rating by sector. Sources: Bloomberg Finance LLP, FTSE Russell, MSCI, Refinitiv, S&P Capital IQ, Wolfe Research Luo's QES*

7.4.3. *Factor exposure*

ESG scores are expected to be positively correlated to size, as larger companies typically are also more heavily invested in sustainability reporting and setting up ESG standards. Alessandrini et al. (2020a) confirm this hypothesis. There are many academic studies that suggest that sustainable companies tend to be more profitable and enjoy lower cost of capital. Therefore, we also expect a negative relationship with value and a positive correlation to profitability. Lastly, given the strong growth in sustainable investing, significant assets have shifted to ESG-friendly companies; therefore, it is also possible that the ESG rating can be (positively) correlated to price momentum.

In this section, we compute the cross-sectional correlation between ESG metrics and our set of 15 common factors. Given the time-varying nature of companies' business and operational profiles, we trace the time-series distribution of the correlation coefficients using boxplots.

The final headline ESG score clearly shows significant factor tilts. As shown in Figure 7.9(A), in the US, the largest exposures are size and illiquidity. Highly rated companies tend to be large-cap (as measured by the large positive exposure to the size factor) and liquid (as represented by the negative exposure to the illiquidity factor). The large-cap tilt is as expected, given that larger companies tend to have more resources committed to environmental, social and governance initiatives and have more extensive disclosures than small-cap firms.

Furthermore, it is evident that better ESG-rated firms are likely to grow (i.e. negative exposure to the book-to-market factor), be more profitable (in terms of ROE or return on equity) and become lower risk (i.e. negative exposure to the beta factor). Higher ESG rated companies enjoy high multiples based on book value of equity, but not on other metrics (e.g. earnings or EBITDA/EV).

The results are similar in other regions – especially on size (large cap tilt) and liquidity (more liquid). Outside of the US, highly rated ESG companies are more likely to enjoy higher dividend yields, but the growth exposure is more muted. In EM, the positive correlations with dividend yield and ROE factors are particularly noticeable (see Figure 7.9(D)).

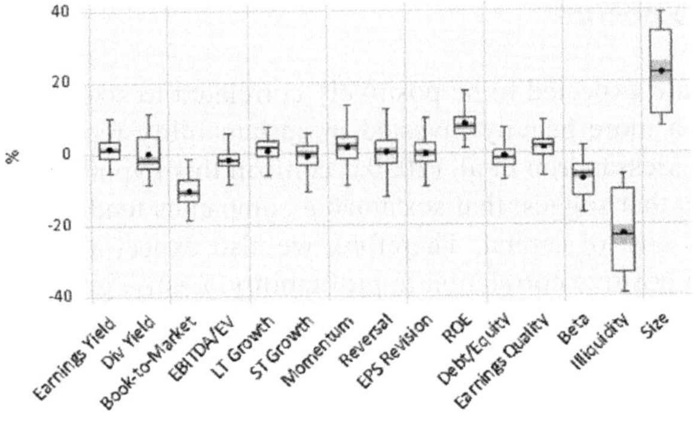

a) US

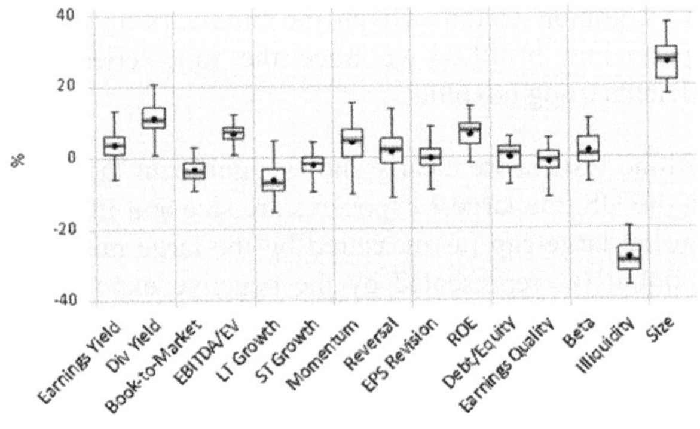

b) Europe

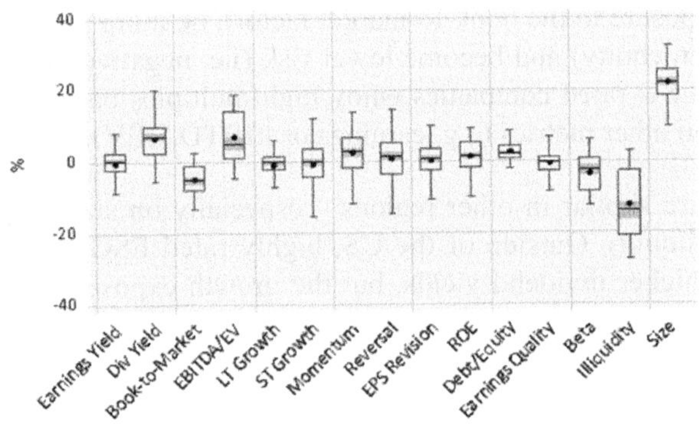

c) Other developed markets

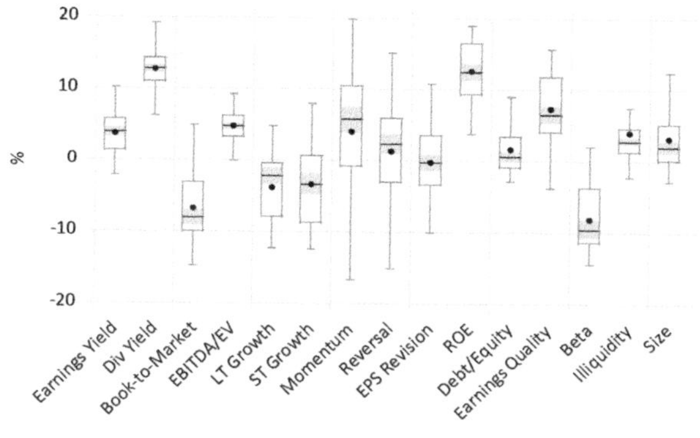

d) EM

Figure 7.9. *MSCI ESG rating's factor exposure (cross-sectional correlation). Sources: Bloomberg Finance LLP, FTSE Russell, MSCI, Refinitiv, S&P Capital IQ, Wolfe Research Luo's QES*

7.4.3.1. *Environmental, social and governance*

In the US and Europe, all three pillars still demonstrate the same large cap and positive liquidity exposures (see Figure 7.10(A) and (B)). Interestingly, firms with better environmental and social track records enjoy higher price-to-book multiples (or lower book-to-market yield) in almost all regions. More profitable companies (as measured by ROE) are generally rated higher on all three pillars in almost all regions.

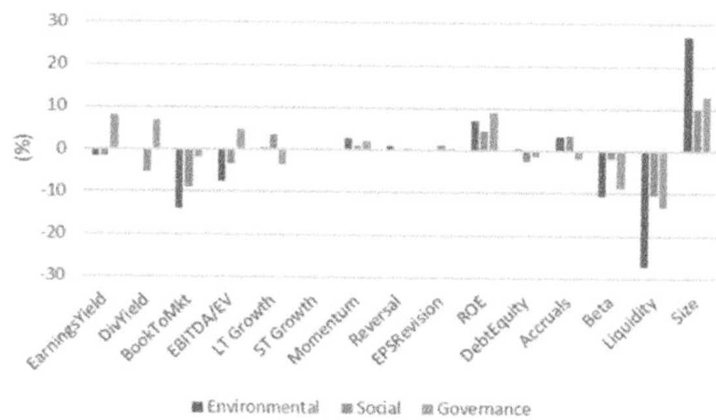

a) US

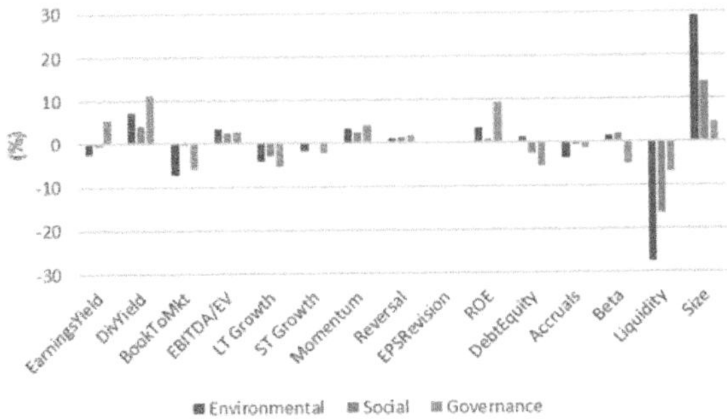

b) Europe

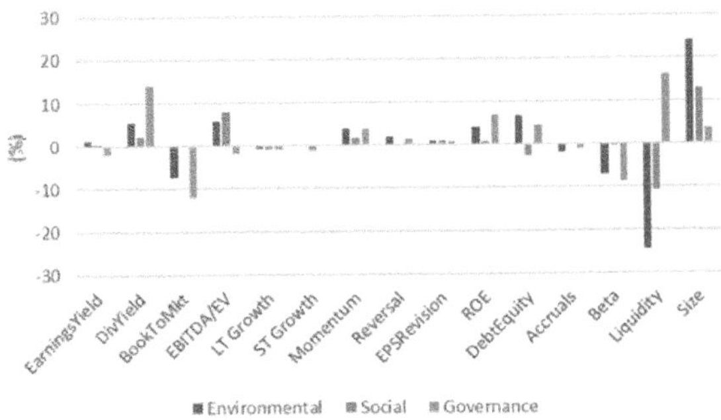

c) Other developed markets

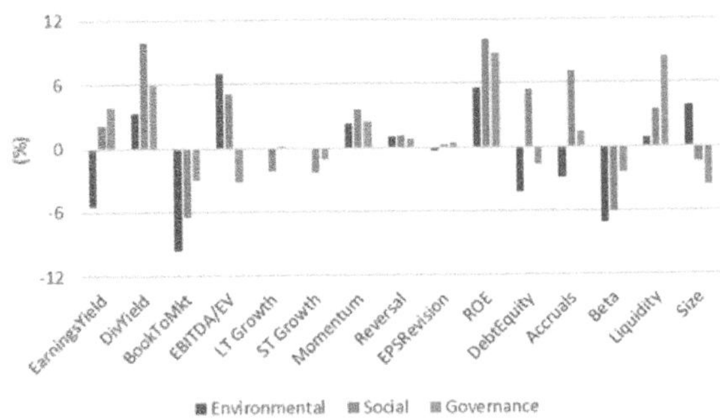

d) EM

Figure 7.10. *Average factor correlation by three pillars. Sources: Bloomberg Finance LLP, FTSE Russell, MSCI, Refinitiv, S&P Capital IQ, Wolfe Research Luo's QES*

7.5. ESG as stock-selection factors

The relationship between ESG and investment performance has been at the center of debate among policymakers, investors and academics for years. If firms rated higher on the ESG side also deliver better stock returns, that would truly allow investors to achieve the goal of "doing well by doing good". Friede, et al. (2015) summarize over 2,000 academic studies on the performance impact of ESG. Boffo, et al. (2020) provide a recent meta study on the impact of ESG on fund performance. Atz et al. (2020) is another recent survey of 1,141 papers and 27 meta-reviews. The general conclusion is that ESG investing is comparable or preferable to conventional investing. The alpha content (or stock return predictive power) from ESG data is perceived as minimal. However, the vast majority of academic studies is limited to the aggregated level of ESG ratings. Few research papers have dived into the granular levels of detailed information that comprises the final ESG score.

Academic evidence on whether ESG data can be used as stock-selection factors is mixed. Dorfleitner et al. (2013) find significant alpha from Asset4 ESG rating (now part of Refinitiv ESG). On the contrary, Breedt et al. (2019) find that ESG is not a unique factor, because return-forecasting information is already captured by other known factors – particularly, size and low beta. On the other hand, the authors agree that adding an ESG overlay to existing factor portfolios does not hurt returns. Nagy et al. (2016) and Kaiser and Schaller (2019) suggest that, although the level of ESG rating may not lead to outperformance, an ESG momentum strategy (i.e. stocks with improving ESG ratings) can be profitable. Chan et al. (2020) find that factor portfolios constructed by jointly optimizing factor exposure with ESG and carbon outcomes result in similar performance as conventional factor strategies. The authors also show two examples on green intangible value and corporate culture quality that enhance traditional value and quality factors, respectively.

In this section, we pursue the idea of whether ESG data can be used as another set of alternative data and more importantly, whether there are stock-selection signals that can be developed using corporate sustainability information. We focus on MSCI ESG data – not only the headline rating, but also hundreds of key issues and metrics that underlie the final score.

7.5.1. *Aggregated ESG rating and the three pillars*

First, we will start from the aggregate ESG ratings. Recall that MSCI provides three aggregate scores:

– ESG rating as represented by letter grades from AAA to CCC in seven categories, which can be converted to numerical scores (e.g. AAA=7, AA=6, …, and CCC=1);

– final ESG score (normalized score within each of the pre-defined industries);

– non-industry-adjusted score.

Next, we have the three pillars – environmental pillar score, social pillar score and governance pillar score. For each pillar, there is also a corresponding weight assigned, reflecting MSCI's opinion on the materiality for a given company. The weight assigned to a given key issue and pillar tends to be more industry-specific than company-specific. Therefore, we do not expect the various weight items to have a strong predictive power of stock returns.

7.5.2. *Revenue, country and industry adjustment*

Given the significant bias on size, country and industry for most ESG metrics, it is prudent to adjust for these risk factors. Therefore, for each of the above nine metrics, we further compute two variants, adjusted for each company's revenue, country and industry. More importantly, we do not want to reward (or penalize) a company's environmental, social or governance practices simply because of its size. Adjusting a company's rating based on revenue is particularly relevant for environmental factors. For example, between two electric utility companies, the larger one typically produces more carbon emissions, even if it employs environmentally friendlier technologies. Revenue-adjusted carbon emissions essentially measure a company's emission intensity.

In the first case, we perform the following cross-sectional regression for each of our region and industry universes. Recall that we divide the global equity space into four geographic regions (US, Europe, other developed markets and EM), and then within each region, we further split the universe by our 26 industries. Therefore, in total, we have $4 \times 26 = 104$ investment universes:

$$OriginalESP_{i,t} = \beta_{0,t} + \beta_{1,t} \log(Revenue_{i,t}) + \varepsilon_{i,t}$$

where $OriginalESP_{i,t}$ is the original MSCI ESG score for company i at time t, $\beta_{0,t}$ and $\beta_{1,t}$ are regression coefficients, $Revenue_{i,t}$ is the last 12-month revenue for company i at time t, and $\varepsilon_{i,t}$ is the residual from the regression, which is also our revenue-adjusted ESG score.

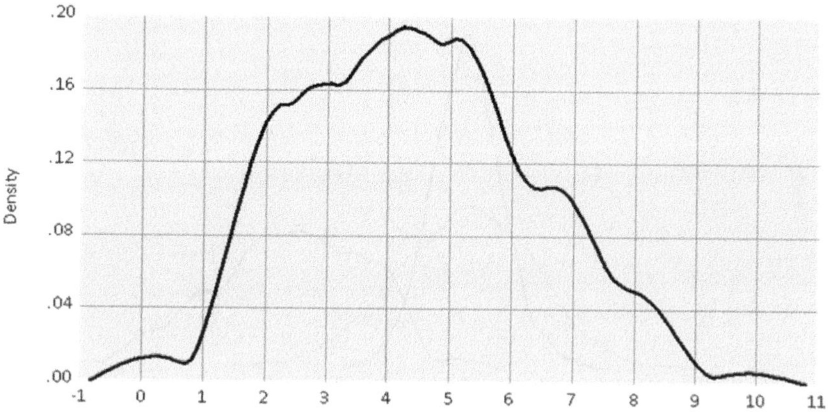

a) Final ESG score

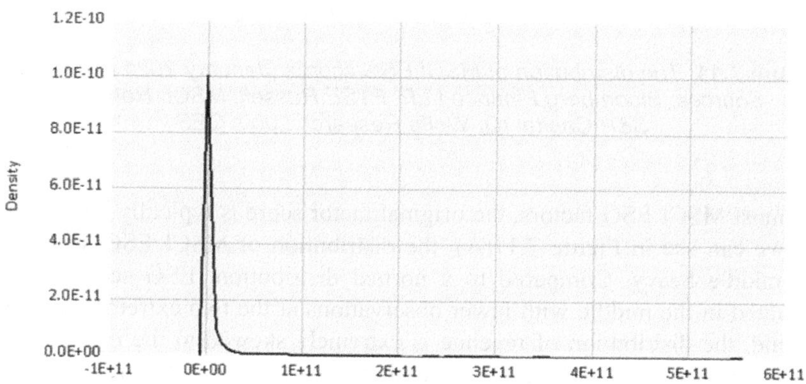

b) Revenue

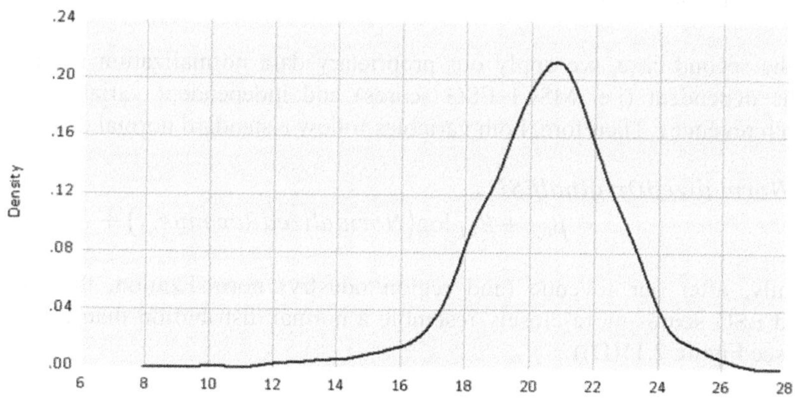

c) Log (revenue)

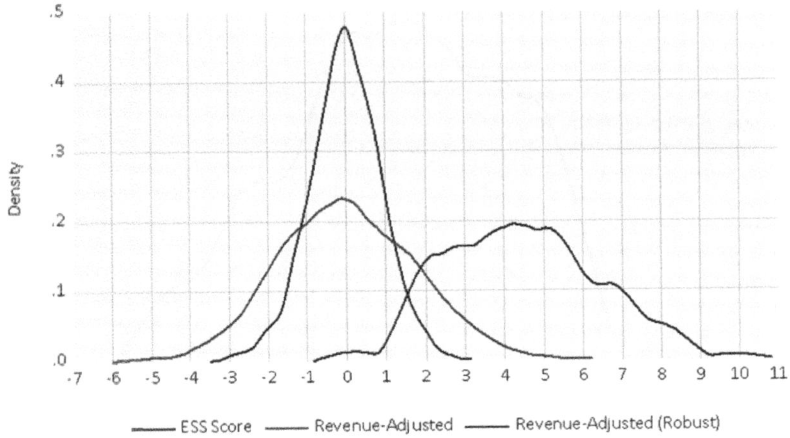

d) Revenue-adjusted final ESG score

Figure 7.11. *The distribution of MSCI ESG scores (January 2021 in the US). Sources: Bloomberg Finance LLP, FTSE Russell, MSCI, Refinitiv, S&P Capital IQ, Wolfe Research Luo's QES*

For most MSCI ESG factors, the original factor score is typically between 0 and 10. As we can see in Figure 7.11(A), the distribution of MSCI ESG score (in the US) is middle heavy. Compared to a normal distribution, ESG scores are more concentrated in the middle with fewer observations at the two extreme tails. On the other hand, the distribution of revenue is extremely skewed to the right, i.e. a few companies have extremely large sales numbers (see Figure 7.11(B)). After taking a logarithm, the distribution of revenue is now much better behaved with fewer outliers (see Figure 7.11(C)).

In the second case, we apply our proprietary data normalization technique to both the dependent (i.e. MSCI ESG scores) and independent variable (i.e. last 12-month revenue). Therefore, both variables follow a standard normal distribution:

$$NormalizedOriginalESP_{i,t} = \beta_{0,t} + \beta_{1,t} \log(NormalizedRevenue_{i,t}) + \varepsilon_{i,t}$$

Finally, after our revenue (and region/industry) normalization, the two new adjusted ESG scores more closely resemble a normal distribution than the original factor (see Figure 7.11(D)).

7.5.3. *Other adjustment*

For some ESG metrics, we further adjust for other company characteristics, such as total assets and number of employees. For some key issues within the social pillar, total assets and number of employees are more relevant than revenue.

7.5.4. *ESG momentum*

Nagy et al. (2016) and Kaiser and Schaller (2019) suggest that, although the original ESG factors have little predictive power of future stock returns, ESG momentum strategies (i.e. changes in ESG scores) show more promising performance. Therefore, for each of the above ESG factors, we further compute a year-over-year change or momentum version.

7.5.5. *Performance of aggregate ESG and three pillar scores*

Let us start by examining a simple example – the final MSCI ESG score, adjusted for revenue, region and industry. As shown in Figure 7.12(A), coverage was more for the S&P 500/MSCI USA prior to 2012, but has since improved substantially – nearly 80% companies in the Russell 3000 index have ESG ratings now. In the long-term, the signal has delivered reasonable performance, with a Sharpe ratio[3] of 0.36. The payoff pattern is also reasonably linear and monotonic (see Figure 7.12(B)). Furthermore, most of the excess return was generated after 2017, when ESG investing started to take off in the US and the performance of traditional factors plunged substantially (see Figure 7.12(C)). Lastly, given the low frequency nature of ESG data, signal decay is also modest, lasting for about half a year (see Figure 7.12(D)). The results are certainly encouraging and confirm the academic argument of "doing well by doing good".

In Europe, the suite of ESG and pillar scores are even more encouraging. As shown in Figure 7.13(A), the headline ESG score delivers a Sharpe ratio of 0.6x, easily beating many common stock-selection factors. Like what we saw in the US, performance is particularly strong in recent years. In addition to the aggregate ESG score, all three pillars also produce solid Sharpe ratios. Performance of the social and governance dimensions is peculiarly noticeable (see Figure 7.13(C) and (D)).

3 Sharpe ratio is computed on a long/short quintile portfolio formed on the signal, with a monthly rebalance.

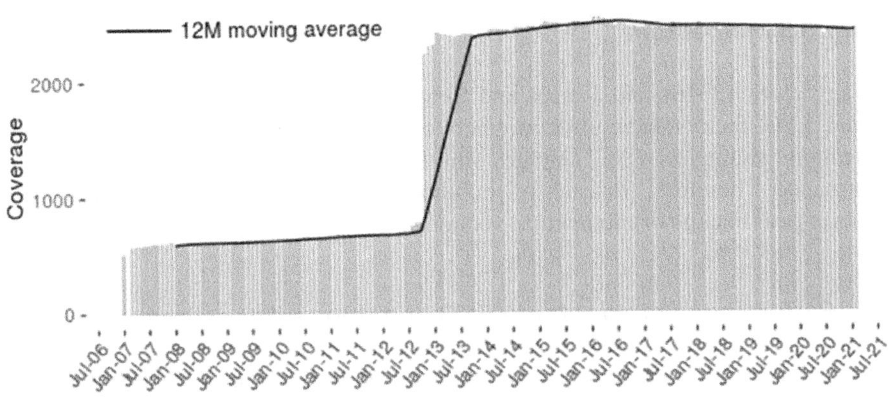

a) Coverage

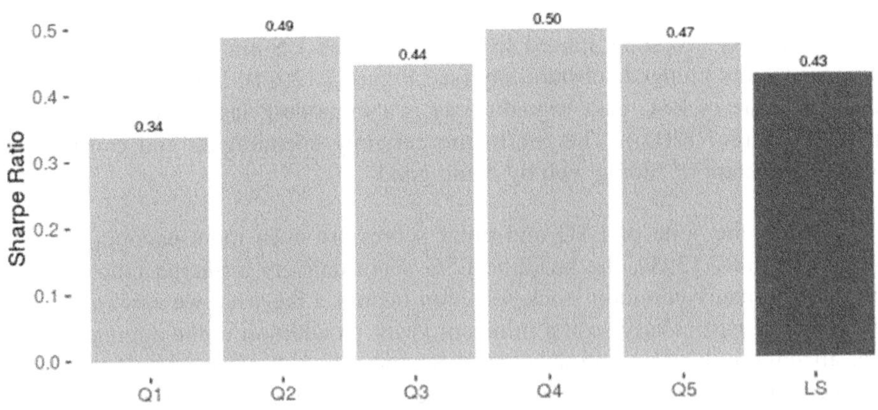

b) Sharpe ratio, by quintiles

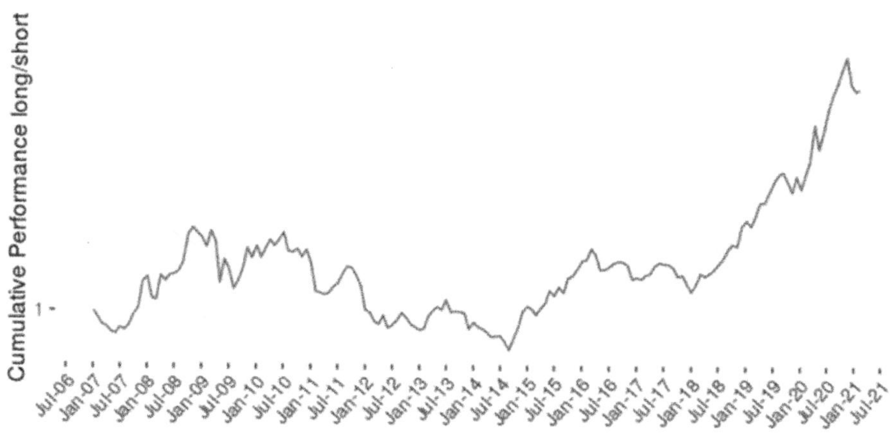

c) Cumulative performance (L/S quintile portfolio)

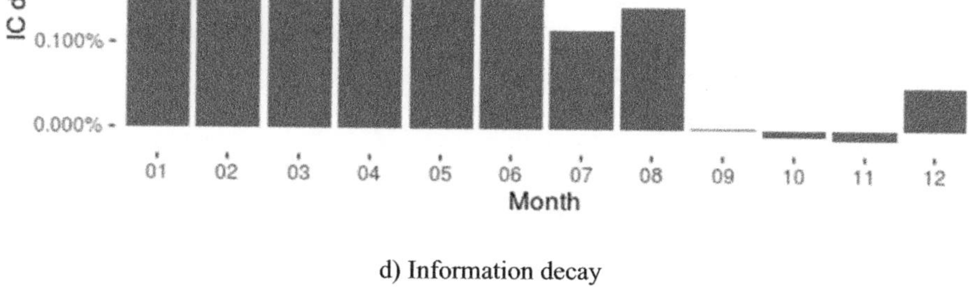

d) Information decay

Figure 7.12. *Revenue-adjusted (normalized) ESG score, US. Sources: Bloomberg Finance LLP, FTSE Russell, MSCI, Refinitiv, S&P Capital IQ, Wolfe Research Luo's QES*

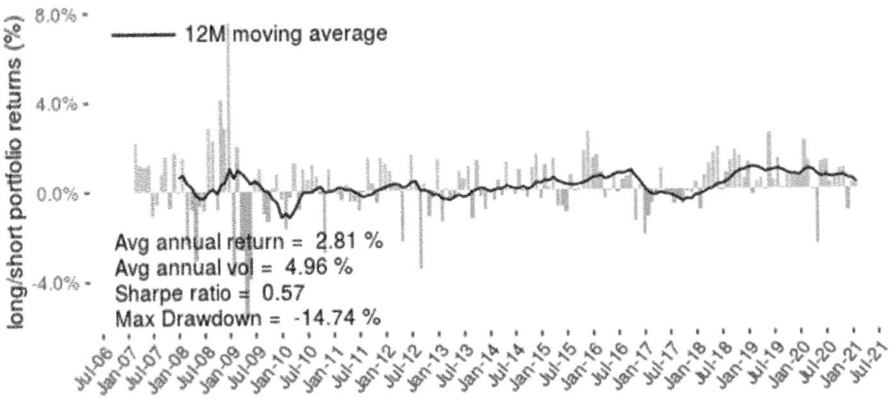

a) Final score

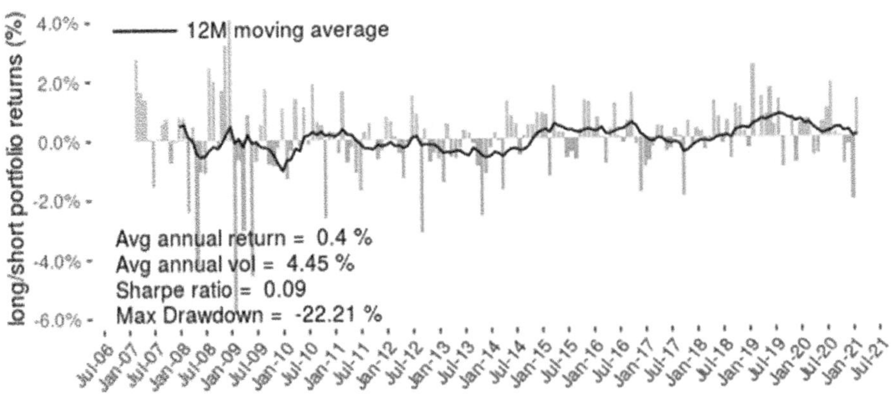

b) Environmental pillar score

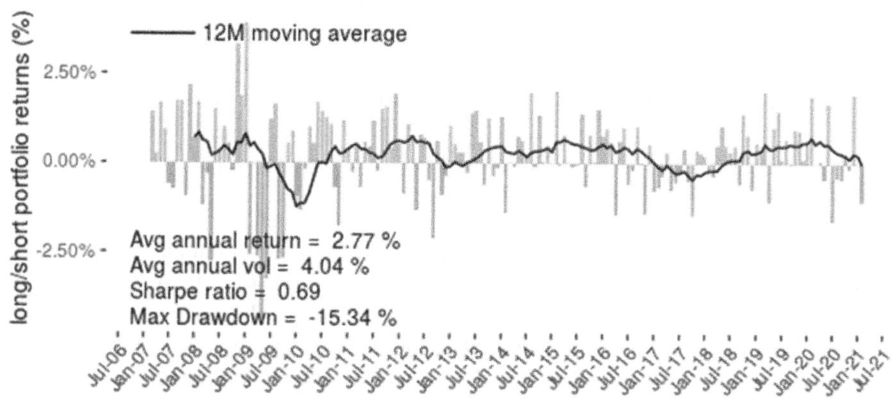

c) Social pillar score

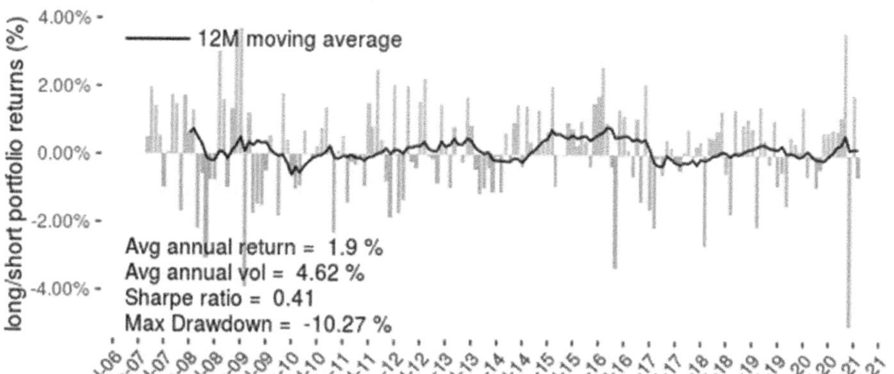

d) Governance pillar score

Figure 7.13. *Revenue-adjusted (normalized) ESG score, Europe. Sources: Bloomberg Finance LLP, FTSE Russell, MSCI, Refinitiv, S&P Capital IQ, Wolfe Research Luo's QES*

7.6. Environmental factors

Due to space limit, in this chapter, we focus on the environmental dimension. For the environmental pillar, as shown in Figure 7.5, there are 13 key issues. MSCI provides four factors for most of the 13 key issues – an overall score, exposure score, management score and weight. For each score, we perform the same two revenue/region/sector adjustments, as discussed in the previous section. Then, we compute the year-over-year change (or momentum) variant for all of the above.

7.6.1. *Zooming into clean technology*

One of the most promising ways to tackle climate change is the rapid development of clean technology. As shown in Figure 7.14(A), there are over 600 companies in the US with exposures to clean technology. The ones receiving more favorable scores have also outperformed their peers (see Figure 7.14(B)). A strategy that invests in clean tech winners has delivered consistent performance, especially since 2019 (see Figure 7.14(C)). The predictive power of the signal lasts for about eight months (see Figure 7.14(D)).

Financial returns offered by clean tech opportunities are even strong in Europe (see Figure 7.14(E)) and other developed markets (see Figure 7.14(F)).

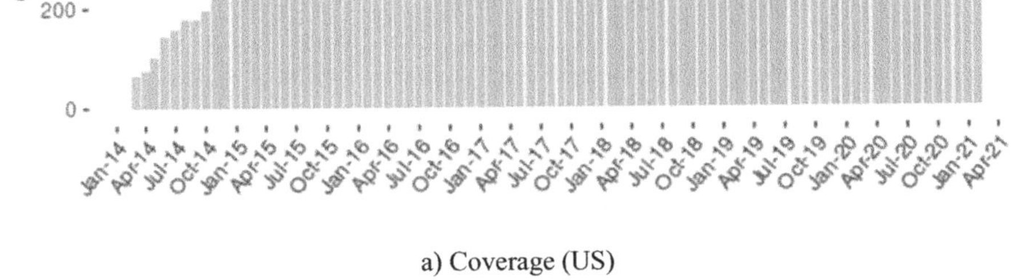

a) Coverage (US)

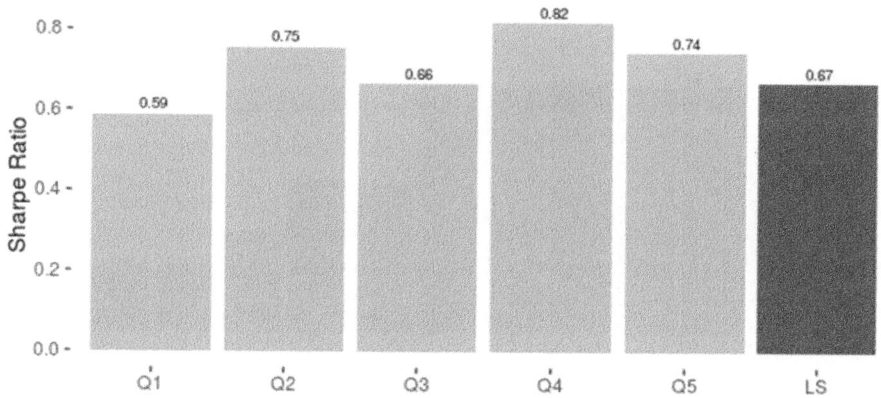

b) Sharpe ratio, by quintiles (US)

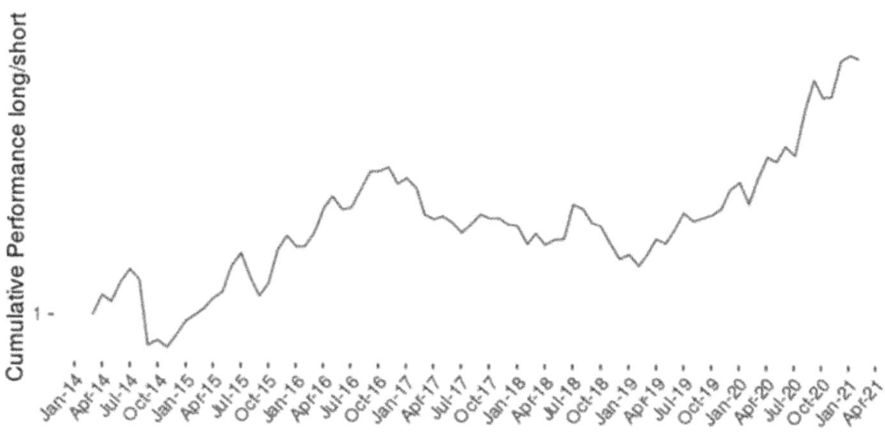

c) Cumulative performance (L/S quintile portfolio), US

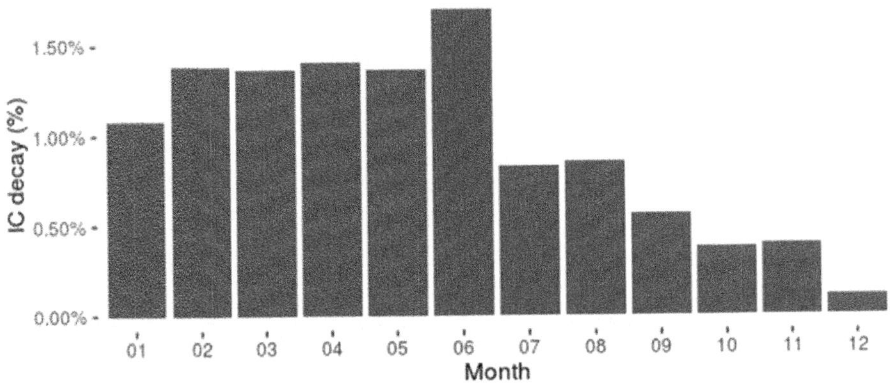

d) Information decay, US

e) Cumulative performance, Europe

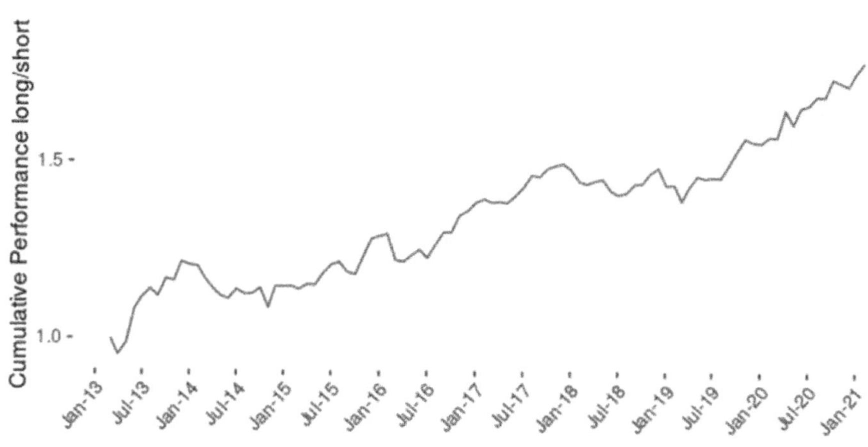

f) Cumulative performance, other DM

Figure 7.14. *Opportunity in clean technology. Sources: Bloomberg Finance LLP, FTSE Russell, MSCI, Refinitiv, S&P Capital IQ, Wolfe Research Luo's QES*

Some of the most exciting developments in clean tech has been in emerging markets, especially China. As shown in Figure 7.15(A), there are considerably more companies with exposures to clean tech in EM than in other regions. Long-term performance of the signal is modest, with a Sharpe ratio of 0.3x, but performance in recent years (since 2019) has been exceptionally strong (see Figure 7.15(B)).

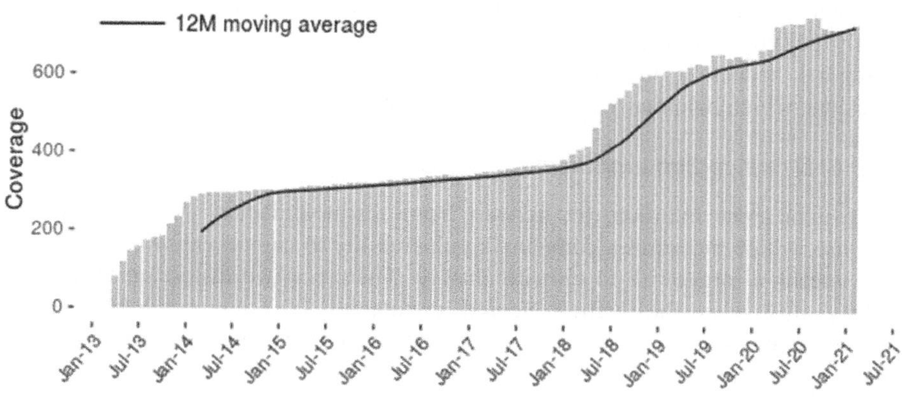

a) Coverage

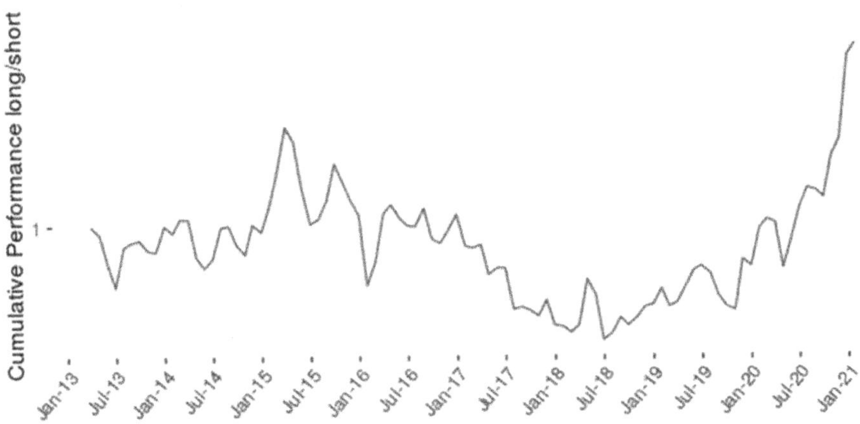

b) Cumulative performance (L/S quintile portfolio)

Figure 7.15. *Opportunity in clean technology (EM). Sources: Bloomberg Finance LLP, FTSE Russell, MSCI, Refinitiv, S&P Capital IQ, Wolfe Research Luo's QES*

7.6.2. *Carbon emissions along the supply chain*

As the developed world started to realize the negative impact of environmental issues, companies are penalized by ESG rating downgrades. It is well documented in academic research that firms with low ESG ratings face high cost of capital. In addition to the different approaches presented in the previous section, another way for individual companies to reduce carbon emissions is to "export" environmental problems to developing/low-cost countries. In this case, firms outsource their heavily pollutive productions to other countries. Consequently, companies report substantial reduction in direct GHG (Greenhouse Gas) emissions and other pollutants, which potentially boosts their ESG ratings.

7.6.2.1. *Direct carbon emissions*

To gain some intuition, let us start from a simple factor – carbon emission exposure, which measures a company's vulnerability to carbon emissions. A higher score means a larger risk to climate change. Therefore, we expect an investment strategy that buys the highest scored companies (i.e. high carbon emission exposures) and shorts the lowest ranked firms to produce *negative* returns in the long run. Empirical analysis confirms our hypothesis for all four regions.

Performance of the carbon emission factor has been particularly strong in recent years as investors are becoming increasingly aware of environmental issues and penalizing the heavy polluters.

a) US

b) Europe

c) Other developed markets

d) EM

Figure 7.16. *Carbon emission exposure factor, cumulative performance (long/short quintile portfolio). Sources: Bloomberg Finance LLP, FTSE Russell, MSCI, Refinitiv, S&P Capital IQ, Wolfe Research Luo's QES*

7.6.2.2. *GHG protocol*

GHG protocol – a partnership between WRI (World Resources Institute) and WBCSD (the World Business Council for Sustainable Development) – establishes global standards and guidance to measure and manage GHG emissions from private and public sectors. GHG protocol works with governments, industry associations, NGOs (Non-Profit Organizations) and businesses to develop GHG accounting and reporting principles. Carbon emission and climate change, however, are environmental issues of a global nature. Therefore, "exporting" emissions to other countries makes no difference from a worldwide perspective. Even worse, it requires further transportation to import the finished products back to home countries, which leads to more energy consumption and pollution.

GHG protocol attempts to mitigate such problems by introducing a three-tier reporting requirement (see GHG protocol (2004)):

– Scope 1 GHG emissions are direct emissions from sources that are owned or controlled by a company, for example, emissions from the production process.

– Scope 2 GHG emissions account for emissions from GHG of purchased electricity consumed by the company.

– Scope 3 GHG emissions include all other indirect emissions resulting from the company's activities. Examples of scope 3 activities are "extraction and production of purchased materials; transportation of purchased fuels; and use of sold products and services".

GHG protocol requires companies to separately account for and report on scopes 1 and 2 at a minimum. Scopes 1 and 2 are defined in a way so that the same emission is not double counted by two or more companies. Scope 3 emissions are not mandatory and far more difficult to quantify. As a result, most companies still do not disclose.

7.6.2.3. *Raw material sourcing*

The "direct carbon emissions" key issue only measures a company's direct environmental impact. One way to quantify the impact of indirect emissions is to incorporate raw material sourcing standards. MSCI defines raw material sourcing as the following:

> This key issue evaluates the extent to which companies are exposed to risks of damaging their brand value by sourcing or utilizing raw materials with high environmental concerns. Companies that have policies to source materials with lower environmental impact and participate in initiatives to reduce environmental impact of raw

materials production score higher on this key issue. Companies that do not utilize sustainably produced raw materials and set no targets for use of such materials in the future score lower.

Given the difficulty of measuring supply chain activities, signal coverage is more limited (see Figure 7.17(A)). The small sample also leads to some outliers, but performance is still relatively monotonic (see Figure 7.17(B)). Performance has been consistent throughout history and particularly strong since 2018 (see Figure 7.17(C)). Lastly, signal decay is extremely slow, lasting beyond a year, which makes it especially attractive to long-term investors (see Figure 7.17(D)).

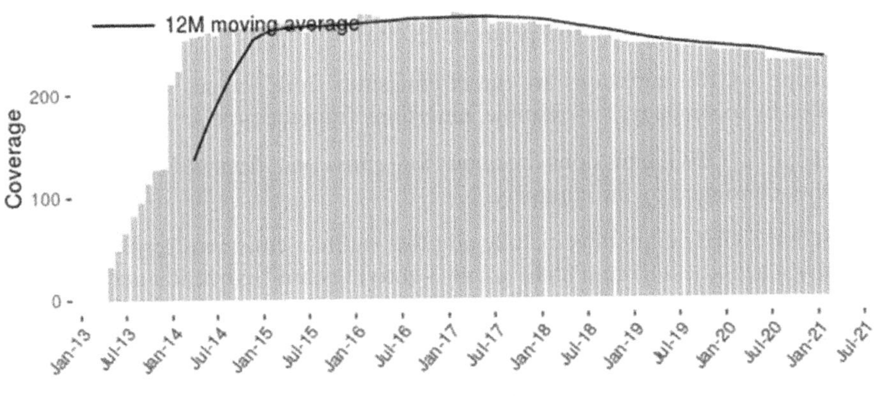

a) Coverage

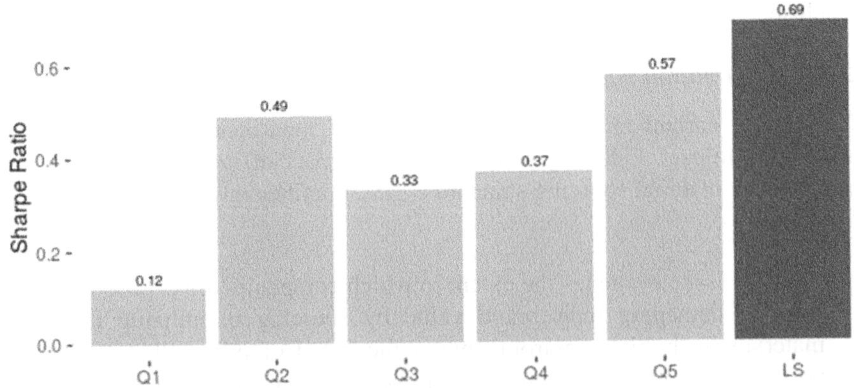

b) Sharpe ratio, by quintiles

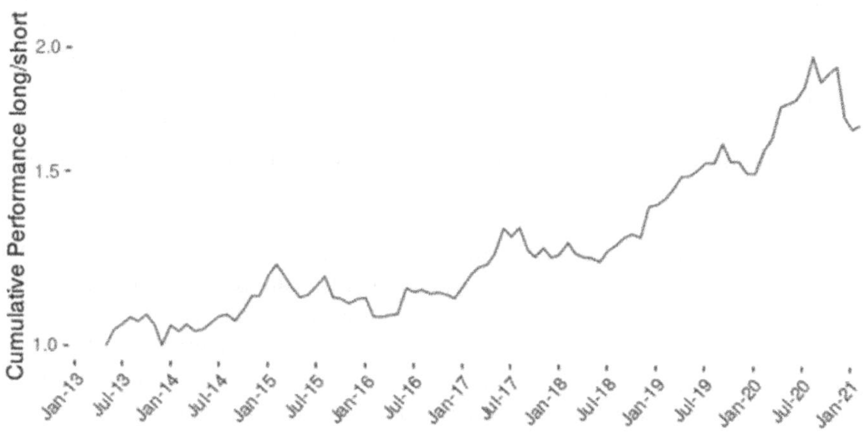

c) Cumulative performance (L/S quintile portfolio)

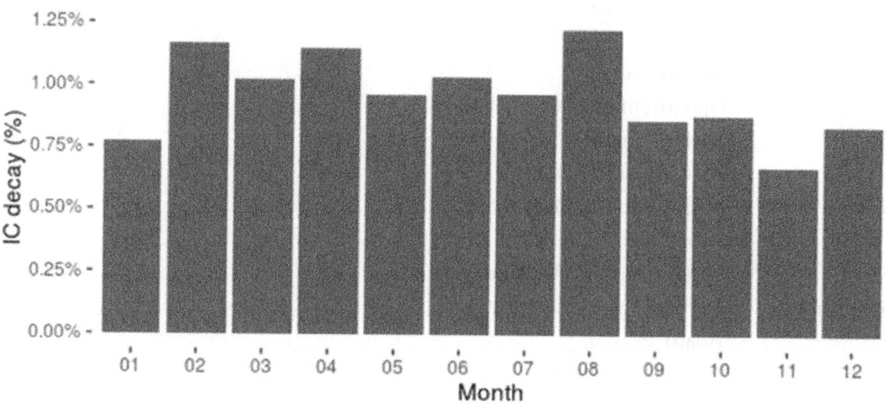

d) Information decay

Figure 7.17. *Raw material sourcing score (US). Sources: Bloomberg Finance LLP, FTSE Russell, MSCI, Refinitiv, S&P Capital IQ, Wolfe Research Luo's QES*

7.7. ESG signals are additive to traditional stock-selection factors

As detailed in the previous sections, both broad ESG scores and "green" alpha signals have significant predictive power of future stock returns. Performance has been particularly strong in recent years, as investors become increasingly aware of the impact of sustainability. On the other hand, the performance of traditional stock-selection factors (or smart beta factors) continues to shrink (see Luo et al. (2019)). Consequently, incorporating ESG signals into the investment process should boost performance meaningfully for active portfolio managers. Similarly, we also expect strong diversification benefit and performance improvement for asset owners by adding ESG portfolios to their asset mix. In this section, we will demonstrate the incremental value-add from ESG factors in details.

For illustration purposes, we will continue to use the same 15 style factors shown in Figure 7.3, along with the various ESG signals introduced in the previous sections. For the convenience of our readers, we summarize the list of ESG metrics in Figure 7.18. Please note that neither the list of 15 common factors nor the set of ESG metrics represent the best-performing signals. Rather, they are chosen for their simplicity, economic intuition and representation of certain investment styles.

Pillar	ESG Metric	US	Europe	Other DM	EM
Aggregate	Final ESG Score	X	X	X	X
	Environmental		X	X	X
	Social	X	X	X	X
	Governance	X	X	X	X
Environmental	Opportunities in Clean Tech	X	X	X	X
	Carbon Emissions	X	X	X	X
	Raw Material Sourcing	X	X		X
Social	Workplace Health & Safety	X	X	X	X
	Labor Management	X	X	X	X
	Supply Chain Labor Standard	X	X	X	
Governance	Corruption & Instability	X	X	X	X
	Board		X		
	Accounting Practice	X	X		X

Figure 7.18. *List of ESG metrics, by region. Sources: MSCI, Wolfe Research Luo's QES*

7.7.1. *Performance comparison with traditional stock-selection factors*

As shown in Figure 7.19(A), the performance of common style factors has retreated considerably since 2017 and plummeted in 2020. Value and quality factors suffered from the largest losses in the 2020 risk rally. On the other hand, ESG metrics have shown strong resilience, especially in recent years. In 2020, on average, the 15 common factors delivered a negative Sharpe ratio of -0.6x versus an average Sharpe ratio of 0.8x for the ESG metrics.

In Europe (see Figure 7.19(B)), other developed markets (see Figure 7.19(C)) and EM (see Figure 7.19(D)), the broad pictures resemble the US, i.e. considerable decline of traditional factors and the robustness of ESG signals.

Globally, within the traditional factor family, value, quality and risk factors suffered from the deepest performance downgrades, while analyst sentiment and growth styles survived reasonably well in 2020. Among the three ESG pillars, the environmental dimension enjoyed the best performance pick-up in recent years.

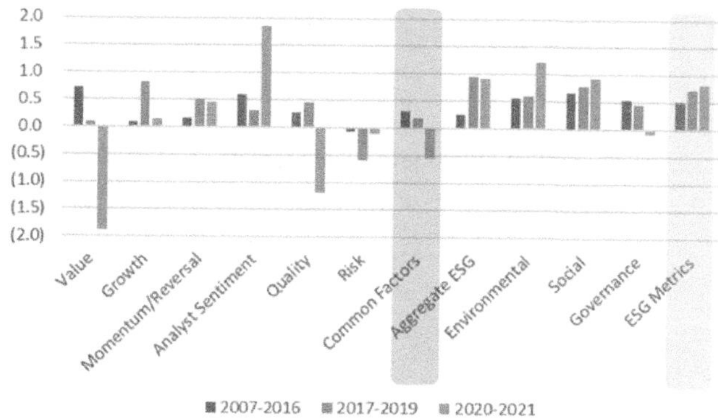

a) US

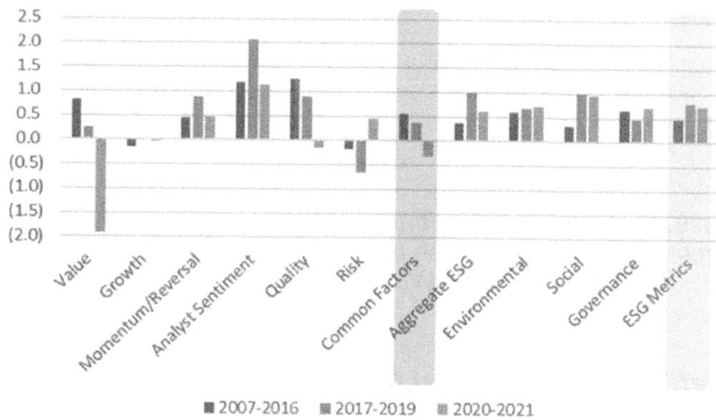

b) Europe

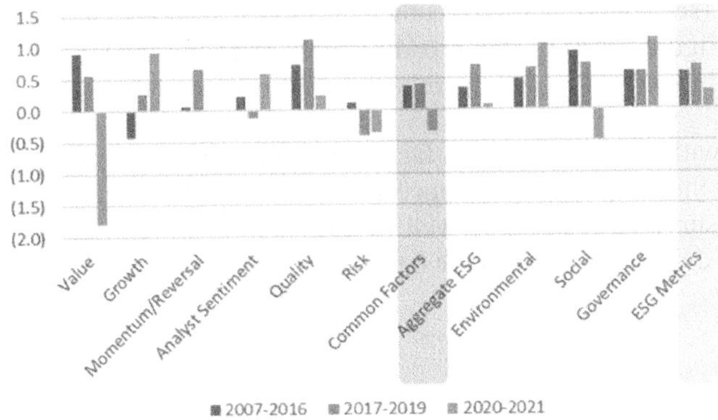

c) Other developed markets

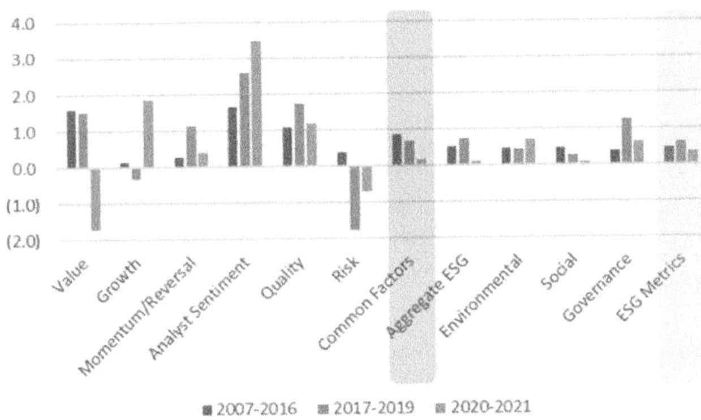

d) EM

Figure 7.19. *Performance comparison (Sharpe ratio, L/S quintile portfolio). Sources: Bloomberg Finance LLP, FTSE Russell, MSCI, Refinitiv, S&P Capital IQ, Wolfe Research Luo's QES*

7.7.2. *Correlation with traditional factors*

As discussed in the previous section, most ESG metrics had modest performance prior to 2017, but saw strong performance improvement in recent years. The trend is the opposite for traditional style factors, which experienced steep performance downgrades since 2017. Therefore, we would expect these two sets of factors to be uncorrelated. Indeed, as shown in Figure 7.20, correlation[4] between common factors and ESG metrics varies from -18% in other developed markets to 20% in Europe.

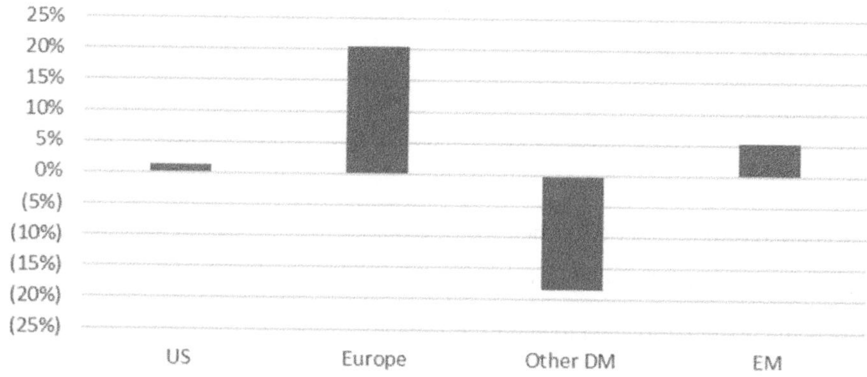

Figure 7.20. *Correlation between common factors and ESG metrics. Sources: Bloomberg Finance LLP, FTSE Russell, MSCI, Refinitiv, S&P Capital IQ, Wolfe Research Luo's QES*

7.7.3. The diversification benefit offered by ESG factors

To demonstrate the diversification benefit offered by ESG metrics, in this section, we use a series of asset allocation examples. We construct three asset allocation portfolios:

– smart beta strategy invests in the 15 style factors (long/short quintile portfolios);

– ESG strategy invests in the ~10 ESG metrics (long/short quintile portfolios);

– all-inclusive strategy invests in both smart beta factors and ESG metrics.

To construct our asset allocation strategy, we apply a risk parity optimization. Essentially, at each month end, we allocate an equal risk budget to all eligible assets, via the following optimization:

4 Correlation is computed on two portfolios. The first portfolio invests in the 15 common factor long/short quintile baskets – equally weighted, while the second portfolio invests in the ESG-based signals.

$$\omega_{i,t} = argmin_{\omega} \sum_{i=1}^{K} \sum_{j=1}^{K} \left[\omega_{i,t} COV\left(r_{i,t}, r_{p,t}\right) - \omega_{j,t} COV\left(r_{j,t}, r_{p,t}\right)\right]^2$$

Subject to:

$\sum_{i=1}^{K} \omega_{i,t} = 1$, and

$\omega_{i,t} \geq 0.$

Here:

$r_{i,t}$ is the return of asset i at time t;

$r_{p,t}$ is the return of the portfolio at time t;

$\omega_{i,t}$ is the weight allocated to asset i at time t.

It has shown in numerous academic and practitioner's research that risk parity allocation tends to produce more balanced portfolios that outperform conventional 60-40% (i.e. 60% equity/40% bond), mean-variance optimization and 1/N (i.e. equally weighting) schemes (see Qian (2011), Jurczenko et al. (2015) and Luo et al. (2017)).

7.7.3.1. *US*

As shown in Figure 7.21(A), although the risk parity portfolio constructed using common style factors (the average of the 15 factors as shown in Figure 7.3) delivers higher Sharpe ratio before 2017, the ESG portfolio clearly outperformed in recent years. More importantly, incorporating ESG metrics in the asset allocation framework boosts performance throughout the entire history. At the beginning of the sample, ESG metrics accounted for less than 20% of our final portfolio. However, the weight allocated to ESG factors has since increased gradually and currently stands at around 42% (see Figure 7.21(B)).

Among the 15 common style factors, the weights of earning revision and illiquidity have expanded at the expense of long-term growth and earning quality (see Figure 7.21(C)). For ESG metrics, governance, carbon emission and accounting practice have gained weights, while the importance of the headline ESG signal and labor management has shrunk (see Figure 7.21(D)).

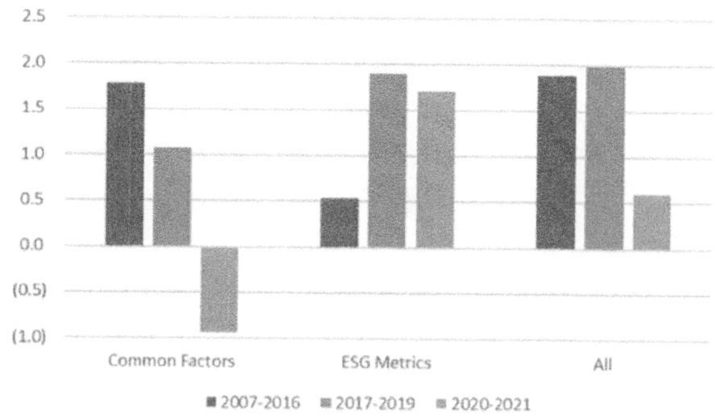

a) Sharpe ratio

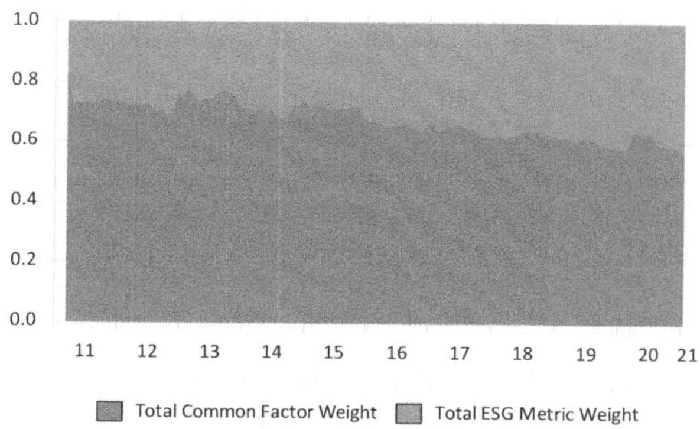

b) Common factors versus ESG metrics

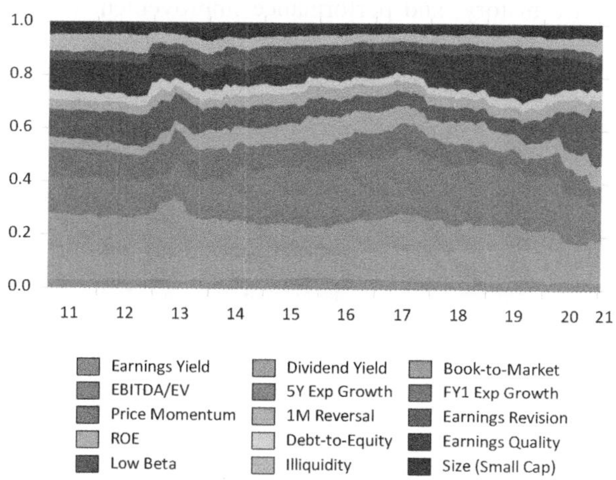

c) Common factor weight

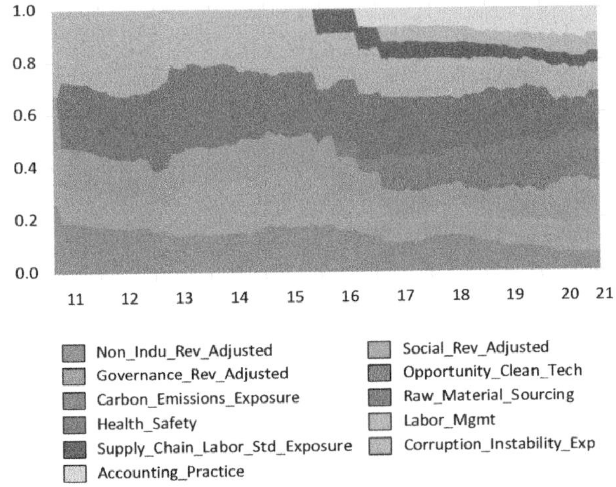

d) ESG signal weight

Figure 7.21. *Risk parity allocation, US. Sources: Bloomberg Finance LLP, FTSE Russell, MSCI, Refinitiv, S&P Capital IQ, Wolfe Research Luo's QES*

7.7.3.2. *Other regions*

Outside of the US, we observe similar patterns. The performance of smart beta portfolios has plunged in recent years (see Figure 7.22), while ESG signals remain robust. Adding ESG factors in factor allocation lifts the Sharpe ratio substantially throughout the entire history, and performance improvement is particularly strong after 2017.

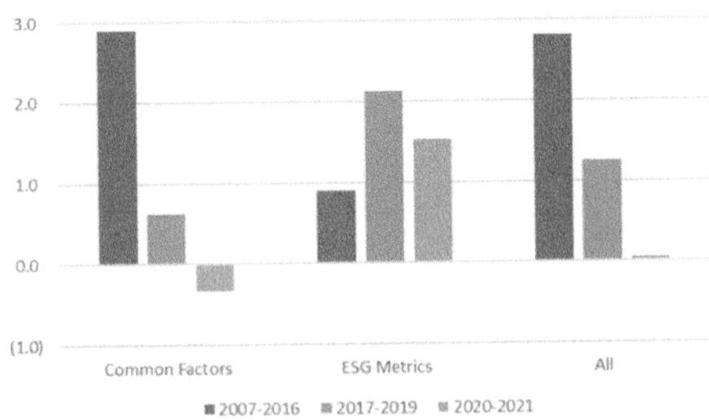

a) US

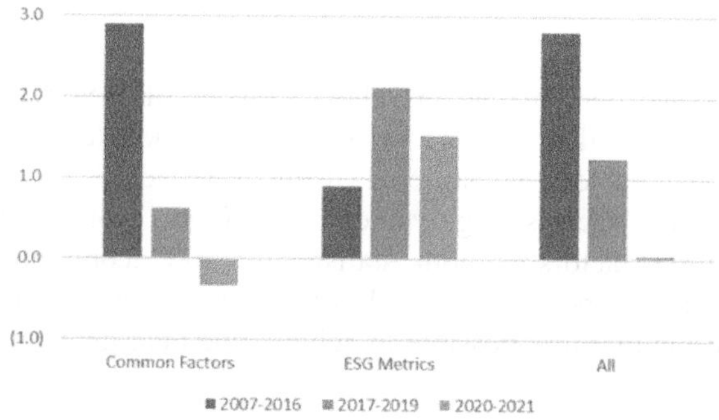

b) Europe

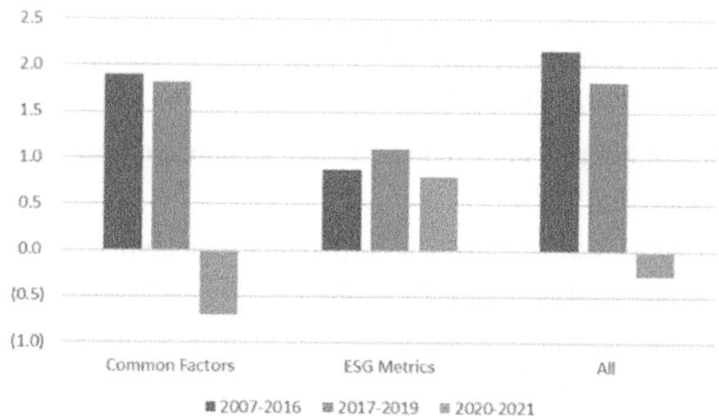

c) Other developed markets

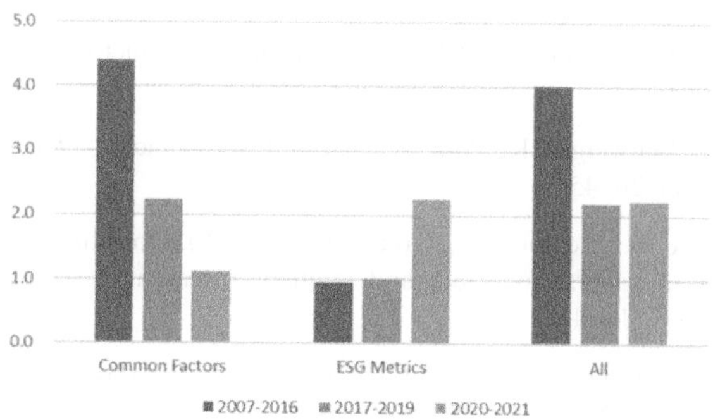

d) EM

Figure 7.22. *ESG versus common factors, global. Sources: Bloomberg Finance LLP, FTSE Russell, MSCI, Refinitiv, S&P Capital IQ, Wolfe Research Luo's QES*

7.8. Conclusion

In this chapter, we demonstrate the characteristics of MSCI ESG rating data. More importantly, we detail how to develop stock-selection factors using environmental data. Factors constructed using ESG data are uncorrelated to traditional risk premia factors (e.g. value, momentum, profitability). Investment performance of ESG factors is both statistically and economically significant, especially in recent years, as sustainability investing has gained tremendous acceptance. We argue that it is possible to achieve the dual goals of "doing well by doing good".

7.9. References

Alessandrini, F. and Jondeau, E. (2020). ESG investing: From sin stocks to smart beta. *Journal of Portfolio Management*, 46(3), 75–94.

Amihu, Y. (2002). Illiquidity and stock returns: Cross-section and time-series effects. *Journal of Financial Markets*, 5, 31–56.

Atz, U., Van Holt, T., Liu, Z.Z. (2020). Do corporate sustainability and sustainable finance generate better financial performance? A review and meta-analysis. *SSRN*, Working paper.

Berg, F., Kölbel, J.F., Rigobon, R. (2020). Aggregate confusion: The divergence of ESG ratings. *SSRN*, Working paper.

Boffo, R., Marshall, C., Patalano, R. (2020). *ESG Investing: Practices, Progress and Challenges*. OECD, Paris.

Breedt, A., Ciliberti, S., Seager, P. (2018). Is ESG an equity factor or just an investment guide? Working Paper.

Chan, Y., Hogan, K., Schwaiger, K., Ang, A. (2020). ESG in factors. *Journal of Impact and ESG Investing*, 1(1), 26–45.

Chatterj, A.K., Durand, R., Levine, D.I., Touboul, S. (2016). Do ratings of firms converge? Implications for managers, investors and strategy researchers. *Strategic Management Journal*, 37(8), 1597–1614.

Christensen, D., Serafeim, G., Sikochi, A. (2019). Why is corporate virtue in the eye of the beholder? The cast of ESG ratings. Working paper.

Dorfleitner, G., Utz, S., Wimmer, M. (2013). Where and when does it pay to be good? A global long-term analysis of ESG investing. *SSRN*, Working paper.

Escrig-Olmedo, E., Fernandez-Izquierdo, M.A., Ferrero-Ferrero, I., Rivera-Lirio, J.M., Munoz-Torres, M.J. (2019). Rating the raters: Evaluating how EST rating agencies integrate sustainability principles. *Sustainability*, 11(3).

Friede, G., Busch, T., Bassen, A. (2015). ESG and financial performance: Aggregated evidence from more than 2000 empirical studies. *Journal of Sustainable Finance & Investment*, 5(4), 210–233.

Gibson, R., Krueger, P., Riang, N., Schmidt, P.S. (2020). ESG rating disagreement and stock returns. European Corporate Governance Institute Finance, Working paper no. 651/2020.

Jurczenko, E., Michel, T., Teiletche, J. (2015). A unified framework for risk-based investing. *Journal of Investment Strategies*, 4, 1–29.

Kaiser, L. and Schaller, F. (2019). Environmentally (un-)friendly portfolio construction. *Journal of Investment Consulting*, 19(1), 43–52.

Liang, H. and Renneboog, L. (2017). On the foundations of corporate social responsibility. *Journal of Finance*, 72(2), 853–910.

Luo, Y., Wang, S., Jussa, J. (2017). Signal research and multifactor models. Report, Wolfe Research Luo's QES.

Luo, Y., Zhong, J., Alvarez, M. (2019). The future of active management. Report, Wolfe Research Luo's QES.

Nagy, Z., Kassam, A., Lee, L.E. (2016). Can ESG add alpha? An analysis of ESG tilt and momentum strategies. Report, MSCI.

Qian, E. (2011). Risk parity and diversification. *Journal of Investing*, 20(1), 119–127.

8

Enhancing Environment-driven Portfolios with Traditional Factors

This chapter investigates the efficacy of combining traditional factors with environmental data when building optimal equity portfolios. Our contribution departs from the traditional literature by focusing on allocations designed to adjust benchmark policies. We allow compositions to be embedded in a general factor framework in which firm characteristics are the main drivers of the portfolio weights. Our results suggest that it is feasible to improve the environmental score of a portfolio without it being too detrimental to its out-of-sample performance. However, pure sustainable attributes alone do not allow us to fulfill this objective: they need to be enhanced by non-ESG predictors to deliver their full potential.

8.1. Introduction

Investors are increasingly concerned about the environmental footprint of their equity portfolios. Consequently, their demand for "*green*" assets is surging exponentially and studies on the relationship between sustainability and performance have blossomed (see Coqueret (2021) for equities and Flammer (2021) for bonds). This has led money managers to include new objectives and indicators when composing their allocations, and the most widespread criteria are the so-called environment, social and governance pillars.

Investors willing to allocate according to environmental considerations (or ESG criteria more generally) face a plethora of integration methods, which we split in two, according to their technical sophistication: methods based on simple heuristic

Chapter written by Guillaume COQUERET, Christian MORGENSTERN, James KELLY, Sascha STIERNEGRIP, Johannes FREY-SKÖTT and Björn ÖSTERBERG.

1 Environmental concerns are often considered along with their social and governance counterparts, in the so-called ESG trichotomy.

methods (such as for instance sorting, screening, best-in-class weighting, etc.) and methods that rely on some form of optimization with an additional sustainable component (reward or constraint).

Both families of methods rely essentially on ESG data, and the second is also driven by the distribution of stock returns. Quite recently, researchers have sought to add factor flavors to their portfolio construction recipes. One tempting direction is to enhance the ways traditional factors are defined by integrating sustainability criteria. An overwhelming majority of contributions work with ESG criteria rather than focusing on the environmental pillar solely. For instance, Fan and Michalski (2020) use ESG sorting and screening to boost quality and momentum factors (see also Cai et al. (2021) for other factors). Likewise, Chan et al. (2020) seek to improve the definition of value and quality factors using sustainable signals. In Kaiser (2020), it is the traditional value, growth and momentum portfolios that are ameliorated with ESG tilts. Finally, Mohanty et al. (2021) propose to alter the factor portfolios by manually increasing the weights of high ESG stocks in order to boost the portfolios' ESG score, which, according to the authors, results in higher alphas.

More generally, a guided tour of simple ESG integration methods is provided in Bender et al. (2017) and Henriksson et al. (2019). Nevertheless, the state-of-the-art contribution is Alessandrini and Jondeau (2021), in which the authors propose a very complex optimization program wherein the ESG score of a portfolio is maximized under a batch of constraints, including tracking error, turnover control, as well as regional, industry and factor exposure. The exhaustive nature of the proposed scheme makes it impossible to interpret however, and the weights, while optimal, emanate from a black box.

The present chapter proposes to exploit traditional firm attributes such as market capitalization, price-to-book ratio or volatility in order to enhance a portfolio optimization scheme that includes a reward for the environmental score of the portfolio. In doing so, the estimation of the first moments of expected returns is both explicit in closed form, and more accurate out-of-sample. This allows us to reasonably improve the environmental footprint of the portfolio without incurring a financial cost. While our analysis focuses on the E pillar, they also hold for ESG criteria more generally. This comes from the relatively high correlation between the three pillars in our dataset.

Nevertheless, we find that the E and ESG constraints are binding and that an increased uptick in ESG exposure will come with a deterioration in returns. While it is possible to improve the sustainability exposure of an allocation without significantly

2 We refer to Chapters 4 and 5 of Coqueret (2021) for an exhaustive list of references.

3 Henceforth, we will refer to "*attributes*", "*characteristics*" or "*factors*" interchangeably.

curtailing its profitability, there is a threshold beyond which too much E or ESG may become overly detrimental.

More generally, we note that governance factors have been found to produce signals which exhibit positive excess returns (see, for example, Bloomberg Professional Service (2020) or Coqueret (2021)). This is intuitive, as governance attributes are the most directly linked to the financial performance of firms, both from a return and risk perspective. Investors have also enhanced equity quality portfolios with governance factors successfully, as there is a natural economic linkage between the two. Environmental characteristics are very much the focus of regulators, non-governmental and governmental organizations, investors and asset owners; however, they also have a much weaker link to the financial performance of companies, in particular in the short term. For this reason, we are not surprised that we do not find that environmental factors improve portfolio returns. We do note that this may change in the medium to long term as the problem of stranded assets increasingly comes into focus for investors, and as flows push the price of green firms up (van der Beck 2021). Notably, Mercure et al. (2021) discuss the estimated impact of different policy goals and the implication for the energy mix on a sector basis, which is used to model stranded assets on a national level.

The remainder of this chapter is structured as follows. In section 8.2, we lay out the optimization framework and outline the factor integration. Section 8.3 presents our baseline empirical tests, while section 8.4 runs robustness checks and sensitivity analyses. Finally, section 8.5 concludes.

8.2. Framework

This section details the theoretical setting of the chapter by proposing two types of ESG-driven optimization: the traditional approach first, followed by an alternative that relies on firm characteristics to estimate the expected returns and covariance matrices.

8.2.1. *ESG overlays: the classic overlay*

Suppose we are given a target portfolio $\boldsymbol{p}$ which is known (or expected) to perform well. On top of this portfolio, the agent wishes to add a layer $\boldsymbol{z}$ so that the total allocation is $\boldsymbol{w} = \boldsymbol{p} + \boldsymbol{z}$. There are N assets (stocks – indexed by n) in the universe which are characterized by the following attributes: a vector of expected returns $\boldsymbol{\mu}$, a covariance matrix (of returns) $\boldsymbol{\Sigma}$ and a vector of observable sustainability metric, $\boldsymbol{g}$ (say, a synthetic environmental score, or a ratio of emissions divided by sales, for instance).

We assume that there is no disagreement in sustainability metrics, though that is not always the case (see Dimson et al. (2020) and Avramov et al. (2022) for theoretical and empirical perspectives on the matter). The objective (utility) function of the investor is the following:

$$U = \boldsymbol{w}'\boldsymbol{\mu} - \frac{\gamma}{2}\boldsymbol{w}'\boldsymbol{\Sigma}\boldsymbol{w} + \delta \boldsymbol{w}'\boldsymbol{g} \qquad [8.1]$$

which is a form that is commonplace in the literature (see Pedersen et al. 2021). We use the convention that $\boldsymbol{v}'$ is the transpose of vector $\boldsymbol{v}$ and the same applies to matrices. In equation [8.1], the usual mean-variance utility is complemented with a sustainable score and the positive scalar δ marks the intensity of the preference towards the green component. In practice, it may not be possible to leave the overlay portfolio $\boldsymbol{z}$ unconstrained – typically because of turnover or leverage considerations. Thus, the feasible optimization program reads

$$\max_{\boldsymbol{z}} \left\{ (\boldsymbol{p}+\boldsymbol{z})'\boldsymbol{\mu} - \frac{\gamma}{2}(\boldsymbol{p}+\boldsymbol{z})'\boldsymbol{\Sigma}(\boldsymbol{p}+\boldsymbol{z}) + \delta(\boldsymbol{p}+\boldsymbol{z})'\boldsymbol{g} \right\}, \quad \text{s.t. } \boldsymbol{z}'\boldsymbol{z} \le 2\kappa, \qquad [8.2]$$

where 2κ denotes the upper bound of the constraint on the leverage of the overlay portfolio (the 2-scaling factor is a notational convention that simplifies the final expression below). The solution to this program is

$$\boldsymbol{z}_o^* = (\gamma\boldsymbol{\Sigma} + \lambda \boldsymbol{I})^{-1} \left(\underbrace{\boldsymbol{\mu}}_{\text{perf.}} + \underbrace{\delta \boldsymbol{g}}_{\text{ESG}} - \underbrace{\gamma\boldsymbol{\Sigma}\boldsymbol{p}}_{\text{risk}} \right), \qquad [8.3]$$

where we have used horizontal brackets to highlight the role of the three components in the portfolio. The "*o*" subscript stands for "*overlay*". The constant λ is chosen to saturate the leverage constraint. As λ increases to infinity, or, equivalently, as κ decreases to zero, the magnitude of the weights in $\boldsymbol{z}$ shrinks to zero. While the overlay strategy is by nature designed as a complement to long-only benchmarks, it can easily accommodate long–short portfolios. Notably, if the target portfolio $\boldsymbol{p}$ has zero weights, the overlay becomes a standalone long–short strategy.

It is worth noticing that the Lagrangian of the program [8.2] has the same parametric form as that of

$$\max_{\boldsymbol{z}} \left\{ (\boldsymbol{p}+\boldsymbol{z})'\boldsymbol{\mu} - \frac{\gamma}{2}(\boldsymbol{p}+\boldsymbol{z})'\boldsymbol{\Sigma}(\boldsymbol{p}+\boldsymbol{z}) \right\}, \quad \text{s.t. } \begin{cases} \boldsymbol{z}'\boldsymbol{z} \le 2\kappa \\ (\boldsymbol{p}+\boldsymbol{z})'\boldsymbol{g} \ge g^* \end{cases}, \qquad [8.4]$$

4 The Lagrangian function $\mathcal{L} = (\boldsymbol{p}+\boldsymbol{z})'\boldsymbol{\mu} - \frac{\gamma}{2}(\boldsymbol{p}+\boldsymbol{z})'\boldsymbol{\Sigma}(\boldsymbol{p}+\boldsymbol{z}) + \delta(\boldsymbol{p}+\boldsymbol{z})'\boldsymbol{g} - \lambda(\boldsymbol{z}'\boldsymbol{z}/2 - \kappa)$ has a gradient

$$\frac{\partial \mathcal{L}}{\partial \boldsymbol{z}} = \boldsymbol{\mu} - \gamma\boldsymbol{\Sigma}(\boldsymbol{p}+\boldsymbol{z}) + \delta\boldsymbol{g} - \lambda\boldsymbol{z},$$

and the first-order condition imposes the form in equation [8.3].

i.e. when we replace the green reward inside the objective function by a simple green constraint which aims to obtain a minimum weighted green score of g^*. This shows that tilting the utility function towards greenness, or imposing a sustainable constraint are the two sides of the same coin in ESG-driven mean-variance investing.

The optimization programs described above impose quadratic constraints ($\boldsymbol{z}'\boldsymbol{z} \leq 2\kappa$) because they allow closed-form solutions. One implication is that under normal circumstances, all weights will be non-zero. Nevertheless, it can be desired to resort to the more selective constraint $|z_n| \leq \kappa$. In this case, the optimization can be written in a LASSO-like form (Tibshirani 1996) and when the penalization is strong enough, stocks start to disappear from the portfolio (i.e. they are assigned a zero weight). Thus, the optimization acts as a filter. For instance, when tracking a long-only target, if the ESG score is highly rewarded, a strong LASSO-type constraint would rule out those stocks with ESG scores that are unfavorable.

8.2.2. *The factor embedding – the factor overlay*

Now, let us assume that the portfolio policy is linearly driven by firm attributes, as in Brandt et al. (2009). Each one of the N firms is characterized by K indicators (or features), which we write $x_{t,n}^{(k)}$, where t is the index of time, n is the index of the firm, and k is the number of the characteristic. The common characteristics in the literature are firm size, the book-to-market ratio and past performance (momentum). Just like for factor models, relying on a limited number of characteristics will help alleviate the problem that arise because of dimension issues. When there are thousands of assets, but dozens of monthly returns or hundreds of daily returns, the estimation of the covariance structure is notoriously noisy (see Chopra and Ziemba (1993) and Kan and Zhou (2007)).

We henceforth index quantities by a time index for notational clarity. The K characteristics are decomposed into two groups, one consisting of K_S fields related to sustainability and K_O which are orthogonal thereto (so that $K_S + K_O = K$). Moreover, we write $\boldsymbol{X}_t$ for the time-t ($N \times K$) matrix of characteristics, so that the overlay is expressed as

$$\boldsymbol{z}_t = \boldsymbol{X}_t\boldsymbol{\theta}_t, \qquad [8.5]$$

where $\boldsymbol{\theta}_t = [\theta_t^{(1)} \dots \theta_t^{(K)}]'$ is the column vector which drives the allocation process. Typically, a positive (*resp.* negative) value for $\theta_t^{(k)}$ implies that the investor believes that the characteristic number k is positively (*resp.* negatively) linked to future returns.

In this case, the Lagrangian function of [8.2] translates to

$$\mathcal{L}_t = (\boldsymbol{p}_t + \boldsymbol{X}_t\boldsymbol{\theta}_t)'\boldsymbol{\mu}_t - \frac{\gamma}{2}(\boldsymbol{p}_t + \boldsymbol{X}_t\boldsymbol{\theta}_t)'\boldsymbol{\Sigma}_t(\boldsymbol{p}_t + \boldsymbol{X}_t\boldsymbol{\theta}_t)$$
$$+ (\boldsymbol{p}_t + \boldsymbol{X}_t\boldsymbol{\theta}_t)'\boldsymbol{X}_t^G\boldsymbol{\delta} - \lambda(\boldsymbol{\theta}_t'\boldsymbol{X}_t'\boldsymbol{X}_t\boldsymbol{\theta}_t - \kappa),$$

where $\boldsymbol{X}_t^G$ is the restriction of $\boldsymbol{X}_t$ to the columns related to the sustainable fields, and $\boldsymbol{\delta}$ is a K_S-long column vector that defines the preference towards these fields, as well as the overall relative importance of sustainability in the utility of the investor. The first-order conditions require

$$\frac{\partial \mathcal{L}_t}{\partial \boldsymbol{\theta}_t} = \boldsymbol{X}_t'\boldsymbol{\mu}_t - \gamma \boldsymbol{X}_t'\Sigma_t(\boldsymbol{p}_t + \boldsymbol{X}_t\boldsymbol{\theta}_t) + \boldsymbol{X}_t'\boldsymbol{X}_t^G\boldsymbol{\delta} - \lambda \boldsymbol{X}_t'\boldsymbol{X}_t\boldsymbol{\theta}_t = \boldsymbol{0}_K,$$

i.e.

$$\boldsymbol{\theta}_t = \left(\gamma \boldsymbol{X}_t'\Sigma_t\boldsymbol{X}_t + \lambda \boldsymbol{X}_t'\boldsymbol{X}_t\right)^{-1} \boldsymbol{X}_t' \left(\boldsymbol{\mu}_t - \gamma \Sigma_t \boldsymbol{p}_t + \boldsymbol{X}_t^G\boldsymbol{\delta}\right),$$

which is a form similar to that of equation [8.3]. As in Proposition 2.1 in Ammann et al. (2016), the elements nonetheless need to be estimated. In particular, without any additional regularization, the inverted matrix may be ill-defined, especially if estimation samples are small. If we assume, as in this latter paper, that utility is a simple quadratic function over realized portfolio returns, we have

$$\hat{\boldsymbol{\theta}}_t = \left(\gamma \hat{\boldsymbol{\Sigma}}_t + \lambda \boldsymbol{X}_t'\boldsymbol{X}_t\right)^{-1} \left(\hat{\boldsymbol{\mu}}_t - \gamma \hat{\boldsymbol{\sigma}}_t + \hat{\boldsymbol{g}}_t\right), \qquad [8.6]$$

where

$$\hat{\boldsymbol{\Sigma}}_t = \tau^{-1} \sum_{s=t-\tau}^{t-1} \boldsymbol{X}_s'\boldsymbol{r}_{s+1}\boldsymbol{r}_{s+1}'\boldsymbol{X}_s, \qquad \hat{\boldsymbol{\sigma}}_t = \tau^{-1} \sum_{s=t-\tau}^{t-1} \boldsymbol{X}_s'\boldsymbol{r}_{s+1}\boldsymbol{r}_{s+1}'\boldsymbol{p}_s,$$

$$\hat{\boldsymbol{\mu}}_t = \tau^{-1} \sum_{s=t-\tau}^{t-1} \boldsymbol{X}_s'\boldsymbol{r}_{s+1}, \qquad \hat{\boldsymbol{g}}_t = \boldsymbol{X}_t'\boldsymbol{X}_t^G\boldsymbol{\delta},$$

where τ is the estimation period (or chronological sample depth) and $\boldsymbol{r}_{s+1}$ are the vectors of realized returns. Working with firm characteristics provides a double advantage. First, it considerably shrinks the dimension in the estimation of the covariance matrix. This is valuable because work with short rolling datasets, and at the same time the universe consists of several hundred assets, implying that the sample covariance matrices are strongly non-singular. Second, the vector of expected returns is driven by characteristics, which is a much more efficient choice than taking *past* returns as proxy (see Brandt et al. (2009), Hjalmarsson and Manchev (2012), Ammann et al. (2016) for references on the performance of characteristic-driven portfolios).

Consequently, each term in equation [8.6] has a clear interpretation. $\gamma\hat{\Sigma}_t$ handles the global risk reduction of the portfolio via the covariances in between assets. $\lambda \boldsymbol{X}_t'\boldsymbol{X}_t$

5 Precisely, $u(r) = r - \gamma r^2/2$, with r being the realized portfolio return. The optimization is run on the average utility.

6 Of course, this obstacle can be circumvented by resorting to more sophisticated estimators.

fine-tunes the leverage of the overlay, i.e. the slack of the adjustment with respect to the benchmark. $\hat{\mu}_t$ drives the expected returns, i.e. the pure performance part of the allocation. $\gamma\hat{\sigma}_t$ reduces the risk by taking into account the correlation between the return of the overlay and that of the benchmark. Finally, $\hat{g}_t$, via the vector $\boldsymbol{\delta}$, gives more or less weight to the sustainability (environmental) component.

This decomposition and the constants therein (which are parameters) allow the investor to simply adjust them according to their preferences and beliefs. In addition, it is feasible to add a final term which would handle turnover and transaction costs (see Coqueret and Guida 2020). However, the basic assumption we make is that benchmarks are already slowly moving and have negligible asset rotation. Therefore, tuning the leverage of the overlay is almost equivalent to adjusting the transaction cost of the integrated strategy.

8.3. Empirical tests

In this section, we test the two strategies defined by equation [8.3] on the one hand, and by equations [8.5] and [8.6] on the other. The first is a standard environmental-adjusted optimal solution. The second seeks the same goal, but operates on firm characteristics.

8.3.1. *Data and protocol*

The investment universe consists of US firms that disclose ESG data. As pointed out in numerous studies, disclosure is key. In order to allocate according to ESG criteria, and to the E pillar in particular, the first filter imposes the availability of these criteria. Most major data providers have now acquired specialized companies, or developed their own in-house solutions. While our primary focus in the environmental score of the firms, we also gather all three pillars computed by *Sustainalytics*. The latter are accessible via *Bloomberg*. The data starts in February 2014 and ends in July 2021. In the sample, the minimum number of firms is 495 and the maximum is 573.

For a given month, each firm is characterized by five common attributes, which we list in Table 8.1.

We also show the evolution of the distributions of these raw variables in Figure 8.11 in Appendix 8.6. The three ESG pillars are computed as ranks in the cross-section of stocks, so that the evolution of quantiles is constant through time, which is why the corresponding time-series are not plotted. Among the other variables, the time-series of quantiles is also relatively stable, except for momentum, which is notoriously volatile (as much as rolling annual returns can be).

Factor	Definition	Bloomberg field	Academic references
Size	Market capitalization	*CUR_MKT_CAP*	Banz (1981) Fama and French (1992, 1993)
Value	Price-to-book ratio	*PX_TO_BOOK_RATIO*	Graham et al. (1934) Fama and French (1992, 1993)
Momentum	t-12m to t-1m return	From *PX_LAST*	Jegadeesh and Titman (1993) Carhart (1997)
Investment	Growth of assets	*ASSET_GROWTH*	Fama and French (2015)
Profitability	Operating profitability	*PROF_MARGIN*	Fama and French (2015)

Table 8.1. *List of traditional factors used in the chapter*

In Figure 8.1, we plot some correlation statistics. It is shown in the left panel that the highest correlations are obtained among the three ESG pillars and the next most significant link is between market capitalization and the E pillar. The link between ESG scores and firm size is for instance investigated in Drempetic *et al.* (2019). In the right panel of Figure 8.1, we for instance see that for a given date, the ESG indicators are not much correlated (in absolute) with subsequent monthly returns. However, at the stock level, they seem to be more important drivers of return dispersion.

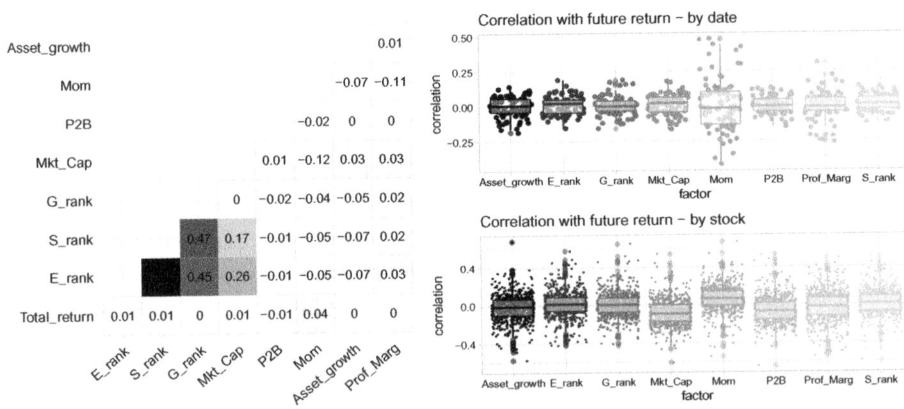

Figure 8.1. *Correlations. We plot the full sample correlation between the variables in the left panel. In the right panel, we show jitter and box plots of covariances between the variables and the one month ahead return. In the top graph, correlations are computed on a date-by-date basis and then displayed (there are 81 months in the sample). In the bottom graph, correlations are evaluated on a stock-by-stock basis, and then shown – there are 672 different firms in the sample*

Before being plugged into equation [8.6], the raw factor variables are all normalized (or ranked), so that their distribution is uniform in the $[-0.5, 0.5]$ interval.

This has become standard in the characteristic-driven asset pricing literature (Kelly et al. 2019; Freyberger et al. 2020). The rationale behind this choice is to propose both comparable scales in predictors as well as invariant distributions, which are preferable from a forecasting perspective (in machine learning, the issue of changing predictors is referred to as *covariate shift* – see Sugiyama and Kawanabe (2012)).

One important component of our approach is the choice of the benchmark portfolio for the overlay. It is not obvious, but we settle on the equally weighted policy. Not only is it simple to compute, but it is also harder to outperform in the long run (see DeMiguel et al. 2009; Plyakha et al. 2015), which will set the hurdle a little higher. Capitalization-weighted portfolios (with weights proportional to market capitalization) will nonetheless be considered for robustness check purposes.

In our baseline backtests, rolling samples of 12 months are used. Investment decisions are based on *past* data and returns are measured purely out-of-sample. Furthermore, the parameters we work with are the following. λ is a multiple of 100, from 100 to 1,000 (after initial tests, these are reasonable values). $\gamma \in \{1, 5, 10\}$, which allows for several levels of risk preferences. $\boldsymbol{\delta} \in \{0.005, 0.05, 0.5\} \times (1, 0, 0)$: i.e. the environmental pillar is the only one that matters. Finally, in the traditional overlay strategy (equation [8.3]), $\boldsymbol{\mu}$ is the sample average of returns, while $\boldsymbol{\Sigma}$ is the regularized covariance estimator from Ledoit and Wolf (2004).

The performance metrics we report are the following. First, the **average return**: $\bar{r} = T^{-1}\sum_{t=1}^{T} r_t$, where r_t are the out-of-sample returns of the portfolio. Next, the **transaction cost-adjusted Sharpe ratio**: $\frac{\bar{r}-0.003\times\text{Turn}}{\hat{\sigma}}$, where $\hat{\sigma}$ is the standard deviation of returns and Turn is the turnover of the portfolio computed as in Brandt et al. (2009). The cost per trade of 30 basis points is conservative, and we use a slightly smaller figure than in Goto and Xu (2015) (50 bps), because trading starts posterior to 2014 (transaction fees decrease over time). We also report the **average ESG score**: $\bar{g} = T^{-1}\sum_{t=1}^{T} g_t$, where g_t is the time-t weighted average ESG score of the portfolio. Finally, the **average realized utility** is $\bar{u} = \bar{r} - \frac{\gamma}{2}\hat{\sigma}^2 + \delta\bar{g}$.

8.3.2. *Baseline results*

In Figure 8.2, we plot the average returns (left panel) and transaction cost-adjusted Sharpe ratio of the overlay strategies, along with those of the equally weighted benchmark (horizontal black dashed line).

The results call for four conclusions. First, the shrinkage intensity on the x-axis has a different impact, depending on the environmental tilt (δ). When the latter is large ($\delta = 0.5$), the portfolio makes strong environmental bets which, in our sample, turn out to be beneficial for returns but strongly detrimental for risk (left panel) because we see that the Sharpe ratio (right panel) is inferior to the benchmark. When it is smaller, the environment score is less meaningful and the return benefits depend on risk aversion, but Sharpe ratios remain below that of the benchmark. Note that λ, the shrinkage intensity, is a major parameter in both schemes as it defines the magnitude of the adjustment that is allowed to the overlay.

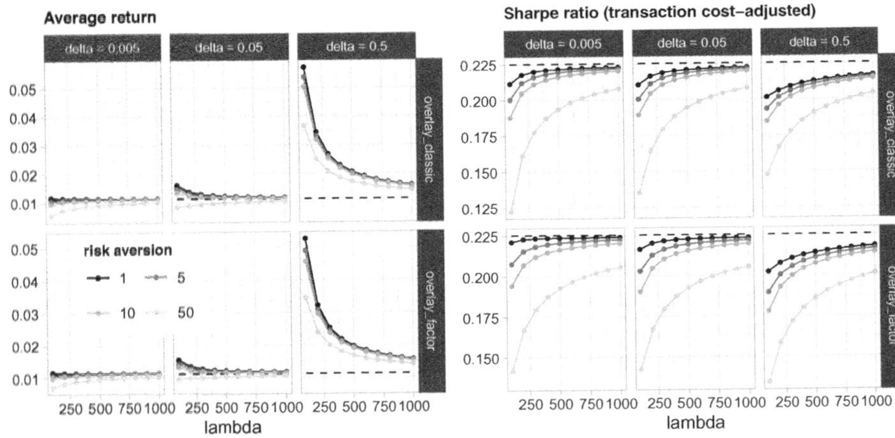

Figure 8.2. *Average return and transaction cost-adjusted Sharpe ratio. We plot both metrics as a function of the shrinkage intensity λ. The top graphs pertain to the classic optimization scheme* [8.3], *while the plots below relate to the factor approach given by* [8.6]. *Risk aversion is shown with gray-scale*

Second, the asymptotic value (for large λ), disappointingly, converges to the benchmark, meaning that no strategy is able to outperform it in terms of the Sharpe ratio, even though for λ large enough (say, $\lambda \geq 500$), the discrepancy becomes marginal for the smallest value of risk aversion. Third, the importance of risk aversion matters sharply for the Sharpe ratio, but can be determinant for returns as well and in any case, low risk aversion seems preferable.

The fourth point relates to the impact of the environmental constraint or target. The results suggest that when the tilt is too pronounced, the policy makes bets that are clearly unrelated to the traditional indicators with possibly large leverage when λ is small. Unsurprisingly, a strong exposure to environmental stocks comes at a certain cost in terms of risk-adjusted performance.

In Figure 8.3, we show the sensitivity of the average environmental score and the realized utility of the portfolios with respect to λ. In the latter case, we looked at what would happen if other ESG components were taken into account. Thus, we tested four configurations for $\boldsymbol{\delta}$, namely $(1, 1, 1)$ (global ESG) , $(3, 0, 0)$ (pure E, the baseline), $(0, 3, 0)$ (pure S) and $0, 0, 3$ (pure G).

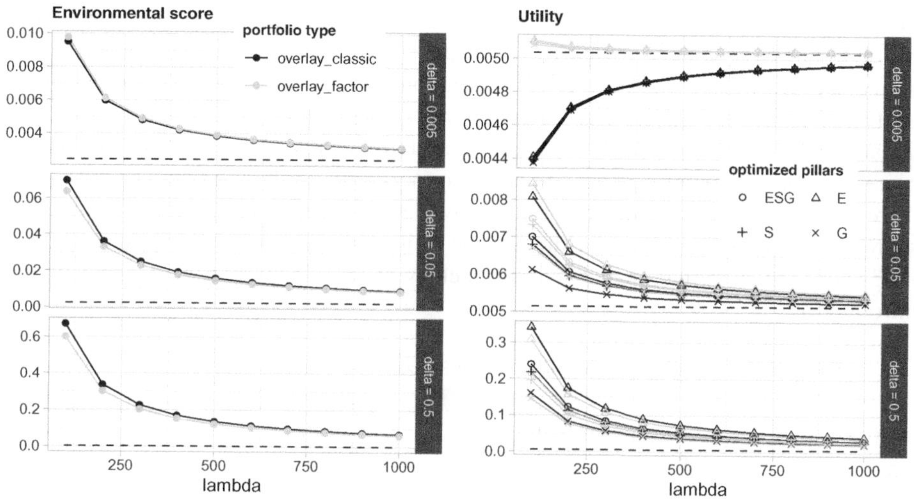

Figure 8.3. *Sustainability scores and realized utility. We plot the average ESG score of the portfolio (left panel), along with the realized utility (right panel). In the right panel, the ESG component is represented with four different shapes, marking the difference between global ESG, pure E, pure S, or pure G. Finally, two colors show the difference between the two approaches: standard (light gray, based on equation* [8.3]*) versus factor-driven (dark gray, based on equation* [8.6]*). The dashed black line marks the values for the equally weighted benchmark*

In the left panel, it is shown that the environmental score decreases with λ. When the latter increases, the optimization has less room to favor the E-score and progressively converges to that of the benchmark. Thus, moderate values of λ (around $\lambda = 400$) improve the environmental exposure of the portfolio. For low levels of risk aversion, this is done with almost no cost in terms of performance (see the previous figure). The two overlay formulae yield similar outcomes but we note the changes in the scale of the y-axis when δ increases: the strongest changes come with large values of δ.

In the right panel of Figure 8.3, we observe that factor-based portfolios are either very close or superior to traditional overlays in terms of the realized utility. Notably, the former always improve the benchmark, thanks to their enhanced ESG tilt which is reflected in the utility. Finally, these conclusions are not altered by the choice of the ESG pillar which drives the sustainable component of the utility.

A last batch of visualizations is presented in Figure 8.4. The left panel compares the TC-adjusted Sharpe ratio of the traditional overlay equation [8.3] with that of the factor-based adjustments from equation [8.6], across all combinations of parameters. The black line shows the points for which both solutions have equal Sharpe ratios.

Straightforwardly, we conclude that factor-based overlays are undoubtedly superior, which is not easy to see in Figure 8.2. In the right panel, we show the ESG versus financial performance trade-off. It shows that, on average and among all the strategies we tested, an increase in ESG score comes with a decrease in Sharpe ratio. While the debate on the cost of sustainability is not settled (see Coqueret 2021), our conclusions point towards a damaging effect and not an opportunity. In the left panel ($\delta = 0.005$), there are some points at the top that show a significant improvement in the ESG score with only a minor loss in the Sharpe ratio. However, the corresponding gains in E-score are marginal, which is consistent with the discussion at the end of the introduction.

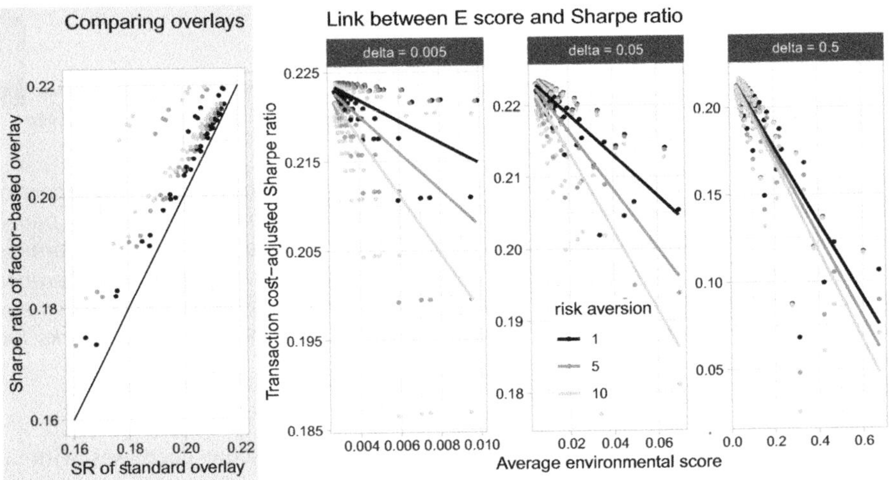

Figure 8.4. *Comparing overlays and the environment–Sharpe ratio trade-off. For given sets of parameters, we plot the TC-adjusted Sharpe ratio of the two overlays: the simple one (x-axis, based on [8.3]) versus the factor-based (y-axis, based on [8.6]). The right panel shows the E-score versus Sharpe ratio trade-off for all baseline strategies. Shades of gray code the level of risk aversion. Linear models for each level of risk aversion are fitted (straight lines)*

8.3.3. *Statistical significance*

In this section, we evaluate the discrepancy in performance between the benchmark and its overlays from a testing perspective. To take risk into account, we resort to the test on the equality of Sharpe ratios introduced in Wright et al. (2014) which improves those of Ledoit and Wolf (2008) and Leung and Wong (2008). As shown in Figure 8.2, overlays are almost always outperformed, so the question here is: is this relative inferiority really significant?

A first unreported set of results yielded p-values that were unexpectedly high, meaning that benchmarks and overlays had relatively comparable performances. One reason for this is the small sample that we use (six years of monthly data). Statistical tests are quite sensitive to sample sizes (see Halsey et al. 2015), so we design a more conservative procedure inspired from bootstrapping. We fix $\lambda = 200$ and at each date, we test four portfolios, based on the three pillars or on each pillar individually. This yields four times more returns that are used for the computation of the t-statistics – thereby mechanically lowering the p-values.

The corresponding results are shown in Figure 8.5. The interpretation is as follows. The p-values are often used as a decision proxy to reject the null. For instance, if the p-value is below a given threshold, for example, 5%, it is considered that the null should be rejected, i.e. that the two portfolios have different Sharpe ratios. In Figure 8.5, all p-values are above 5%, which suggests that the benchmark and overlays have Sharpe ratios that are not statistically different. However, as the sustainability component becomes more important (δ increases, i.e. the gray becomes lighter), the discrepancy between the two augments. In this case, when the environmental exposure becomes too pronounced, the loss in performance is such that the difference in Sharpe ratio shifts into significant territory. The overlays are then strongly outperformed by the benchmark.

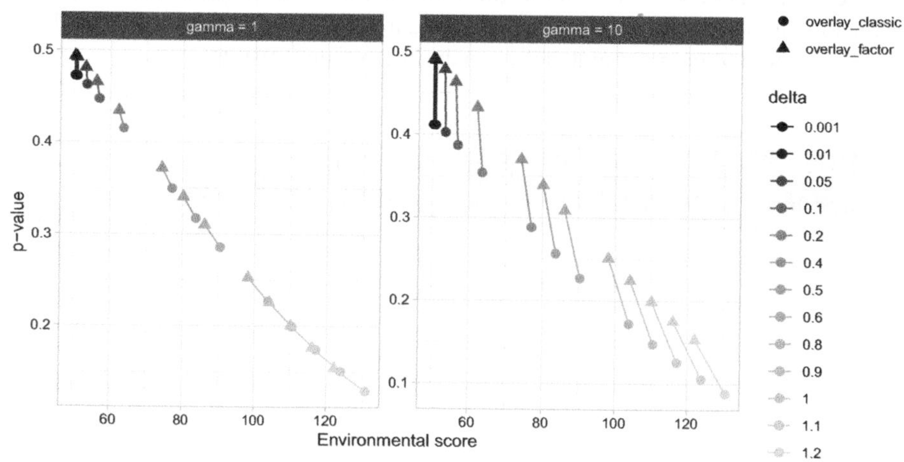

Figure 8.5. *Comparing Sharpe ratios along environmental gains. For $\lambda = 200$ and 12 month rolling samples, we plot the p-values of the test for which the null hypothesis is that both strategies (benchmark and overlay) have the same Sharpe ratio. The p-values are shown as a function of the average E score of the portfolios. Each color codes a value for the importance of the E component, as measured by δ. The upper points pertain to the factor-based overlays, while the lower points relate to the traditional overlays. The left panel corresponds to low risk aversion, while the right one deals with moderately high risk aversion*

There is little difference between the two overlay methods, as shown by the proximity of the points with the same colors (linked with a line). However, the fact that the factor-based overlay is always above proves that it is more competitive (from a Sharpe ratio standpoint, as inferred from Figure 8.4). The most noteworthy findings pertain to the scale of the x axis. It starts at 50, which is the green score (out of 100) of the equally weighted benchmark. Thus, as δ increases and the portfolios become greener, their relative performance deteriorates. However, this loss in performance is *not* significant. It seems possible to significantly increase the ESG exposure without affecting the Sharpe ratio too much. Note that scores above 100 imply some form of leverage: shorting some low ESG firms in order to buy more of the high ESG companies.

In sum, the combination of all results presented above proves that it is possible to generate a performance that is similar to the benchmark while at the same time considerably improving the green exposure of the portfolio via the small adjustments stemming from the overlay.

8.3.4. *Sector exposure*

For completeness, we now examine the sectorial tilts of the portfolios and overlays. We use the Global Industry Classification Standard (GICS Level 1/Sectors) of MSCI because it is not too granular and yields easily interpretable results.

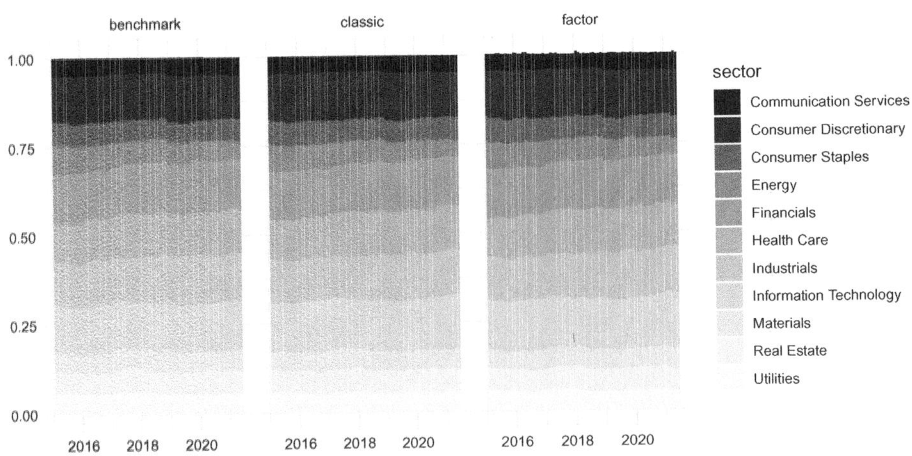

Figure 8.6. *Sector exposure. We show the sectorial decomposition of the portfolios for $\gamma = 5$, $\lambda = 200$, $\delta = 0.1$ and a sample size of 12 months. The traditional overlay is displayed in the middle panel, while the factor-based overlay is shown to its right (the benchmark being the left plot)*

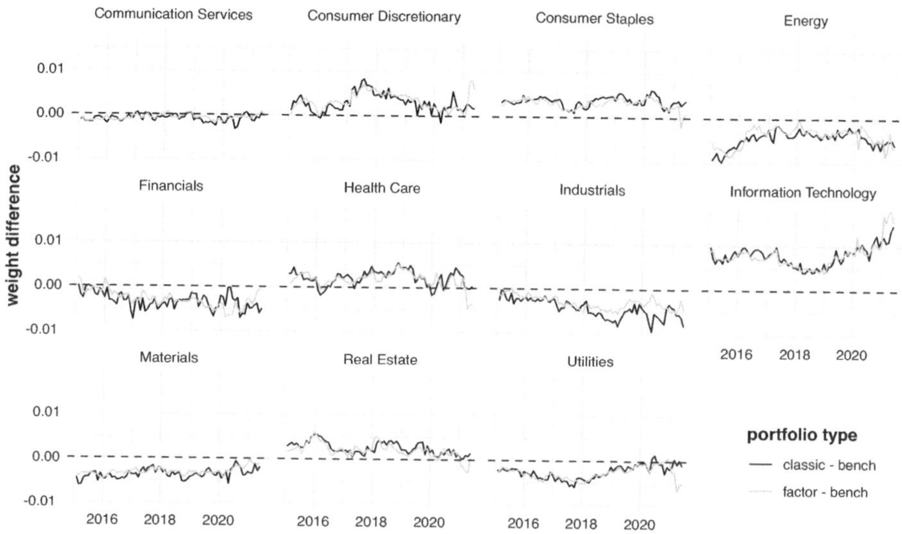

Figure 8.7. *Net sector exposure versus benchmark. We show the net sectorial decomposition of the portfolios for* $\gamma = 5$, $\lambda = 200$, $\delta = 0.1$ *and a sample size of 12 months. It is computed as the sector weight of the optimized portfolio minus that of the benchmark*

Figure 8.6 provides the evolution of the sector exposures (aggregate sector weights) for the portfolios over time. Figure 8.7 provides a different view of this data, considering only the deviations of the weights of each sector in the overlays minus the weights of the benchmark. We note the deviations from the benchmark sector weights are small and moreover the deviations are consistent for both the classic and factor portfolios, which we would expect as the main sector tilt will come from the environmental scores rather than factors.

Moreover, the tilts are in line with our prior expectations, i.e. sectors that have a significant environmental footprint are under-weighted (e.g. energy or industrials). In contrast, sectors such as health care or information technology, which have less of an environmental impact as businesses, have larger weights in the overlays.

8.3.5. *Transfer coefficients*

Following Clarke (2020), we define transfer coefficients as the correlation (in the cross-section) between portfolio weights $\boldsymbol{w}_t$ and the environmental score of firms $\boldsymbol{g}_t$: $\mathrm{t}_s = \mathrm{cor}(\boldsymbol{w}_s, \boldsymbol{g}_s)$.

Figure 8.8 provides the transfer coefficients over time for both the classic and the factor portfolios. As γ increases, we see that transfer coefficients are shifted slightly lower, which is expected as the risk is more emphasized and hence the correlation of the signal to the portfolio shrinks. The fact that the factor portfolio has a higher transfer coefficient than the classic portfolio is consistent with the fact that the E pillar has a lower correlation with total returns. However, we observe a 25% correlation to the market capitalization attribute. The volatility of the transfer coefficient for the factor portfolio is driven by the highly time-varying correlation of the cross-sectional correlation of the factor signals and the E pillar score.

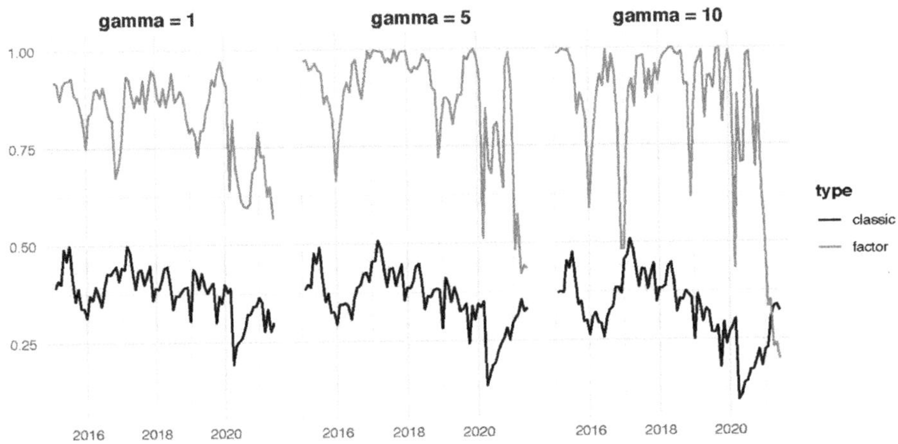

Figure 8.8. *Transfer coefficients. We plot the transfer coefficients* t_s *of the portfolios for* $\gamma \in \{1, 5, 10\}$, $\lambda = 200$, $\delta = 0.01$ *and a sample size of 12 months*

8.4. Robustness checks

To test the robustness of our results, we replicate our backtests with changes in two directions. First, we analyze the sensitivity to sample size and in a second part, we choose an alternative benchmark.

8.4.1. *The sample size*

Our baseline results rely on rolling 12 month samples. Given the relatively short length of our complete dataset, it is impractical to increase it because it would shrink the chronological depth of the backtest. However, it is feasible to reduce it, so as to make allocations more reactive to recent data. To achieve this, we can vary the parameter τ, as defined in equation [8.6]. In Figure 8.9, we plot the difference in the average return between the overlays and the benchmark. The optimization is performed on the environmental pillar with $\delta = 0.05$.

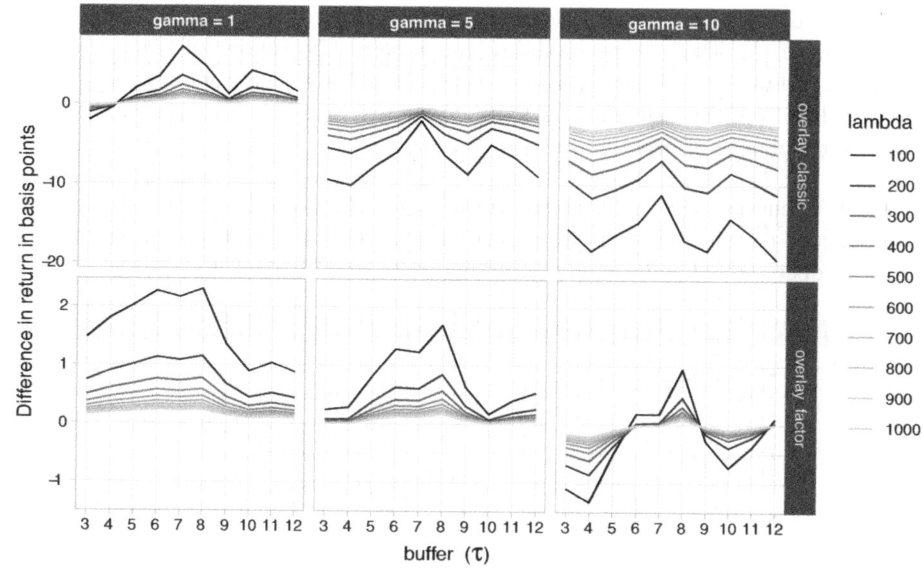

Figure 8.9. *Impact of sample size (τ). We plot the difference (in basis points) in average return between the overlay strategies and their benchmark (a positive value meaning that the overlay beats the benchmark). Three risk aversion values are tested $\gamma = 1$ (left), $\gamma = 5$ (center) and $\gamma = 10$ (right). The traditional overlay is displayed on the upper graphs, while the factor-based overlay is shown in the bottom plots. The optimization is performed on the environmental pillar with $\delta = 0.05$*

A first result is the confirmation of the superiority of the factor overlay. Looking at the y axis scales shows the magnitude of under-performance yielded by the simple overlay strategy. The fact that there is at most a 2-basis point discrepancy between the factor overlay and the benchmark is further proof that tilting the allocation towards green stocks is *not* too costly. It also underlines that the sample size has at most a marginal impact. For low levels of risk aversion, the overlay even beats the benchmark for most values of rolling samples. Average returns are higher for low risk aversion, which was expected, as the focus is less on variance reduction.

8.4.2. *A more passive benchmark*

Up to now, the core portfolio that drives the allocation was the equally weighted (EW) portfolio. In the financial industry, capitalization-weighted (CW) indices (like the S&P500) often serve as yardstick for equity performance. It is thus important to assess if the baseline results hold if the benchmark has weights that are proportional to the market capitalization of the stocks.

In Figure 8.10, we plot the average returns of overlays and benchmarks as a function of λ. The rolling samples encompass 12 months of data and the environmental intensity is $\delta = 0.05$. We observe that both curves are relatively close (the y-scale is compact) and generate similar patterns, meaning that our methodology is robust to a change of benchmark.

The CW curves are slightly above those of the EW portfolios, which is likely a by-product of the recent over-performance of large stocks. Traditionally, the reverse is true, as documented in the asset pricing literature (see Plyakha et al. 2015).

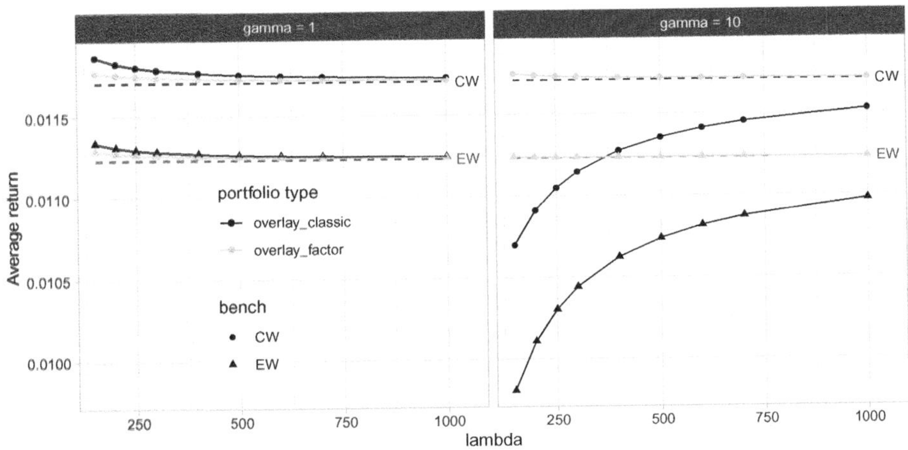

Figure 8.10. *The cap-weighted benchmark. We plot the average return between the overlay strategies and their benchmark. Two risk aversion values are tested* $\gamma = 1$ *(left) and* $\gamma = 10$ *(right). The optimization is performed on the environmental pillar with* $\delta = 0.05$. *EW stands for "equally-weighted", while CW pertains to "capitalization-weighted" portfolios. Benchmark values are shown at the top with dashed lines*

8.5. Conclusion

This chapter provides new evidence that sustainable investing is neither necessarily very costly from a financial standpoint, nor extremely useful to generate abnormal profits. Our methodology uses traditional firm characteristics as drivers of estimators in portfolio optimization, while, at the same time, the utility function incorporates a flexible sustainable component based on the environmental footprint of firms. Because the estimation of the expected returns and covariance matrix is performed via their link to characteristics, the portfolios are able to increase their ESG score, while at the same time delivering a level of risk-adjusted performance which is similar to their benchmarks. However, portfolios driven by returns and ESG data only are not able to bridge the gap in performance, which highlights the incremental importance of non-ESG information in the crafting of the allocation. Therefore, sustainable overlays are most effective when they combine ESG goals with traditional factors. In Figure 8.5, we showed that the environmental score can be increased by approximately 140% without a statistically significant change in the Sharpe ratio.

We nevertheless underline that our results are contingent on the data at our disposal. The divergence in ESG ratings is a notorious issue in the field (Berg et al. 2022), and changes in data providers are of course likely to alter performance (Billio et al. 2020). That being said, the combination of sustainability criteria with more fundamental variables undoubtedly seems like a promising area for portfolio managers and researchers in asset pricing more generally.

8.6. Appendix: Distribution of variables

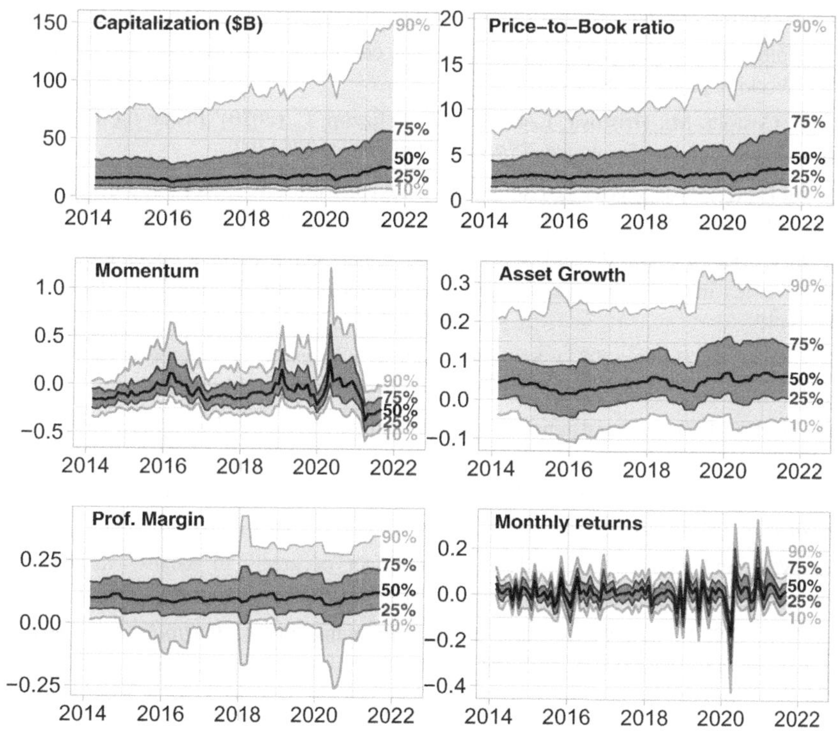

Figure 8.11. *Evolution of the distribution of variables. We plot the evolution of the 10%, 25%, 50%, 75% and 90% quantiles of the market capitalization of firms in our sample (top left panel). The price-to-book ratio, past 12 month returns, asset growth and profitability margin are also displayed (see Table 8.1 for their definitions). Finally, the quantiles of the monthly total returns of assets (i.e. including dividends) are shown in the bottom right panel*

8.7. References

Alessandrini, F. and Jondeau, E. (2021). Optimal strategies for ESG portfolios. *Journal of Portfolio Management*, 47(6), 114–138.

Ammann, M., Coqueret, G., Schade, J.-P. (2016). Characteristics-based portfolio choice with leverage constraints. *Journal of Banking & Finance*, 70, 23–37.

Avramov, D., Cheng, S., Lioui, A., Tarelli, A. (2022). Sustainable investing with ESG rating uncertainty. *Journal of Financial Economics*, 145(2), 642–664.

Banz, R.W. (1981). The relationship between return and market value of common stocks. *Journal of Financial Economics*, 9(1), 3–18.

van der Beck, P. (2021). Flow-driven ESG returns. *SSRN*, Working paper 3929359.

Bender, J., Sun, X., Wang, T. (2017). Thematic indexing, meet smart beta! Merging ESG into factor portfolios. *Journal of Index Investing*, 8(3), 89–101.

Berg, F., Kölbel, J.F., Rigobon, R. (2022). Aggregate confusion: The divergence of ESG ratings. *SSRN*, Working paper 3438533.

Billio, M., Costola, M., Hristova, I., Latino, C., Pelizzon, L. (2020). Inside the ESG ratings: (Dis)agreement and performance. *SSRN*, Working paper 3764493.

Bloomberg Professional Service (2020). Seeking signals from ESG data [Online]. Available at: https://www.bloomberg.com/professional/blog/seeking-signals-from-esg-data/.

Brandt, M.W., Santa-Clara, P., Valkanov, R. (2009). Parametric portfolio policies: Exploiting characteristics in the cross-section of equity returns. *Review of Financial Studies*, 22(9), 3411–3447.

Cai, L., Cooper, R., He, D. (2021). Socially responsible investing and factor investing, is there an opportunity cost? *Journal of Portfolio Management*, 48(2), 181–197.

Carhart, M.M. (1997). On persistence in mutual fund performance. *Journal of Finance*, 52(1), 57–82.

Chan, Y., Hogan, K., Schwaiger, K., Ang, A. (2020). ESG in factors. *Journal of Impact and ESG Investing*, 1(1), 26–45.

Chopra, V.K. and Ziemba, W.T. (1993). The effect of errors in means, variances, and covariances on optimal portfolio choice. *Journal of Portfolio Management*, 19(2), 6.

Clarke, R.G. (2020). Analysis of active portfolio management. *Portfolio Management in Practice, Volume 3: Equity Portfolio Management*, 141.

Coqueret, G. (2021). *Perspectives in Sustainable Equity Investing*. CRC Hall, Boca Raton, FL.

Coqueret, G. and Guida, T. (2020). *Machine Learning for Factor Investing: R Version*. CRC Press, Boca Raton, FL.

DeMiguel, V., Garlappi, L., Uppal, R. (2009). Optimal versus naive diversification: How inefficient is the 1/N portfolio strategy? *Review of Financial Studies*, 22(5), 1915–1953.

Dimson, E., Marsh, P., Staunton, M. (2020). Divergent ESG ratings. *Journal of Portfolio Management*, 47(1), 75–87.

Drempetic, S., Klein, C., Zwergel, B. (2019). The influence of firm size on the ESG score: Corporate sustainability ratings under review. *Journal of Business Ethics*, 167, 1–28.

Fama, E.F. and French, K.R. (1992). The cross-section of expected stock returns. *Journal of Finance*, 47(2), 427–465.

Fama, E.F. and French, K.R. (1993). Common risk factors in the returns on stocks and bonds. *Journal of Financial Economics*, 33(1), 3–56.

Fama, E.F. and French, K.R. (2015). A five-factor asset pricing model. *Journal of Financial Economics*, 116(1), 1–22.

Fan, J.H. and Michalski, L. (2020). Sustainable factor investing: Where doing well meets doing good. *International Review of Economics & Finance*, 70, 230–256.

Flammer, C. (2021). Corporate green bonds. *Journal of Financial Economics*, 142(2), 499–516.

Freyberger, J., Neuhierl, A., Weber, M. (2020). Dissecting characteristics nonparametrically. *Review of Financial Studies*, 33(5), 2326–2377.

Goto, S. and Xu, Y. (2015). Improving mean variance optimization through sparse hedging restrictions. *Journal of Financial and Quantitative Analysis*, 50(6), 1415–1441.

Graham, B., Dodd, D.L.F., Cottle, S. (1934). *Security Analysis*. McGraw-Hill, New York.

Halsey, L.G., Curran-Everett, D., Vowler, S.L., Drummond, G.B. (2015). The fickle P-value generates irreproducible results. *Nature*, 12(3), 179–185.

Henriksson, R., Livnat, J., Pfeifer, P., Stumpp, M. (2019). Integrating ESG in portfolio construction. *Journal of Portfolio Management*, 45(4), 67–81.

Hjalmarsson, E. and Manchev, P. (2012). Characteristic-based mean-variance portfolio choice. *Journal of Banking & Finance*, 36(5), 1392–1401.

Jegadeesh, N. and Titman, S. (1993). Returns to buying winners and selling losers: Implications for stock market efficiency. *Journal of Finance*, 48(1), 65–91.

Kaiser, L. (2020). ESG integration: Value, growth and momentum. *Journal of Asset Management*, 21(1), 32–51.

Kan, R. and Zhou, G. (2007). Optimal portfolio choice with parameter uncertainty. *Journal of Financial and Quantitative Analysis*, 42(3), 621–656.

Kelly, B.T., Pruitt, S., Su, Y. (2019). Characteristics are covariances: A unified model of risk and return. *Journal of Financial Economics*, 134(3), 501–524.

Ledoit, O. and Wolf, M. (2004). A well-conditioned estimator for large-dimensional covariance matrices. *Journal of Multivariate Analysis*, 88(2), 365–411.

Ledoit, O. and Wolf, M. (2008). Robust performance hypothesis testing with the Sharpe ratio. *Journal of Empirical Finance*, 15(5), 850–859.

Leung, P.-L. and Wong, W.-K. (2008). On testing the equality of multiple Sharpe ratios, with application on the evaluation of iShares. *Journal of Risk*, 10(3), 15.

Mercure, J.F., Salas, P., Vercoulen, P., Semieniuk, G., Lam, A., Pollitt, H., Holden, P.B., Vakilifard, N., Chewpreecha, U., Edwards, N.R. et al. (2021). Reframing incentives for climate policy action. *Nature Energy*, 6(12), 1133–1143.

Mohanty, S.S., Mohanty, O., Ivanof, M. (2021). Alpha enhancement in global equity markets with ESG overlay on factor-based investment strategies. *Risk Management*, 23(3), 213–242.

Pedersen, L.H., Fitzgibbons, S., Pomorski, L. (2021). Responsible investing: The ESG-efficient frontier. *Journal of Financial Economics*, 142(2), 572–597.

Plyakha, Y., Uppal, R., Vilkov, G. (2015). Why do equal-weighted portfolios outperform value-weighted portfolios? *SSRN*, Working paper 2724535.

Sugiyama, M. and Kawanabe, M. (2012). *Machine Learning in Non-stationary Environments: Introduction to Covariate Shift Adaptation*. MIT Press, Cambridge, MA.

Tibshirani, R. (1996). Regression shrinkage and selection via the lasso. *Journal of the Royal Statistical Society: Series B (Methodological)*, 58(1), 267–288.

Wright, J., Yam, S., Pang Yung, S. (2014). A test for the equality of multiple Sharpe ratios. *Journal of Risk*, 16(4), 3–21.

9

Enhancing the Accuracy of Firm Valuation with Multiples Using Carbon Emissions

Carbon emissions are nowadays an essential driver of the value of a firm. We are the first to analyze the potential of carbon emission data in enhancing the accuracy of firm valuations using the similar public company methodology with multiples. Using carbon emissions can improve firm valuation accuracy in two separate ways. First, we construct multiples based on carbon emissions (CEM) which are able to estimate firm values. Moreover, second, we create more precise peer groups by including carbon emissions (CEPG) in the composition process. To gain deeper insights, we are conducting further analyses, for example, by measuring the accuracy of carbon emission peer groups and carbon emission multiples at valuing carbon-intensive or carbon-inefficient firms. We extend our study by looking at firms in countries with carbon pricing or taking ESG and SDG concerns into account. Overall, we find that CEPG improves the accuracy of firm valuations in more than three-quarters of all cases, whereas CEM have limited use. Therefore, we recommend that analysts, asset managers and investors include carbon emission data in their peer group composition.

In the latest global risk perception survey, climate change mitigation and adaptation failure are ranked first by their impact (World Economic Forum 2021). The awareness of these climate-related financial risks has grown recently, especially since the Paris COP 21 (International Monetary Fund 2019). In addition, most countries have committed to emission mitigation and are introducing respective

Chapter written by Martin NERLINGER.

I am grateful for the helpful comments and suggestions by Marco Wilkens, Andrea Jacob and Maximilian Görgen. I am responsible for all errors. This research did not receive any specific grant from funding agencies in the public, commercial or not-for-profit sectors. Declarations of interest: none.

policies, for example, carbon pricing or carbon taxes (World Bank Group 2019). These recent developments impact firms depending on their carbon emissions and can lead to a significant change in their firm value. Motivated by the concerns of investors, asset managers, regulators and standard setters about the risk of an accelerating climate change on firm value, we assess possible applications of carbon emission data to improve the accuracy of firm valuation.

We are the first to analyze the valuation potential of carbon emission data using one of the most used firm-valuation approaches: multiples based on the similar public company method. Using carbon emissions can help construct more accurate multiples for firm valuation in two ways. First, we create carbon emission-based multiples (carbon emission multiples, CEM) and evaluate their firm valuation accuracy. Moreover, second, we identify and compose a more suitable peer group (carbon emission peer group, CEPG) for firm valuations using carbon emissions as a classification criterion. Our results can be used by practitioners, for example, analysts, asset managers and investors, to improve the accuracy of their firm valuation approaches.

Our fundamental multiples approach is as follows. First, we construct multiples based on different financial and non-financial indicators. Next, we compose peer groups of similar firms based on their characteristics. Now, we can form a self-excluding average multiple within each peer group for each firm. We use the average multiple of a peer group to estimate a firm's value. Afterward, we can compare the valuation error by subtracting the estimated from the observed firm value. Subsequently, we evaluate the valuation accuracy of a multiple by computing four different error measures: logarithmic error, absolute logarithmic error, overvaluation percentage and absolute logarithmic error percentage. We aggregate these error measures in four different ways across all firms and peer groups using either the arithmetic mean, the median, the geometric mean or the harmonic mean. After this procedure, we can evaluate the firm valuation accuracy of each financial and non-financial multiple as well as each peer group composing.

Throughout all our analyses, we use common financial and accounting data from Refinitiv Datastream and Worldscope to compile a global firm data sample of more than 27,000 firms yearly. We add carbon emission data from three major data providers, Refinitiv ESG, CDP and Sustainalytics, to broaden our coverage and address known biases within the data collection approaches.

We use carbon emissions to construct CEM and evaluate their firm valuation accuracy in our first analyses. For this purpose, we use scope 1 and scope 2 carbon emissions in relation to either the firm's equity or entity value. We find that CEM have, on average, higher absolute logarithmic errors and a lower percentage of absolute logarithmic errors below 15%. In contrast, in most cases, they have lower logarithmic errors, which points to a significantly higher percentage of undervaluations. This observation is reinforced by a higher percentage of undervaluations in all cases.

In our second analysis, we examine the valuation potential of carbon emission data by including them in the peer group composition procedure. Firms with similar carbon emissions are affected to a similar extent by investor behavior, for example, divestment, as well as additional risks from climate policies, for example, carbon tax. We find that we can increase the accuracy of a firm valuation significantly using CEPG. We measure a higher accuracy of the estimated firm value using the logarithmic error and especially the absolute logarithmic error as our error measure. Our results hold across all four error aggregation methods as well as for both equity and entity multiples. Measured by the percentage of overvaluation, we receive a mixed picture, but more precise valuations are achieved in half of the cases. However, the percentage of absolute logarithmic error of multiples below 15% is slightly higher if we include carbon emissions.

If we combine our analyses by constructing CEM using CEPG, we observe only a limited improvement in the firm valuation accuracy in comparison with financial multiples in all cases across all aggregation measures as well as for both equity and entity multiples. We find that our results are mostly driven by a more suitable composition of peer groups rather than the use of CEM.

To gain further insights, we carry out numerous analyses. Our main findings even hold for important subgroups. We find that we were able to increase the accuracy of firm valuations, especially for firms with reported carbon emission data available. A similar picture emerges if we look at carbon-intensive industries. Here, the inclusion of carbon emissions into the composition of peer groups leads to more precise firm valuations. However, we cannot find similar results when considering carbon-intensive firms.

In order to include the current climate policy development, we also carried out two analyses in which we look at firms that are located in countries with a carbon pricing initiative. In the first case, we cover countries with a national carbon pricing initiative. In the second case, we also take regional carbon pricing initiatives into account. We show that an improvement in the accuracy of firm valuations can be achieved in both CEM and CEPG cases.

If, on the other hand, we take into account the recent increasing relevance of the climate topic for asset management, we conclude that more precise firm valuations are obtained for relatively fewer firms even before and for relatively more firms since 2010. Considering the differences between various geographical regions, we observe only minor differences. The highest CEM and CEPG firm valuation accuracy is achieved in Asia, but if we combine the two approaches, the best results are achieved in Oceania. Overall, 16 out of 18 geographical cases show a 50% or higher firm valuation accuracy.

In a further analysis, we are taking ESG and SDG issues into account. We find that the firm valuation accuracy increases, especially for firms with low ESG ratings. This is due to the fact that ESG ratings also include many carbon-related components such as carbon efficiency or carbon emissions. However, beyond that, ESG ratings are also expressing the general adaptability to ESG issues, which is particularly weak at low ratings and thus affects firm value in particular. We find similar effects for firms that are located in countries with a high SDG 13 (Climate Action) performance. In this case, the country's SDG performance indicates that it is actively involved in combatting climate change and therefore has, for example, reduced its carbon emissions to meet its NDCs.

Overall, our study shows that it is worthwhile to use carbon emissions for the composition of suitable peer groups (CEPG) and, in certain cases, also for the construction of multiples (CEM). Analysts, asset managers and investors can improve their firm valuation accuracy by using CEPG. Our chapter contributes to studies that analyze the impact of carbon emissions on firm value and to studies on the valuation of firms, especially those that use multiples.

A fast-growing strand of literature is analyzing the impact of carbon emissions on firm value. Carbon emissions and carbon disclosure have a significant positive effect on the value of a firm (Matsumura et al. 2014). In particular, mandatory carbon disclosure has a significant impact on the market valuation of a firm and the overall market efficiency. It increases market liquidity (higher trading volume) and lowers information asymmetries (lower bid-ask spreads) for carbon-intensive firms (Krueger 2015; Liesen et al. 2017). Furthermore, investors and firms do care about carbon emission disclosures (Brammer and Pavelin 2006; Griffin et al. 2017).

From a risk perspective, high carbon emissions add additional risk to a firm, thereby impacting its value. Environmental friendliness and environmental risks, of which carbon emissions are a huge part, have a significant impact on firm value (Fernando et al. 2017). The risk of emitting carbon as a firm can also be measured by analyzing, for example, extreme weather events, which show a lowering firm value effect for carbon-intensive firms (Berkman et al. 2021). Moreover, carbon emissions increase a firm's tail risk (Ilhan et al. 2019). In addition, firms with high carbon emission intensity and high carbon risk have lower stock returns (Bolton and Kacperczyk 2021; Görgen et al. 2021). However, if firms receive free carbon certificates, they profit from this "free lunch" and have comparable higher returns (Oestreich and Tsiakas 2015).

A firm valuation can be conducted via various equity and entity approaches that are mainly divided into being market-based or fundamental-based. One of the most used firm valuation market-based approaches in practice is the similar public company method (Asquith et al. 2005; Koller et al. 2005; Pinto et al. 2019). This method is grounded in the belief of Jevon's law of one price (Jevons 1879) so that no identical good can be traded at different prices in efficient markets. Due to the large number of different ways to apply multiples for firm valuation, we follow best practices in applying multiples for firm valuation purposes (Plenborg and Pimentel 2016).

The literature focuses on the valuation accuracy of multiples as the main criterion to prove their usefulness. Most studies analyze either the composition of suitable peer groups, the aggregation of multiples within a peer group or the most useful variables to compose multiples. Studies focusing on a suitable peer group composition (Eberhart 2001, 2004) should control for differences within peer groups (Henschke and Homburg 2009) to identify similar firms. Early studies advise composing peer groups using a firm's industry classification (Alford 1992) and its characteristics (Bhojraj and Lee 2002). New studies propose to integrate insights about similar Internet-searched firms (Lee et al. 2015) to compile optimal peer groups.

To aggregate multiples within a peer group, several authors have suggested correcting for related biases using different aggregation methods, for example, mean, median, geometric mean or harmonic mean (Cooper and Lambertides 2014; Dittmann and Maug 2008). Furthermore, we use several error measures to obtain detailed knowledge of distortions within our results: logarithmic error, absolute logarithmic error, overvaluation percentage and absolute logarithmic error percentage (Chullen et al. 2015). To additionally improve the accuracy of a multiple-based firm valuation, a study suggests using a combination of multiples (Yoo 2006).

The selection of financial indicators for multiples is part of some studies. Many multiples are based on either cash flows or earnings (Chen et al. 2014), which contain different valuation information (Liu et al. 2007). Other studies construct valid multiples based on historical balance sheet key indicators, for example, asset, sales and earning values (Lie and Lie 2002; Yin et al. 2018), analyst forecasts, for example, forward earnings (Liu et al. 2002), or different cash flow measures, for example, operating cash flow (Liu et al. 2007; Koller et al. 2015).

The range of applications with multiples is large. Multiples are used, for example, for the assessment of the value of initial public offerings (IPOs) (Kim and

Ritter 1999; Deloof et al. 2009). Different studies focus on the valuation accuracy of multiples in specific geographical areas, for example, European or Asian countries (Herrmann and Richter 2003; Schreiner and Spremann 2007; Sehgal and Pandey 2010) or sector-specific areas, for example, Internet firms (Trueman et al. 2000). Furthermore, multiples are used to predict future returns (An et al. 2010) or to test the value impact of the majority or minority ownership (Graham and Lefanowicz 1999).

The remainder of our analysis is structured as follows: section 9.1 presents the used carbon and financial data. Section 9.2 explains the applied construction methodology to obtain suitable peer groups and multiples. Section 9.3 provides the results using CEM to determine firm value. Section 9.4 presents the findings of the usefulness of CEPG. Section 9.5 combines both CEM and CEPG. Section 9.6 provides further insights based on additional robustness analyses. Section 9.7 concludes the chapter.

9.1. Data

Following common global sample construction approaches of papers like Ince and Porter (2006), Hou et al. (2011) and Schmidt et al. (2019), we compile yearly global financial data from Refinitiv Datastream. We apply common screening techniques introduced in Ince and Porter (2006) and exclude all firms that are not identified as equity (e.g., ADRs) or which are not primarily listed. Moreover, we include only firms that account for approximately 99.5% of a country's market capitalization to reduce liquidity biases. This leaves us a global firm data sample of 27,667 unique firms for our sample period from 2002 to 2019.

To avoid common data biases related to carbon data, for example, self-reporting bias, we also take three different carbon data providers with different data collection and quality approaches into account (Busch et al. 2018). Furthermore, this merge enables us to lengthen our analysis in time as well as broaden our coverage of global firms.

9.1.1. *Carbon data*

With respect to the latest literature dealing with data issues (Busch et al. 2018), for example, data collection processes and self-reporting biases in carbon data, we merge three major data providers: CDP, Refinitiv ESG and Sustainalytics. Each database is used in many publications covering environmental, carbon and climate topics, for example, CDP (Ioannou et al. 2016; Görgen et al. 2020; Ilhan et al. 2021); Refinitiv Asset 4/ESG (Dyck et al. 2019; Görgen et al. 2020; Gibson et al. 2019); and Sustainalytics (Verheyden et al. 2016; Engle et al. 2020; Gibson et al. 2021).

Our data comprises two dimensions of reported carbon emissions: scope 1 (direct emissions) and scope 2 (indirect emissions). Due to the lack of high-quality data for scope 3 (indirect emissions within the value chain of a firm), we leave them out of our analysis. Furthermore, we extend our analysis beyond pure carbon emissions toward carbon intensity by dividing all carbon emissions by net sales.

9.1.2. *Financial data*

We use common financial data from Refinitiv Datastream and Worldscope. From the P&L, we use net sales and revenues, earnings, net, and gross income, EBIT and EBITDA to construct respective multiples. We build balance sheet multiples using the following variables: common equity, total assets and total capital. The third group of multiples uses cash flow data, including net cash flow from operating activities and cash dividends paid. We extend our dataset using estimates from I/B/E/S, for example, 1- and 2-year analyst forecasts for earnings, net sales and EBITDA to compile forward-looking multiples. Equity value is measured by price and entity value by its respective value. To sort firms according to their size, we use market values. We identify the sector of a firm using the Thomson Reuters Business Classification (TRBC).

9.2. Multiple construction methodology

To determine the value of a firm using multiples, we start by identifying similar firms and composing peer groups. Subsequently, we construct multiples and calculate the self-excluding average multiple for each peer group firm. Afterward, we are able to determine a value for a firm using these multiples and the respective reported variable of the firm. Therefore, we can determine the firm valuation accuracy by comparing the reported value of a firm with its estimated multiple-based value.

Table 9.1 shows all financial and carbon multiples used within this analysis. We assume that carbon emission data can be used to construct either equity or entity-based multiples. Therefore, we use them in both cases.

	(1) Non-financial	(2) P&L	(3) Balance sheet	(4) Cash flow	(5) Forward-looking
Equity value	*P/CE I*	*P/SA*	*P/CEQ*	*P/OCF*	*P/E 1*
	P/CE II	*P/EBT*		*P/D*	*P/E 2*
	P/CE I + II	*P/E*			
Entity value	*EV/CE I*	*EV/SA*	*EV/TA*	*EV/OCF*	*EV/SA 1*
	EV/CE II	*EV/GI*	*EV/TC*		*EV/SA 2*
	EV/CE I + II	*EV/EBIT*			*EV/EBITDA 1*
		EV/EBITDA			*EV/EBITDA 2*

This table provides an overview of the main calculated multiples. They are either equity or entity value-based and can be classified into five categories: (1) carbon emission multiples (CEM) are based on carbon emission scope 1 (CE I), scope 2 (CE II) and scopes 1 and 2 combined (CE I + II). (2) Multiples using key figures from the profit and loss statement consist of net sales or revenues (SA), pre-tax income (EBT), net income (E), gross income (GI), earnings before interest and taxes (EBIT) and earnings before interest, taxes, depreciation and amortization (EBITDA). (3) Balance sheet multiples are composed of common equity (CEQ), total assets (TA) or total capital (TC). (4) Net cash flow from operating activities (OCF) and paid cash dividends (D) are used to compile cash flow multiples. (5) Forward-looking multiples are made of analyst one- and two-year forecasts of earnings (E 1 and E 2), net sales or revenues (SA 1 and SA 2) and earnings before interest, taxes, depreciation and amortization (EBITDA 1 and EBITDA 2).

Table 9.1. *Overview of the main calculated multiples*

9.2.1. *Identifying and composing suitable peer group*

We are using several firm characteristics to identify a similar public firm, which is suitable for a respective peer group. First, we use the business classification of a firm as the main peer group criteria. In untabulated results, we extend our peer group composition, adding further common fundamental characteristics, for example, size or profitability (Lie and Lie 2002). We include carbon emissions in our second analysis to compose CEPG. Within each peer group, there must be at least 10 different firms so that we obtain meaningful results that are less driven by the extreme values of individual firms.

9.2.2. *Constructing and aggregating multiples*

The used multiples are commonly formed as follows in our analyses:

$$m_p = \frac{fv_p}{rv_p} \qquad [9.1]$$

A multiple m of a peer group firm p is a fraction with the numerator fv as the firm value and the denominator rv as the reference value

$$m_p \, rv_t = \widehat{fv}_t \qquad [9.2]$$

By multiplying the multiple of the peer group m_p with the reference value rv_t of the target firm, we receive an estimated firm value $\widehat{fv}_t$.

Following Chullen, Kaltenbrunner and Schwetzler (2015), we use four different aggregation measures to improve the valuation accuracy of our peer group multiples: arithmetic mean $(\bar{m}_p{}^A)$, median $(\bar{m}_p{}^M)$, harmonic mean $(\bar{m}_p{}^H)$ and geometric mean $(\bar{m}_p{}^G)$. These measures are defined as:

$$\bar{m}_p{}^A = \sum_{i=1}^{n} \left[\frac{fv_i}{rv_i}\right] \frac{1}{n} \qquad [9.3]$$

$$\bar{m}_p{}^M = inf\left\{\frac{fv_i}{rv_i} : F\left(\frac{fv_i}{rv_i}\right) \geq \frac{1}{2}\right\} \qquad [9.4]$$

$$\bar{m}_p{}^H = \frac{1}{\left[\sum_{i=1}^{n} \left[\frac{1}{\frac{fv_i}{rv_i}}\right]\right] \div n} \qquad [9.5]$$

$$\bar{m}_p{}^G = exp\left\{\frac{1}{n} \sum_{i=1}^{n} ln\left[\frac{fv_i}{rv_i}\right]\right\} \qquad [9.6]$$

In practice, analysts widely use the average of the peers' multiples as the simplest approach to calculate the aggregated multiple. However, this approach lacks robustness towards outliers. Peer group multiples based on mean values suffer from the impact of extreme observations. Both the harmonic mean and median values avoid the impact of extreme values, and empirical evidence documents that both averaging processes perform significantly better than mean values (and geometric means).

In the literature, most studies use the median or the harmonic mean as an aggregate multiple approach (Baker and Ruback 1999; Henschke and Homburg 2009; Liu et al. 2002; Schreiner and Spremann 2007). Only some studies include the geometric mean (Herrmann and Richter 2003). Overall, it is documented in the literature that the harmonic mean is less biased than the arithmetic mean, the geometric mean or the median if percentage errors are used as the error measure for firm valuation accuracy. Using logarithmic errors, however, the harmonic mean is biased downward as much as the arithmetic mean is biased upward, whereas the geometric mean and the median are unbiased. Therefore, we use each aggregation method in its most proven field of application and briefly describe how the others are performing.

9.2.3. *Determining firm valuation errors*

Several error measures are used to judge the accuracy of a firm valuation approach, which all have their benefits and shortcomings (Chullen et al. 2015). Therefore, we use several measures capturing different accuracy dimensions, for example, the percentage of over- or undervaluations, logarithmic-scaled and the absolute logarithmic-scaled errors. These accuracy measures are defined as follows:

$$e_{per} = \frac{\widehat{fv}_t - fv_p}{fv_p} \quad [9.7]$$

$$e_{log} = \ln\left(\frac{\widehat{fv}_t}{fv_p}\right) \quad [9.8]$$

$$e_{alog} = \left|\ln\left(\frac{\widehat{fv}_t}{fv_p}\right)\right| \quad [9.9]$$

In our analyses, we generally report the logarithmic error, the absolute logarithmic error, the overvaluation percentage and the absolute logarithmic error percentage below 15%.

9.3. Constructing new multiples using carbon data

First, we want to study if non-financial but valuation-relevant multiples of comparable firms can be used in a similar manner to common financial ones. According to the aforementioned literature, high carbon emissions and carbon footprints have an impact on the value of a firm. Therefore, we use carbon emissions to construct new multiples and analyze their firm valuation accuracy.

We construct carbon emission multiples (CEM) based on scope 1 (CE I), scope 2 (CE II) and the sum of both scope (CE I and II) emissions for both estimating equity and entity values. As mentioned in the methodology chapter, we are using four different aggregation methods and four different error measures to evaluate the firm valuation accuracy of multiples.

	Multiples	AM	ME	HM	GM
Equity value multiples	*P/CE I*	1.029	0.027	-0.917	0.055
	P/CE II	0.752	0.033	-0.568	0.080
	P/CE I & II	0.752	0.030	-0.691	0.068
	P/SA	0.971	0.081	-0.524	0.120
	P/EBT	0.701	0.087	-0.191	0.183
	P/E	0.709	0.083	-0.200	0.170
	P/CEQ	0.546	0.084	-0.268	0.128
	P/OCF	0.753	0.072	-0.322	0.134
	P/D	0.889	0.086	-0.274	0.171
	P/E 1	0.354	0.072	-0.168	0.088
	P/E 2	0.270	0.056	-0.177	0.046
Entity value multiples	*EV/CE I*	1.270	0.198	-0.507	0.294
	EV/CE II	1.045	0.254	-0.370	0.265
	EV/CE I & II	0.980	0.197	-0.395	0.264
	EV/SA	1.021	0.187	-0.371	0.243
	EV/GI	0.862	0.186	-0.290	0.233
	EV/EBITDA	0.632	0.136	-0.175	0.193
	EV/EBIT	0.705	0.169	-0.167	0.237
	EV/TA	0.588	0.106	-0.411	0.136
	EV/TC	0.616	0.128	-0.339	0.171
	EV/OCF	0.887	0.188	-0.215	0.269
	EV/SA 1	0.588	0.160	-0.304	0.147
	EV/SA 2	0.529	0.135	-0.327	0.124
	EV/EBITDA 1	0.360	0.080	-0.130	0.095
	EV/EBITDA 2	0.294	0.070	-0.172	0.065
ø Overall		0.724	0.116	-0.339	0.159
ø carbon EQ multiples		0.844	0.030	-0.725	0.067
ø fin. EQ multiples		0.649	0.078	-0.265	0.130
ø carbon EV multiples		1.099	0.216	-0.424	0.274
ø fin. EV multiples		0.644	0.140	-0.264	0.174

This table presents 25 multiples constructed. They are either equity or entity value-based and can be classified into five categories: (1) carbon emission multiples (CEM) are based on carbon emission scope 1 (CE I), scope 2 (CE II) and scope 1 and 2 combined (CE I + II). Please check previous tables for the definitions of the other four categories. Each logarithmic error of each multiple is a measure for the multiples valuation accuracy and represents the aggregated error value over all firm valuations. We use four different methods to aggregate multiples across peer groups: average mean (AM), median (ME), harmonic mean (HM) and geometric mean (GM). The analysis is based on yearly values for all firms within the data sample from 2002 to 2019.

Table 9.2. *Carbon emission multiples (CEM) firm valuation accuracy as per logarithmic error*

	Multiples	AM	ME	HM	GM
Equity value multiples	*P/CE I*	1.256	0.849	1.151	0.917
	P/CE II	0.951	0.722	0.848	0.728
	P/CE I & II	1.017	0.749	0.939	0.757
	P/SA	1.101	0.703	0.810	0.716
	P/EBT	0.813	0.500	0.482	0.523
	P/E	0.826	0.494	0.471	0.529
	P/CEQ	0.692	0.534	0.553	0.537
	P/OCF	0.878	0.544	0.592	0.557
	P/D	1.000	0.564	0.571	0.583
	P/E 1	0.471	0.360	0.357	0.366
	P/E 2	0.401	0.329	0.345	0.332
Entity value multiples	*EV/CE I*	1.416	0.865	0.955	0.966
	EV/CE II	1.148	0.833	0.802	0.794
	EV/CE I & II	1.166	0.776	0.825	0.834
	EV/SA	1.127	0.721	0.741	0.733
	EV/GI	0.974	0.665	0.675	0.678
	EV/EBITDA	0.727	0.513	0.500	0.528
	EV/EBIT	0.793	0.531	0.502	0.555
	EV/TA	0.713	0.560	0.642	0.565
	EV/TC	0.737	0.560	0.593	0.572
	EV/OCF	0.984	0.572	0.564	0.604
	EV/SA 1	0.764	0.614	0.651	0.618
	EV/SA 2	0.714	0.588	0.635	0.595
	EV/EBITDA 1	0.495	0.399	0.397	0.415
	EV/EBITDA 2	0.449	0.374	0.389	0.381
ø Overall		0.864	0.597	0.640	0.615
ø carbon EQ multiples		1.075	0.773	0.980	0.801
ø fin. EQ multiples		0.773	0.503	0.523	0.518
ø carbon EV multiples		1.243	0.825	0.861	0.865
ø fin. EV multiples		0.771	0.554	0.572	0.568

This table presents 25 multiples constructed. They are either equity or entity value-based and can be classified into five categories: (1) carbon emission multiples (CEM) are based on carbon emission scope 1 (CE I), scope 2 (CE II) and scope 1 and 2 combined (CE I + II). Please check previous tables for the definitions of the other four categories. Each absolute logarithmic error of each multiple is a measure for the multiples valuation accuracy and represents the aggregated error value over all firm valuations. We use four different methods to aggregate multiples across peer groups: average mean (AM), median (ME), harmonic mean (HM) and geometric mean (GM). The analysis is based on yearly values for all firms within the data sample from 2002 to 2019.

Table 9.3. *Carbon emission multiples (CEM) firm valuation accuracy as per absolute logarithmic error*

Table 9.2 provides us with the results, displaying the logarithmic error for 25 different multiples. The multiples based on both carbon emission scopes lead to the least logarithmic error and are comparable to some financial-based multiples. However, if we consider the higher average logarithmic error of our carbon emission equity and entity multiples, we can only note a limited use of CEM. This is in line with our expectations since carbon emissions have an influence, but not the most important impact, on the value of a firm.

In Table 9.3, we find these results confirmed for the next error measure. Regarding the absolute logarithmic error, we find significantly higher firm valuation errors both for each single carbon emission multiple and for the average across them.

In contrast, when looking at the percentage of overvaluations in Table 9.4, we find a significantly lower value for all CEM. With only 47.7% of overvalued firms, the carbon emission scope 1 multiples have the lowest value across all 25 constructed multiples. This can also be confirmed by Table 9.5, where we present the percentage of absolute logarithmic errors above 15%. On average, we get four percentage points lower errors. This is particularly important, as CEM-based estimated firm values are thus only in a few cases far from the true observed firm value.

Overall, we can therefore conclude that CEMs are suitable for a firm valuation to only a limited extent. However, its valuation error is usually caused by an undervaluation and may therefore compensate for the potential overvaluation of financial multiples.

	Multiples	**AM**	**ME**	**HM**	**GM**
Equity value multiples	*P/CE I*	47.7%	49.7%	23.8%	51.0%
	P/CE II	48.0%	50.9%	29.0%	52.0%
	P/CE I & II	47.9%	50.4%	26.6%	51.1%
	P/SA	80.7%	53.2%	30.8%	54.9%
	P/EBT	80.9%	56.2%	40.5%	61.3%
	P/E	81.2%	56.2%	40.0%	61.2%
	P/CEQ	75.7%	54.7%	37.0%	56.3%
	P/OCF	80.6%	54.4%	34.4%	57.6%
	P/D	83.1%	54.7%	36.8%	58.8%
	P/E 1	73.8%	55.6%	39.5%	57.3%
	P/E 2	71.1%	55.6%	37.7%	55.2%

Entity value multiples	*EV/CE I*	61.2%	56.5%	33.3%	59.2%
	EV/CE II	60.3%	56.2%	36.6%	58.2%
	EV/CE I & II	60.6%	56.1%	34.7%	57.9%
	EV/SA	82.1%	57.4%	35.7%	59.3%
	EV/GI	80.1%	57.8%	38.3%	59.8%
	EV/EBITDA	78.8%	58.8%	42.2%	61.7%
	EV/EBIT	80.2%	60.0%	43.0%	63.2%
	EV/TA	75.8%	56.0%	33.8%	57.0%
	EV/TC	76.8%	57.0%	36.5%	58.7%
	EV/OCF	83.2%	58.8%	40.2%	62.8%
	EV/SA 1	74.6%	56.0%	36.3%	56.4%
	EV/SA 2	73.1%	55.7%	36.0%	55.5%
	EV/EBITDA 1	73.3%	57.2%	41.6%	58.5%
	EV/EBITDA 2	71.5%	56.9%	39.9%	56.9%
ø Overall		72.1%	55.7%	36.2%	57.7%
ø carbon EQ multiples		47.8%	50.4%	26.5%	51.4%
ø fin. EQ multiples		78.4%	55.1%	37.1%	57.8%
ø carbon EV multiples		60.7%	56.3%	34.9%	58.4%
ø fin. EV multiples		77.2%	57.4%	38.5%	59.1%

This table presents 25 multiples constructed. They are either equity or entity value-based and can be classified into five categories: (1) carbon emission multiples (CEM) are based on carbon emission scope 1 (CE I), scope 2 (CE II) and scope 1 and 2 combined (CE I + II). Please check previous tables for the definitions of the other four categories. Each overvaluation percentage of each multiple is a measure for the multiples valuation accuracy and represents the aggregated error value over all firm valuations. We use four different methods to aggregate multiples across peer groups: average mean (AM), median (ME), harmonic mean (HM) and geometric mean (GM). The analysis is based on yearly values for all firms within the data sample from 2002 to 2019.

Table 9.4. *Carbon emission multiples (CEM) firm valuation accuracy as per overvaluation percentage*

	Multiples	AM	ME	HM	GM
Equity value multiples	*P/CE I*	6.0%	9.7%	6.9%	9.4%
	P/CE II	6.6%	11.2%	9.4%	11.1%
	P/CE I & II	6.6%	10.9%	8.2%	10.7%
	P/SA	7.1%	11.8%	10.2%	11.5%
	P/EBT	8.4%	16.9%	17.6%	15.3%
	P/E	8.3%	17.0%	17.7%	15.5%
	P/CEQ	11.5%	15.3%	14.9%	15.0%
	P/OCF	8.2%	15.6%	14.1%	14.9%
	P/D	6.9%	14.6%	14.4%	13.8%
	P/E 1	16.4%	22.9%	22.5%	22.2%
	P/E 2	19.8%	24.8%	23.7%	24.3%
Entity value multiples	*EV/CE I*	6.2%	9.8%	8.6%	9.1%
	EV/CE II	6.9%	10.4%	10.3%	10.2%
	EV/CE I & II	6.4%	10.5%	9.5%	10.0%
	EV/SA	6.7%	11.6%	11.1%	11.2%
	EV/GI	7.6%	12.2%	12.5%	11.7%
	EV/EBITDA	10.0%	15.9%	16.7%	15.1%
	EV/EBIT	8.6%	15.3%	16.5%	14.3%
	EV/TA	10.9%	14.4%	13.1%	14.2%
	EV/TC	10.5%	14.5%	14.0%	14.2%
	EV/OCF	6.8%	14.3%	14.7%	13.4%
	EV/SA 1	10.5%	13.4%	12.8%	13.3%
	EV/SA 2	11.3%	13.9%	13.0%	13.7%
	EV/EBITDA 1	15.9%	20.7%	20.5%	20.2%
	EV/EBITDA 2	17.9%	22.1%	21.2%	21.7%
ø Overall		9.7%	14.8%	14.2%	14.2%
ø carbon EQ multiples		6.4%	10.6%	8.2%	10.4%
ø fin. EQ multiples		10.8%	17.3%	16.9%	16.6%
ø carbon EV multiples		6.5%	10.2%	9.5%	9.8%
ø fin. EV multiples		10.6%	15.3%	15.1%	14.8%

This table presents 25 multiples constructed. They are either equity or entity value-based and can be classified into five categories: (1) carbon emission multiples (CEM) are based on carbon emission scope 1 (CE I), scope 2 (CE II) and scope 1 and 2 combined (CE I + II). Please check previous tables for the definitions of the other four categories. Each absolute logarithmic error percentage above 15% of each multiple is a measure for the multiples valuation accuracy and represents the aggregated error value over all firm valuations. We use four different methods to aggregate multiples across peer groups: average mean (AM), median (ME) harmonic mean (HM) and geometric mean (GM). The analysis is based on yearly values for all firms within the data sample from 2002 to 2019.

Table 9.5. *Carbon emission multiples (CEM) firm valuation accuracy as per absolute log. error percentage*

9.4. Constructing peer groups using carbon data

To determine the potential of carbon emission data in enhancing multiples, we first analyze if the inclusion of carbon emissions into the peer group composition can improve the accuracy of firm valuations. Therefore, we apply the described methodology to construct different equity and entity value-based multiples. Following this, we annually divide all firms into deciles according to their carbon emissions. This classification is now also incorporated into the composition of peer groups towards carbon emission peer groups (CEPG). As a result, within one year and one industry, there are now firms in a peer group that have similar carbon emissions. We also include several other common characteristics to compose peer groups, for example, size or profitability. Our results remain robust.

	Multiples	AM	ME	HM	GM
Equity value multiples	*P/SA*	0.324	0.050	-0.303	0.068
	P/EBT	0.362	0.062	-0.123	0.098
	P/E	0.321	0.052	-0.131	0.088
	P/CEQ	0.322	0.090	-0.077	0.123
	P/OCF	0.341	0.063	-0.175	0.046
	P/D	0.536	0.106	-0.068	0.161
	P/E 1	0.204	0.049	-0.084	0.057
	P/E 2	0.159	0.050	-0.087	0.048
Entity value multiples	*EV/SA*	0.417	0.157	-0.137	0.145
	EV/GI	0.418	0.159	-0.042	0.174
	EV/EBITDA	0.319	0.125	-0.038	0.161
	EV/EBIT	0.423	0.171	-0.015	0.190
	EV/TA	0.337	0.153	-0.053	0.160
	EV/TC	0.375	0.139	-0.019	0.186
	EV/OCF	0.466	0.136	-0.031	0.180
	EV/SA 1	0.358	0.143	-0.084	0.115
	EV/SA 2	0.341	0.160	-0.128	0.095
	EV/EBITDA 1	0.215	0.091	-0.059	0.073
	EV/EBITDA 2	0.173	0.074	-0.072	0.062

ø Overall	0.337	0.107	-0.091	0.117
ø fin. EQ multiples	0.321	0.065	-0.131	0.086
ø fin. EV multiples	0.349	0.137	-0.062	0.140

This table presents 19 multiples, which are either equity or entity value-based and can be classified into five categories: (1) multiples using key figures from the profit and loss statement consisting of net sales or revenues (SA), pre-tax income (EBT), net income (E), gross income (GI), earnings before interest and taxes (EBIT) and earnings before interest, taxes, depreciation and amortization (EBITDA). (2) Balance sheet multiples are composed of common equity (CEQ), total assets (TA) or total capital (TC). (3) Net cash flow from operating activities (OCF) and paid cash dividends (D) are used to compile cash flow multiples. (4) Forward-looking multiples are made of analyst one- and two-year forecasts of earnings (E 1 and E 2), net sales or revenues (SA 1 and SA 2), and earnings before interest, taxes, depreciation and amortization (EBITDA 1 and EBITDA 2). Each logarithmic error of each multiple is a measure for the multiples valuation accuracy and represents the aggregated error value over all firm valuations. We use four different methods to aggregate multiples across peer groups: average mean (AM), median (ME), harmonic mean (HM) and geometric mean (GM). The used peer groups are compiled using carbon emissions in addition. Therefore, they are called carbon emission peer groups (CEPG). The analysis is based on yearly values for all firms within the data sample from 2002 to 2019.

Table 9.6. *Carbon emission peer group (CEPG) firm valuation accuracy as per logarithmic error*

Table 9.6 presents firm valuation accuracy evaluated by the logarithmic error of the estimated firm value using 19 different financial multiples in comparison to the observed firm value. Each multiple is classified as either equity or entity value-based. The table also presents all four possible aggregation methods of firm valuation errors across all firms and years. As an example, the first multiple P/SA, which is based on net sales or revenues and aggregated across peer group firms with the average mean, has a logarithmic error across all firm-year observations within our sample of 0.324. In comparison with the corresponding value of 0.971 shown in Table 9.2, which is just using normal peer groups, we observe that a firm valuation logarithmic error is reduced by two-thirds. The average mean aggregation of multiples helps reduce the logarithmic error of overvaluation in all cases. We achieve similar results for both the median and the geometric mean, but the harmonic mean results in an increased occurrence of significant undervaluation percentages.

In Table 9.7, we now compare the results for the absolute logarithmic error. Taking a look, for example, at the earning-based multiple (P/EBT), we observe an absolute logarithmic error with a value of 0.826. By enhancing the peer group using carbon emissions, this error value is halved to 0.462. This improved firm valuation accuracy can be assessed across all four aggregation methods.

	Multiples	AM	ME	HM	GM
Equity value multiples	*P/SA*	0.554	0.505	0.590	0.510
	P/EBT	0.436	0.360	0.378	0.366
	P/E	0.462	0.347	0.366	0.353
	P/CEQ	0.455	0.402	0.409	0.386
	P/OCF	0.488	0.408	0.440	0.401
	P/D	0.641	0.450	0.438	0.464
	P/E 1	0.345	0.288	0.308	0.296
	P/E 2	0.306	0.263	0.286	0.266
Entity value multiples	*EV/SA*	0.657	0.588	0.592	0.586
	EV/GI	0.629	0.539	0.524	0.531
	EV/EBITDA	0.484	0.405	0.414	0.410
	EV/EBIT	0.570	0.445	0.436	0.457
	EV/TA	0.512	0.447	0.450	0.460
	EV/TC	0.513	0.429	0.418	0.458
	EV/OCF	0.593	0.455	0.438	0.457
	EV/SA 1	0.619	0.556	0.558	0.563
	EV/SA 2	0.601	0.538	0.546	0.538
	EV/EBITDA 1	0.394	0.364	0.361	0.351
	EV/EBITDA 2	0.354	0.325	0.331	0.322
ø Overall		0.506	0.427	0.436	0.430
ø fin. EQ multiples		0.461	0.378	0.402	0.380
ø fin. EV multiples		0.539	0.463	0.461	0.467

This table presents 19 multiples, which are either equity or entity value-based and can be classified into five categories: (1) multiples using key figures from the profit and loss statement consisting of net sales or revenues (SA), pre-tax income (EBT), net income (E), gross income (GI), earnings before interest and taxes (EBIT) and earnings before interest, taxes, depreciation and amortization (EBITDA). (2) Balance sheet multiples are composed of common equity (CEQ), total assets (TA) or total capital (TC). (3) Net cash flow from operating activities (OCF) and paid cash dividends (D) are used to compile cash flow multiples. (4) Forward-looking multiples are made of analyst one- and two-year forecasts of earnings (E 1 and E 2), net sales or revenues (SA 1 and SA 2), and earnings before interest, taxes, depreciation and amortization (EBITDA 1 and EBITDA 2). Each absolute logarithmic error of each multiple is a measure for the multiples valuation accuracy and represents the aggregated error value over all firm valuations. We use four different methods to aggregate multiples across peer groups: average mean (AM), median (ME), harmonic mean (HM) and geometric mean (GM). The used peer groups are compiled using carbon emissions in addition. Therefore, they are called carbon emission peer groups (CEPG). The analysis is based on yearly values for all firms within the data sample from 2002 to 2019.

Table 9.7. *Carbon emission peer groups (CEPG) firm valuation accuracy as per absolute logarithmic error*

	Multiples	AM	ME	HM	GM
Equity value multiples	*P/SA*	67.6%	52.7%	34.0%	51.0%
	P/EBT	73.9%	56.1%	41.9%	58.4%
	P/E	74.3%	56.2%	41.4%	58.5%
	P/CEQ	70.3%	54.8%	40.8%	55.8%
	P/OCF	71.1%	53.8%	37.0%	53.9%
	P/D	76.5%	54.3%	41.1%	59.1%
	P/E 1	69.7%	56.6%	42.6%	56.9%
	P/E 2	67.2%	56.7%	42.2%	55.2%
Entity value multiples	*EV/SA*	70.7%	56.5%	41.4%	56.6%
	EV/GI	71.9%	56.9%	42.3%	57.2%
	EV/EBITDA	73.3%	59.1%	47.9%	60.8%
	EV/EBIT	76.0%	60.5%	49.0%	63.2%
	EV/TA	69.6%	56.8%	43.2%	57.0%
	EV/TC	71.2%	57.2%	45.4%	58.7%
	EV/OCF	76.0%	58.4%	45.5%	60.8%
	EV/SA 1	69.2%	56.5%	42.8%	56.4%
	EV/SA 2	68.3%	56.3%	42.7%	55.8%
	EV/EBITDA 1	68.2%	57.6%	46.9%	57.8%
	EV/EBITDA 2	66.4%	56.8%	45.9%	56.4%
ø Overall		71.1%	56.5%	42.8%	57.3%
ø fin. EQ multiples		71.3%	55.2%	40.1%	56.1%
ø fin. EV multiples		71.0%	57.5%	44.8%	58.2%

This table presents 19 multiples, which are either equity or entity value-based and can be classified into five categories: (1) multiples using key figures from the profit and loss statement consisting of net sales or revenues (SA), pre-tax income (EBT), net income (E), gross income (GI), earnings before interest and taxes (EBIT) and earnings before interest, taxes, depreciation and amortization (EBITDA). (2) Balance sheet multiples are composed of common equity (CEQ), total assets (TA) or total capital (TC). (3) Net cash flow from operating activities (OCF) and paid cash dividends (D) are used to compile cash flow multiples. (4) Forward-looking multiples are made of analyst one- and two-year forecasts of earnings (E 1 and E 2), net sales or revenues (SA 1 and SA 2), and earnings before interest, taxes, depreciation and amortization (EBITDA 1 and EBITDA 2). Each overvaluation percentage of each multiple is a measure for the multiples valuation accuracy and represents the aggregated error value over all firm valuations. We use four different methods to aggregate multiples across peer groups: average mean (AM), median (ME), harmonic mean (HM) and geometric mean (GM). The used peer groups are compiled using carbon emissions in addition. Therefore, they are called carbon emission peer groups (CEPG). The analysis is based on yearly values for all firms within the data sample from 2002 to 2019.

Table 9.8. *Carbon emission peer groups (CEPG) firm valuation accuracy as per overvaluation percentage*

	Multiples	AM	ME	HM	GM
Equity value multiples	*P/SA*	13.9%	14.9%	12.7%	14.7%
	P/EBT	15.4%	21.9%	21.5%	21.2%
	P/E	15.5%	22.5%	21.9%	21.3%
	P/CEQ	15.4%	18.6%	18.3%	18.3%
	P/OCF	15.5%	19.4%	17.1%	19.0%
	P/D	10.2%	17.7%	18.0%	16.6%
	P/E 1	22.0%	27.5%	26.2%	26.7%
	P/E 2	25.4%	29.4%	27.4%	28.2%
Entity value multiples	*EV/SA*	11.9%	13.7%	13.6%	13.7%
	EV/GI	12.3%	15.0%	15.0%	15.0%
	EV/EBITDA	15.2%	19.2%	19.2%	18.5%
	EV/EBIT	12.5%	18.4%	19.1%	17.7%
	EV/TA	14.3%	16.9%	16.1%	16.3%
	EV/TC	14.6%	17.0%	16.7%	16.9%
	EV/OCF	12.2%	17.7%	17.4%	16.8%
	EV/SA 1	12.7%	14.4%	13.7%	14.3%
	EV/SA 2	13.3%	14.5%	14.1%	14.6%
	EV/EBITDA 1	19.9%	22.2%	21.8%	22.1%
	EV/EBITDA 2	21.5%	23.5%	23.3%	23.4%
ø Overall		15.5%	19.2%	18.6%	18.7%
ø fin. EQ multiples		16.7%	21.5%	20.4%	20.8%
ø fin. EV multiples		14.6%	17.5%	17.3%	17.2%

This table presents 19 multiples, which are either equity or entity value-based and can be classified into five categories: (1) multiples using key figures from the profit and loss statement consist of net sales or revenues (SA), pre-tax income (EBT), net income (E), gross income (GI), earnings before interest and taxes (EBIT) and earnings before interest, taxes, depreciation and amortization (EBITDA). (2) Balance sheet multiples are composed of common equity (CEQ), total assets (TA) or total capital (TC). (3) Net cash flow from operating activities (OCF) and paid cash dividends (D) are used to compile cash flow multiples. (4) Forward-looking multiples are made of analyst one- and two-year forecasts of earnings (E 1 and E 2), net sales or revenues (SA 1 and SA 2), and earnings before interest, taxes, depreciation and amortization (EBITDA 1 and EBITDA 2). Each absolute logarithmic error percentage above 15% of each multiple is a measure for the multiples valuation accuracy and represents the aggregated error value over all firm valuations. We use four different methods to aggregate multiples across peer groups: average mean (AM), median (ME), harmonic mean (HM) and geometric mean (GM). The used peer groups are compiled using carbon emissions in addition. Therefore, they are called carbon emission peer groups (CEPG). The analysis is based on yearly values for all firms within the data sample from 2002 to 2019.

Table 9.9. *Carbon emission peer groups (CEPG) firm valuation accuracy as per absolute log error percentage*

To provide further insights into the ratio of over- and undervaluations, we measure the percentage of overvaluations in Table 9.8. For all financial multiples, we get approximately 6% fewer overvaluations for the carbon emission peer groups. This supports the result of Table 9.6 that the use of carbon emissions to compose peer groups helps improve the firm valuation accuracy.

In our last analysis within this chapter, we have a look at the percentage of absolute logarithmic errors below 15% in Table 9.9. In line with the results of Table 9.7, we have an increased percentage across all aggregation methods and multiples. Therefore, the use of CEPG leads to less accurate firm valuations according to this error measure.

Overall, we can conclude that two of the error measures lead to improvements in the firm valuation accuracy, one too mixed and one to a worsening of the results. Furthermore, it can be stated that the use of CEPG leads to significantly fewer overvaluations. The possible undervaluation is, therefore, the biggest problem and must be taken into account.

9.5. Combining carbon emission multiples and carbon emission enhanced peer groups

In the following analyses, we now look at the results when we combine both approaches, CEM and CEPG. Therefore, we construct carbon emission-based multiples and use carbon emission-enhanced peer groups. Our evaluation again considers the four different aggregation methods for multiples and the four alternative firm valuation error measures.

Table 9.10 provides first the results of the logarithmic error analysis. Compared to the CEM from Table 9.2 without using CEPG, we now obtain significantly lower valuation errors. If we now compare these new CEM with the different multiples from Table 9.6, they are more precise but still not the most accurate firm valuation multiple.

This finding is also continued in Table 9.11. We can see that we now have significantly fewer absolute logarithmic errors than before, but they are still higher compared to financial multiples. These results do not vary if we use different aggregation methods.

	Multiples	AM	ME	HM	GM
Equity value multiples	*P/CE I*	0.677	0.031	-0.591	0.027
	P/CE II	0.551	0.013	-0.463	0.031
	P/CE I & II	0.478	0.022	-0.409	0.033
Entity value multiples	*EV/CE I*	0.855	0.215	-0.253	0.251
	EV/CE II	0.777	0.216	-0.142	0.273
	EV/CE I & II	0.657	0.158	-0.244	0.171
ø Overall		0.666	0.109	-0.350	0.131
ø carbon EQ multiples		0.569	0.022	-0.487	0.030
ø carbon EV multiples		0.763	0.196	-0.213	0.232

This table presents six carbon emission multiples (CEM) constructed. They are either equity or entity value-based and can be classified into one category: (1) carbon emission multiples (CEM) are based on carbon emission scope 1 (CE I), scope 2 (CE II) and scopes 1 and 2 combined (CE I + II). Each logarithmic error of each multiple is a measure for the multiples valuation accuracy and represents the aggregated error value over all firm valuations. We use four different methods to aggregate multiples across peer groups: average mean (AM), median (ME), harmonic mean (HM) and geometric mean (GM). The used peer groups are compiled using carbon emissions in addition. Therefore, they are called carbon emission peer groups (CEPG). The analysis is based on yearly values for all firms within the data sample from 2002 to 2019.

Table 9.10. *Carbon emission multiples (CEM) with carbon emission peer groups (CEPG) firm valuation accuracy as per logarithmic error*

	Multiples	AM	ME	HM	GM
Equity value multiples	*P/CE I*	0.964	0.766	0.906	0.752
	P/CE II	0.817	0.679	0.758	0.679
	P/CE I & II	0.756	0.663	0.737	0.668
Entity value multiples	*EV/CE I*	1.105	0.779	0.827	0.801
	EV/CE II	0.972	0.730	0.745	0.754
	EV/CE I & II	0.904	0.729	0.745	0.734
ø Overall		0.920	0.724	0.786	0.731
ø carbon EQ multiples		0.846	0.703	0.800	0.700
ø carbon EV multiples		0.994	0.746	0.772	0.763

This table presents six carbon emission multiples (CEM) constructed. They are either equity or entity value-based and can be classified into one category: (1) carbon emission multiples (CEM) are based on carbon emission scope 1 (CE I), scope 2 (CE II) and scopes 1 and 2 combined (CE I + II). Each absolute logarithmic error of each multiple is a measure for the multiples valuation accuracy and represents the aggregated error value over all firm valuations. We use four different methods to aggregate multiples across peer groups: average mean (AM), median (ME), harmonic mean (HM) and geometric mean (GM). The used peer groups are compiled using carbon emissions in addition. Therefore, they are called carbon emission peer groups (CEPG). The analysis is based on yearly values for all firms within the data sample from 2002 to 2019.

Table 9.11. *Carbon emission multiples (CEM) with carbon emission peer groups (CEPG) firm valuation accuracy as per absolute logarithmic error*

	Multiples	AM	ME	HM	GM
Equity value multiples	*P/CE I*	48.6%	50.2%	29.7%	50.6%
	P/CE II	49.5%	51.5%	32.5%	51.8%
	P/CE I & II	49.3%	50.8%	31.5%	50.6%
Entity value multiples	*EV/CE I*	58.8%	57.0%	39.2%	58.5%
	EV/CE II	58.7%	57.0%	41.0%	58.8%
	EV/CE I & II	59.1%	56.2%	39.2%	56.8%
ø Overall		54.0%	53.8%	35.5%	54.5%
ø carbon EQ multiples		49.1%	50.8%	31.2%	51.0%
ø carbon EV multiples		58.9%	56.7%	39.8%	58.0%

This table presents six carbon emission multiples (CEM) constructed. They are either equity or entity value-based and can be classified into one category: (1) carbon emission multiples (CEM) are based on carbon emission scope 1 (CE I), scope 2 (CE II) and scopes 1 and 2 combined (CE I + II). Each overvaluation percentage of each multiple is a measure for the multiples valuation accuracy and represents the aggregated error value over all firm valuations. We use four different methods to aggregate multiples across peer groups: average mean (AM), median (ME), harmonic mean (HM) and geometric mean (GM). The used peer groups are compiled using carbon emissions in addition. Therefore, they are called carbon emission peer groups (CEPG). The analysis is based on yearly values for all firms within the data sample from 2002 to 2019.

Table 9.12. *Carbon emission multiples (CEM) with carbon emission peer groups (CEPG) firm valuation accuracy as per overvaluation percentage*

In the next step, we again consider the share of overvaluations in Table 9.12. We observe only a little difference, which, in the context of the previous results, suggests that the combination of the two approaches has a low effect on reducing the percentage of undervaluations.

Table 9.13 provides the results for the fourth error measure. In comparison with earlier results, we get slightly higher error values across all aggregation measures.

Overall, it can be concluded that the combination of the two approaches is definitely useful for improving the firm valuation accuracy of CEM. However, CEM are still not able to estimate firm values to the same precision as many financial multiples. For this reason, we continue to view it as particularly worthwhile to use carbon emissions mainly for the composition of the peer groups of financial multiples (CEPG).

	Multiples	AM	ME	HM	GM
Equity value multiples	*P/CE I*	7.9%	10.8%	9.1%	11.0%
	P/CE II	8.6%	12.3%	10.4%	12.1%
	P/CE I & II	8.3%	12.7%	10.3%	12.3%
Entity value multiples	*EV/CE I*	7.8%	10.4%	9.8%	10.2%
	EV/CE II	8.2%	11.1%	11.1%	10.8%
	EV/CE I & II	8.1%	11.4%	10.8%	11.0%
ø Overall		8.2%	11.5%	10.2%	11.2%
ø carbon EQ multiples		8.3%	11.9%	9.9%	11.8%
ø carbon EV multiples		8.1%	11.0%	10.6%	10.7%

This table presents six carbon emission multiples (CEM) constructed. They are either equity or entity value-based and can be classified into one category: (1) carbon emission multiples (CEM) are based on carbon emission scope 1 (CE I), scope 2 (CE II) and scopes 1 and 2 combined (CE I + II). Each absolute logarithmic error percentage above 15% of each multiple is a measure for the multiples valuation accuracy and represents the aggregated error value over all firm valuations. We use four different methods to aggregate multiples across peer groups: average mean (AM), median (ME), harmonic mean (HM) and geometric mean (GM). The used peer groups are compiled using carbon emissions in addition. Therefore, they are called carbon emission peer groups (CEPG). The analysis is based on yearly values for all firms within the data sample from 2002 to 2019.

Table 9.13. *Carbon emission multiples (CEM) with carbon emission peer groups (CEPG) firm valuation accuracy as per absolute log error percentage*

9.6. Robustness

In order to increase the significance of our results regarding the usefulness of CEM and CEPG, we carry out numerous further analyses, which we briefly outline in the following. Table 9.14 presents the results of all different cases.

Each case is evaluated whether it increases the valuation accuracy either using CEPG, CEM, or both. Both equity and entity multiples, as well as the four aggregation measures and the four error measures, are used for the evaluation. Therefore, a total of 32 values per procedure are considered for the evaluation of the firm valuation accuracy. We indicate in the table the percentage of the 32 values in which an improvement of the firm valuation accuracy is found.

Case	Peer group	Multiples	Peer group and multiples	Case	Peer group	Multiples	Peer group and multiples
(1) All firms	*81.3%*	*37.5%*	*37.5%*	*(11) USA*	*75.0%*	*37.5%*	*37.5%*
(2) Carbon data available	*87.5%*	*37.5%*	*37.5%*	*(12) Europe*	*65.6%*	*37.5%*	*25.0%*
(3) Carbon-intensive industries	*87.5%*	*28.1%*	*31.3%*	*(13) Americas*	*75.0%*	*37.5%*	*37.5%*
(4) Carbon-intensive firms	*37.5%*	*31.3%*	*31.3%*	*(14) Asia*	*81.3%*	*40.6%*	*37.5%*
(5) Carbon-efficient firms	*78.1%*	*34.4%*	*31.3%*	*(15) Oceania*	*68.8%*	*31.3%*	*43.8%*
(6) Carbon-inefficient firms	*34.4%*	*34.4%*	*31.3%*	*(16) Africa*	*53.1%*	*37.5%*	*37.5%*
(7) National carbon pricing	*75.0%*	*34.4%*	*31.3%*	*(17) High ESG firms*	*43.8%*	*28.1%*	*15.6%*
(8) Any carbon pricing	*81.3%*	*37.5%*	*37.5%*	*(18) Low ESG firms*	*65.6%*	*25.0%*	*34.4%*
(9) Since 2010	*71.9%*	*31.3%*	*31.3%*	*(19) High SDG 13 countries*	*81.3%*	*34.4%*	*31.3%*
(10) Before 2010	*87.5%*	*25.0%*	*37.5%*	*(20) Low SDG 13 countries*	*75.0%*	*31.3%*	*12.5%*

This table provides an overview of all firm valuation accuracy results. Each case is evaluated whether it increases the valuation accuracy either using CEPG, CEM, or both. Both equity and entity multiples, as well as the four aggregation measures and the four valuation accuracy measures, are used for the evaluation. Therefore, a total of 32 values per procedure are considered for the evaluation of the firm valuation accuracy. The table indicates the percentage of these 32 cases in which an improvement of the firm valuation accuracy is found.

Table 9.14. *Overview of all cases and their firm valuation accuracy results*

In the first step, we provide the results of our last analyses (case 1). In the second step (case 2), we consider only those firms that have carbon emissions available in at least 50% of our sample period. This ensures that the selected firms regularly report their carbon emissions. As a result, it can be assumed that the quality of these emission data is significantly higher. We get slightly better results using CEPG than in our evaluation of all firms (case 1), while the results for CEM and the combination of both remain basically the same.

In the next case, we look at firms that are either part of carbon-intensive industries (case 3) or are emitting high carbon emissions (case 4). Carbon-intensive industries include firms from the following business sectors according to the Thomson Reuters Business Classification (TRBC): Energy–fossil fuels, mineral resources, transportation, automobiles and auto parts, and utilities. Carbon-intensive firms are defined as those firms which are during a year among the 25% largest carbon emitters. Overall, the firms of carbon-intensive industries can be valued accurately, and their firm valuation accuracy is improved to a similar extent by our two approaches CEM and CEPG. If we look at the carbon-intensive firms, we can see that they can also be valued accurately overall, i.e. we observe low valuation errors. However, both CEM and CEPG lead to a more precise firm value accuracy in only one-third of the used aggregation methods and error measures, which is lower than in most other cases.

Usually, it is not only the absolute amount of carbon emissions that plays a role in the valuation of a firm. The carbon efficiency, measured as carbon intensity, is also considered in two cases for carbon-efficient and carbon-inefficient firms (cases 5 and 6). Our results remain very consistent for both firm groups so that we assume that both indicators can be considered in a similar manner when valuing firms based on their carbon intensity.

In a further case, we assume that the existence of carbon pricing initiatives (CPI) within a country leads to an increased impact of carbon emissions on a firm value. For example, the introduction of a carbon emission certificate system leads to increased costs for firms. Therefore, we use in one case (case 7) only firms of countries with a national CPI. Furthermore, in another case (case 8), we loosen this condition by including all countries with either a national or a regional CPI. We are referring to the carbon pricing dashboard of the World Bank to detect currently active CPIs (Dolphin et al. 2020; World Bank Group 2019). The accuracy of firm valuations increases, in particular, if we use CEPG.

Next, we consider whether our results depend on the topicality of the climate crisis debate (cases 9 and 10). It can be assumed that only in recent years the interest in carbon emissions has increased by a huge amount. This leads to the assumption that carbon emissions have become more relevant in terms of value impact nowadays (Engle et al. 2020). Therefore, we split our sample into two separate time periods: before and since 2010. The choice of the year 2010 was chosen because, for example, we observe a huge increase of available carbon emission data within all three of our data providers. In line with our expectations, we have been getting better firm valuation results since 2010, but the use of our methods is also useful before 2010.

Afterward, we examine our results for differences across geographical regions. We analyze the following regions individually: USA (case 11), Europe (case 12), Americas (case 13), Asia (case 14), Oceania (case 15), and Africa (case 16). However, we cannot see any major differences between the different regions, and in all cases, CEPG is worth using to improve the firm valuation accuracy.

In addition to the previous analyses, we are now including ESG information about individual firms (cases 17 and 18). We assume that the quality of the non-financial reporting and thus also of the carbon emissions are higher for ESG firms, as well as that the influence of emissions on the firm value is strongly moderated by the ESG rating. For our analysis, we use the Refinitiv ESG score. Overall, we find that for firms with high ESG ratings, CEPG is the most useful approach to improve firm value accuracy. In contrast, low ESG firm values are much less accurately assessed.

In our last analysis, we want to analyze whether the SDG 13 (Climate Action) performance of a country has an influence on the carbon emissions and firm value relationship. Such an influence should be reflected by more accurate firm value estimations (cases 19 and 20). We are using data from the Sustainable Development Report (Sachs et al. 2019) to analyze the top 25 and the worst 50 SDG 13 performing countries separately. We find only slight improvements for the top 25 countries.

Overall, we can state on the results of the 20 different cases that the use of CEPG can be recommended in all cases to improve the accuracy of firm valuations using multiples.

9.7. Recommendation for using carbon emissions for multiples and further research

Based on the results of our numerous analyses, we find a limited potential of estimating firm values with CEM. However, we can suggest the use of CEPG in most cases. The inclusion of carbon emissions to compose peer groups leads to an increase in the firm valuation accuracy in more than three-quarters of our cases.

In order to further improve the composition of peer groups, propensity score matching can be useful. Beyond common financial indicators, carbon emissions can be included to identify similar firms for valuation purposes. Further studies may focus on including the illiquidity discount (Damodaran 2005; Officer 2007; Pratt and Niculita 2008) or the control premium (Petersen et al. 2006; Pratt and Niculita 2008; Betton et al. 2009) into the valuation approach. It may also be promising to study the valuation potential of multiples comprising ESG ratings or scores, even if the ratings of different providers differ considerably (Gibson et al. 2021).

Overall, we recommend including carbon emissions into the composition of peer groups based on our results. Our approach yields consistent, efficient and accurate firm valuations for asset managers and investors to improve their investment decision-making. It also increases the accuracy of analysts' firm valuation estimations, especially for carbon emission-impacted firms. Finally, it helps capital market participants, regulators and firms to better understand the inclusion of carbon emission information into the valuation process of a firm.

9.8. References

Alford, A.W. (1992). The effect of the set of comparable firms on the accuracy of the price-earnings valuation method. *Journal of Accounting Research*, 30(1), 94. https://doi.org/10.2307/2491093.

An, J., Bhojraj, S., Ng, D.T. (2010). Warranted multiples and future returns. *Journal of Accounting, Auditing & Finance*, 25(2), 143–169. https://doi.org/10.1177/0148558X1002500201.

Asquith, P., Mikhail, M.B., Au, A.S. (2005). Information content of equity analyst reports. *Journal of Financial Economics*, 75(2), 245–282. https://doi.org/10.1016/j.jfineco.2004.01.002.

Baker, M. and Ruback, R.S. (1999). Estimating industry multiples. Working paper. https://www.hbs.edu/faculty/Pages/item.aspx?num=1435.

Berkman, H., Jona, J., Soderstrom, N.S. (2021). Firm-specific climate risk and market valuation. Working paper, Advance online publication. https://doi.org/10.2139/ssrn.2775552.

Betton, S., Eckbo, B.E., Thorburn Karin, S. (2009). Merger negotiations and the toehold puzzle. *Journal of Financial Economics*, 91(2), 158–178. https://doi.org/10.1016/j.jfineco.2008.02.004.

Bhojraj, S. and Lee, C.M.C. (2002). Who is my peer? A valuation-based approach to the selection of comparable firms. *Journal of Accounting Research*, 40(2), 407–439. https://doi.org/10.1111/1475-679X.00054.

Bolton, P. and Kacperczyk, M. (2021). Do investors care about carbon risk? *Journal of Financial Economics*, 142(2), 517–549. https://doi.org/10.1016/j.jfineco.2021.05.008.

Brammer, S. and Pavelin, S. (2006). Voluntary environmental disclosures by large UK companies. *Journal of Business Finance & Accounting*, 33(7–8), 1168–1188. https://doi.org/10.1111/j.1468-5957.2006.00598.x.

Busch, T., Johnson, M., Pioch, T., Kopp, M. (2018). Consistency of corporate carbon emission data. University of Hamburg/WWF Deutschland, Hamburg. https://ec.europa.eu/jrc/sites/default/files/paper_timo_busch.pdf.

Chen, L.H., Folsom, D.M., Paek, W., Sami, H. (2014). Accounting conservatism, earnings persistence, and pricing multiples on earnings. *Accounting Horizons*, 28(2), 233–260. https://doi.org/10.2308/acch-50664.

Chullen, A., Kaltenbrunner, H., Schwetzler, B. (2015). Does consistency improve accuracy in multiple–based valuation? *Journal of Business Economics*, 85(6), 635–662. https://doi.org/10.1007/s11573-015-0768-2.

Cooper, I.A. and Lambertides, N. (2014). Is there a limit to the accuracy of equity valuation using multiples? Working paper. https://doi.org/10.2139/ssrn.2291869.

Damodaran, A. (2005). Marketability and value: Measuring the illiquidity discount. Working paper. https://doi.org/10.2139/ssrn.841484.

Deloof, M., de Maeseneire, W., Inghelbrecht, K. (2009). How do investment banks value initial public offerings (IPOs)? *Journal of Business Finance & Accounting*, 36(1–2), 130–160. https://doi.org/10.1111/j.1468-5957.2008.02117.x.

Dittmann, I. and Maug, E.G. (2008). Biases and error measures: How to compare valuation methods. Working paper. https://doi.org/10.2139/ssrn.947436.

Dolphin, G., Pollitt, M.G., Newbery, D.M. (2020). The political economy of carbon pricing: A panel analysis. *Oxford Economic Papers*, 72(2), 472–500. https://doi.org/10.1093/oep/gpz042.

Dyck, A., Lins, K.V., Roth, L., Wagner, H.F. (2019). Do institutional investors drive corporate social responsibility? International evidence. *Journal of Financial Economics*, 131(3), 693–714. https://doi.org/10.1016/j.jfineco.2018.08.013.

Eberhart, A.C. (2001). Comparable firms and the precision of equity valuations. *Journal of Banking & Finance*, 25(7), 1367–1400. https://doi.org/10.1016/S0378-4266(01)00171-6.

Eberhart, A.C. (2004). Equity valuation using multiples. *The Journal of Investing*, 13(2), 48–54. https://doi.org/10.3905/joi.2004.412306.

Engle, R., Giglio, S., Lee, H., Kelly, B., Stroebel, J. (2020). Hedging climate change news. *The Review of Financial Studies*, 33(3), 1184–1216.

Fernando, C.S., Sharfman, M.P., Uysal, V.B. (2017). Corporate environmental policy and shareholder value: Following the smart money. *Journal of Financial and Quantitative Analysis*, 52(5), 2023–2051. https://doi.org/10.1017/S0022109017000680.

Gibson, R., Krueger, P., Riand, N., Schmidt, P.S. (2021). ESG rating disagreement and stock returns. *Financial Analysts Journal*, 77(4), 104–127.

Görgen, M., Jacob, A., Nerlinger, M., Riordan, R., Rohleder, M., Wilkens, M. (2021). Carbon risk. Working paper. https://dx.doi.org/10.2139/ssrn.2930897.

Graham, R.C. and Lefanowicz, C.E. (1999). Majority and minority ownership of publicly-traded firms: A test of the value of control using market multiples. *Journal of Business Finance & Accounting*, 26(1–2), 171–198. https://doi.org/10.1111/1468-5957.00252.

Griffin, P.A., Lont, D.H., Sun, E.Y. (2017). The relevance to investors of greenhouse gas emission disclosures. *Contemporary Accounting Research*, 34(2), 1265–1297. https://doi.org/10.1111/1911-3846.12298.

Henschke, S. and Homburg, C. (2009). Equity valuation using multiples: Controlling for differences between firms. Working paper. https://doi.org/10.2139/ssrn.1270812.

Herrmann, V. and Richter, F. (2003). Pricing with performance-controlled multiples. *Schmalenbach Business Review*, 55(3), 194–219. https://doi.org/10.1007/BF03396674.

Hou, K., Karolyi, G.A., Kho, B.-C. (2011). What factors drive global stock returns? *The Review of Financial Studies*, 24(8), 2527–2574. https://doi.org/10.1093/rfs/hhr013.

Ilhan, E., Sautner, Z., Vilkov, G. (2021). Carbon tail risk. *The Review of Financial Studies*, 34(3), 1540–1571. https://doi.org/10.1093/rfs/hhaa071.

Ince, O.S. and Porter, R.B. (2006). Individual equity return data from thomson datastream: Handle with care! *Journal of Financial Research*, 29(4), 463–479. https://doi.org/10.1111/j.1475-6803.2006.00189.x.

International Monetary Fund (2019). Global financial stability – Lower for longer. https://www.imf.org/-/media/Files/Publications/GFSR/2019/October/English/text.ashx.

Ioannou, I., Li, S.X., Serafeim, G. (2016). The effect of target difficulty on target completion: The case of reducing carbon emissions. *The Accounting Review*, 91(5), 1467–1492. https://doi.org/10.2308/accr-51307.

Jevons, W.S. (1879). The theory of political economy. http://gbv.eblib.com/patron/FullRecord.aspx?p=1609114.

Kim, M. and Ritter, J.R. (1999). Valuing IPOs. *Journal of Financial Economics*, 53(3), 409–437. https://doi.org/10.1016/S0304-405X(99)00027-6.

Koller, T., Goedhart, M., Wessels, D. (2005). The right role for multiples in valuation. *McKinsey on Finance*, 15, 7–11. https://ssrn.com/abstract=805166.

Koller, T., Goedhart, M., Wessels, D. (2015). *Valuation: Measuring and Managing the Value of Companies*, 6th edition. Wiley, New York.

Krueger, P. (2015). Climate change and firm valuation: Evidence from a quasi-natural experiment. Working paper, Advance online publication. https://doi.org/10.2139/ssrn.2565523.

Lee, C.M.C., Ma, P., Wang, C.C.Y. (2015). Search-based peer firms: Aggregating investor perceptions through internet co-searches. *Journal of Financial Economics*, 116(2), 410–431. https://doi.org/10.1016/j.jfineco.2015.02.003.

Lie, E. and Lie, H.J. (2002). Multiples used to estimate corporate value. *Financial Analysts Journal*, 58(2), 44–54. https://doi.org/10.2469/faj.v58.n2.2522.

Liesen, A., Figge, F., Hoepner, A., Patten, D.M. (2017). Climate change and asset prices: Are corporate carbon disclosure and performance priced appropriately? *Journal of Business Finance & Accounting*, 44(1–2), 35–62. https://doi.org/10.1111/jbfa.12217.

Liu, J., Nissim, D., Thomas, J. (2002). Equity valuation using multiples. *Journal of Accounting Research*, 40(1), 135–172. https://doi.org/10.1111/1475-679X.00042.

Liu, J., Nissim, D., Thomas, J. (2007). Is cash flow king in valuations? *Financial Analysts Journal*, 63(2), 56–68. https://doi.org/10.2469/faj.v63.n2.4522.

Matsumura, E.M., Prakash, R., Vera-Muñoz, S.C. (2014). Firm-value effects of carbon emissions and carbon disclosures. *The Accounting Review*, 89(2), 695–724. https://doi.org/10.2308/accr-50629.

Oestreich, A.M. and Tsiakas, I. (2015). Carbon emissions and stock returns: Evidence from the EU emissions trading scheme. *Journal of Banking & Finance*, 58, 294–308. https://doi.org/10.1016/j.jbankfin.2015.05.005.

Officer, M.S. (2007). The price of corporate liquidity: Acquisition discounts for unlisted targets. *Journal of Financial Economics*, 83(3), 571–598. https://doi.org/10.1016/j.jfineco.2006.01.004.

Petersen, C., Plenborg, T., Scholer, F. (2006). Issues in valuation of privately held firms. *The Journal of Private Equity*, 10(1), 33–48. https://doi.org/10.3905/jpe.2006.667557.

Pinto, J.E., Robinson, T.R., Stowe, J.D. (2019). Equity valuation: A survey of professional practice. *Review of Financial Economics*, 37(2), 219–233. https://doi.org/10.1002/rfe.1040.

Plenborg, T. and Pimentel, R.C. (2016). Best practices in applying multiples for valuation purposes. *The Journal of Private Equity*, 19(3), 55–64. https://doi.org/10.3905/jpe.2016.19.3.055.

Pratt, S.P. and Niculita, A.V. (2008). *Valuing a Business: The Analysis and Appraisal of Closely Held Companies*. McGraw-Hill Education Ltd, New York.

Sachs, J., Schmidt-Traub, G., Kroll, C., Lafortune, G., Fuller, G. (2019). Sustainable development report. Cambridge University Press.

Schmidt, P.S., von Arx, U., Schrimpf, A., Wagner, A.F., Ziegler, A. (2019). Common risk factors in international stock markets. *Financial Markets and Portfolio Management*, 33(3), 213–241. https://doi.org/10.1007/s11408-019-00334-3.

Schreiner, A. and Spremann, K. (2007). Multiples and their valuation accuracy in European equity markets. Working paper. https://doi.org/10.2139/ssrn.957352.

Sehgal, S. and Pandey, A. (2010). Equity-valuation-using-price-multiples-evidence-from-India. *Asian Academy of Management Journal of Accounting and Finance*, 6(1), 89–108.

Trueman, B., Wong, M.F., Zhang, X.-J. (2000). The eyeballs have it: Searching for the value in internet stocks. *Journal of Accounting Research*, 38, 137. https://doi.org/10.2307/2672912.

Verheyden, T., Eccles, R.G., Feiner, A. (2016). ESG for all? The impact of ESG screening on return, risk, and diversification. *Journal of Applied Corporate Finance*, 28(2), 47–55. https://doi.org/10.1111/jacf.12174.

World Bank Group (2019). State and trends of carbon pricing Report. https://doi.org/10.1596/978-1-4648-1435-8.

World Economic Forum (2021). The global risks report 2021, 16th edition. Report, World Economic Forum.

Yin, Y., Peasnell, K., Hunt, H.G. (2018). How do sell-side analysts obtain price-earnings multiples to value firms? *Accounting and Business Research*, 48(1), 108–135. https://doi.org/10.1080/00014788.2016.1230486.

Yoo, Y.K. (2006). The valuation accuracy of equity valuation using a combination of multiples. *Review of Accounting and Finance*, 5(2), 108–123. https://doi.org/10.1108/14757700610668958.

10

Risk Management Challenges in Sustainability Themed Portfolios: An Application to GHG-constrained Portfolios

10.1. Introduction

The past decade has seen greater incorporation of the concept of sustainability into investment portfolios. Sustainability has been one of the key drivers behind the incorporation of environment, social and corporate governance (ESG) criteria to achieve the dual objectives of generating long-term competitive financial returns and positive environmental and societal impacts. Investor demand for strategies that incorporate such criteria is evidenced by assets managed incorporating ESG criteria exceeding $30 trillion on a global basis[1].

Incorporating these criteria into an investment portfolio presents numerous challenges – primarily because of difficulty in both quantifying the attribute question and the associated investment implications. While there are over 160 providers that provide rankings on ESG dimensions – there is little agreement on the relative scores of companies and the evolution of these scores based on the behavior of these companies[2]. From the standpoint of a professional investor however, the

---

Chapter written by Ryan M. BROWN, Harindra DE SILVA and David W. KRIDER.

For a color version of all figures in this chapter, see www.iste.co.uk/jurczenko/climate.zip.

The authors are Portfolio Managers with Allspring Global Investors ("Allspring"). The opinions expressed in this chapter are those of the authors and do not necessarily reflect the views or policy of Allspring or any related company.

1 See (Matos 2020).

2 See (The ESG Mirage 2021) for discussion of what drives ratings changes and differences in ratings between different providers.

poor data quality and the limited availability present both a challenge and an opportunity. Investment decisions, by their very nature, involve making forecasts and risk assessments.

In this chapter, we focus on a particular measure of sustainability related to the "E" in ESG – that is associated with GHG emissions as commonly used in climate themed portfolios. Specifically, we evaluate the risk management implications of explicitly controlling the reported GHG emissions exposure of a portfolio. Direct emissions are unique in the ESG dimension as they are self-reported by companies using a common framework – unlike scores related to less quantifiable measures such as water usage. As with other self-reported data, emissions are reported with a lag and only with annual frequency in accordance with the Financial Stability Board Task Force on Climate-related Financial Disclosure (TCFD)[3]. Decent coverage of carbon data within investable universes are only available for less than a decade, so drawing conclusions on the returns to emissions as an investment characteristic is challenging, especially given the non-stationarity caused by increased governmental and social focus on GHG emissions.

Increased pressure on companies to actively manage and reduce their GHG footprint in accordance with global and local regulations has undoubtedly resulted in growing direct costs being associated with these emissions. The rising price of the global carbon credit market – evidenced by the liquid carbon futures doubling in price in 2021 – is indicative of the fact that high GHG emissions are associated with a real direct economic cost[4] and therefore a likely factor in explaining the cross-section of stock returns. In addition to the direct costs related to emissions, higher GHG companies are increasingly susceptible to transition risk due to shifts in demand for their products. In short, scaled GHG emission intensity is correlated with a systematic economic risk that was not present even a decade ago.

While our research design is focused on emissions, the analysis is broadly applicable to sustainability themed portfolio construction which is characterized by quantifiable risk exposures and dearth of deep historical data on which to calibrate a risk model. As with the specific case of emissions considered here, there are two sources of risk measurement error in such portfolios: (1) identification issues caused by sustainability related exposures that are correlated with other factors and (2)

3 TCFD is a consortium established in 2015 to develop a voluntary, consistent framework for use by financial companies to disclose climate-related financial risk to investors and other stakeholders.

4 See performance of the IHS Markit Global Carbon Index (ticker GLCARB) in USD in 2021. This index tracks the most liquid segment of the tradable carbon credit futures markets.

forecast bias stemming from the fact that sustainability exposures are not directly incorporated into commonly used risk models.

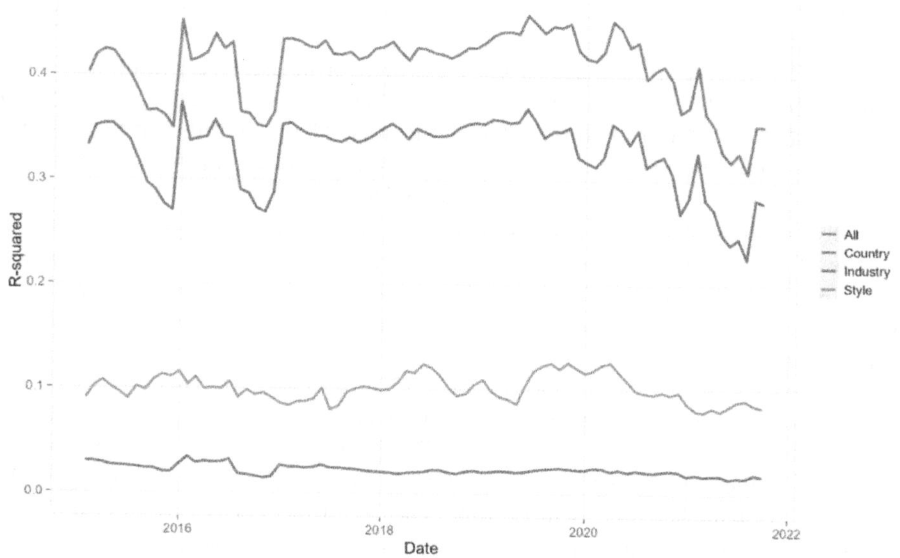

The time span covers 2015:02 to 2021:10; universe is the MSCI World. Scope 1 and Scope 2 emissions sourced from Trucost are divided by firm revenue, standardized each period and regressed on Barra's GEM3L country, industry, and style factors. The cross-sectional R-squared is on the vertical axis, for each set of variables, as well as all factors simultaneously ("All").

Figure 10.1. *R-squared of carbon exposures on various factor controls*

From a risk forecasting standpoint, the correlation between sustainability factors and commonly used risk factors – such as industry, country and style – present a significant identification issue that has the potential to produce a greater than expected forecasting error. This identification issue is further compounded by the fact that the correlation between the factors is time-varying. The extent of this relationship is illustrated in Figure 10.1, where we show the ability of Barra's GEM3L country, industry and style-related risk factors to explain cross-sectional variation in emissions. Industry effects by far have the largest impact, although their impact has declined over the past few years – arguably reflecting the differential response of companies within high emission industries to GHG-related issues[5]. While these correlations and associated identification issues will decrease the precision of any forecast, an investor can draw comfort from the fact that they do not induce a bias in the forecast.

It is difficult to separate these two sources of error in risk forecasting; therefore, we have specifically designed this study to eliminate this first issue by focusing on portfolios that are neutral to commonly used factors. The main thrust of this chapter is to demonstrate the impact of a "missing" factor in the risk model. Because risk models in most cases do not incorporate GHG-related factors, portfolios that explicitly target such factors will likely have higher than forecasted levels of risk because of their exposure to a common factor that is missing from the risk model. This is particularly true in the case of GHG emissions which reflect a systemic risk to companies, certain industries and countries that do not have access to low GHG energy sources.

In this chapter, we attempt to quantify the magnitude of this type error in terms of a bias statistic, and the extent to which investors can manage such forecasting error. We further evaluate the merits of two commonly used risk management techniques to manage and identify this type of error: (1) the return to factor portfolios based on the characteristic in question and (2) the measured exposure to factor-mimicking portfolios. The ability to use such techniques to manage and identify new risks is a major benefit to potential investors challenged by investor demand to incorporate the new sustainability-related factors into their investment framework.

In the next section, we describe the methodology used to evaluate the risk estimation error in portfolios optimized to have low GHG emissions. We then go on to describe the data used in the tests, followed by the results of bias tests and then tests that highlight the source of measurement error using common risk management techniques. The final section summarizes our results and implications for practitioners seeking to actively manage these types of exposures in their portfolios.

10.2. Methodology

Portfolio managers who integrate sustainability into their investment portfolio either explicitly or implicitly undertake an optimization process that combines some

5 Aswani et al. (2022) highlight this identification issue. They show that the documented returns to carbon emissions are primarily due industry and market cap effects.

measure of valuation or risk with an assessment of ESG exposure. The ESG score may be sourced from an external provider, estimated by the manager, or be some combination of these two approaches. While the risk associated with the typical investment characteristics are well-understood and incorporated into the risk models, the same cannot be said for ESG-related factors.

An example of a process where the optimization is explicit would include the MSCI Global Low Carbon Target Index, where the objective is to minimize carbon exposure subject to a tracking error of 30 basis points relative to a parent index[6]. A more heuristic rule-based approach is used in the S&P Global Carbon Efficient index, where companies are over and underweighted by sorting them into deciles within GICS industry groups and reweighting them accordingly[7]. As with other approaches, the goal is to maintain similar levels of risk/return to the benchmark but provide the investor with a meaningful level of carbon reduction. Regardless of the approach used in portfolio construction, an investor is faced with the prospect of quantifying the investment risk associated with additional carbon exposure relative to the benchmark. Both approaches rely on company disclosed GHG data – as does this study. A manager seeking active returns will often also incorporate their own expected return directly into the same optimization process.

We begin our development of the impact of carbon risk with the assumption that each month, returns are generated by the following factor model:

$$r = FB + f_c b_c + \epsilon \qquad [10.1]$$

where F is a matrix of factor exposures – such as country, industry and style exposures, B is a vector of payoffs, f_c is a vector of carbon or climate-related exposures, and b_c is the scalar payoff to the carbon factor. We further assume that the vector f_c is orthogonal to all columns of F.

Due to the mechanics of multiple regression, the b_c parameter can be estimated by running the following modified regression[8]:

$$\left(r - F\hat{B}\right) = f_c \hat{b}_c \qquad [10.2]$$

6 See MSCI Global Low Carbon Target Indexes Methodology (2018).

7 See Spotlight on Japan: How Carbon-Efficient Indices Can Shape the ESG Landscape (2020).

8 The Frisch–Waugh–Lovell theorem states that the coefficient b_c, estimated in a univariate context using partialled-out returns, will produce the same coefficient estimated from the original multivariate regression.

since f_c is assumed to be orthogonal to F. This re-framing is a convenient specification when assessing incremental carbon effects in conjunction with a pre-existing risk model.

Assuming this factor structure, the variance of returns will be:

$$V(r) = F\Omega_F F^T + \sigma_{fc}^2 f_c f_c^T + \Lambda \quad [10.3]$$

The terms Ω_F and σ_{fc}^2 denote the respective factor covariance matrix and the variance to climate payoffs, respectively. We treat these payoff covariances as independent, and justify this assumption later in our portfolios construction. Λ refers to the matrix of specific variances. For a given portfolio weight vector, and estimates of these variances, the corresponding estimated portfolio variance will be:

$$\hat{V}(w^T r) = w^T \left(F \hat{\Omega}_F F^T + \hat{\sigma}_{fc}^2 f_c f_c^T + \hat{\Lambda}\right) w = w^T F \hat{\Omega}_F F^T w + \hat{\sigma}_{fc}^2 \beta_{\text{port,c}} + w^T \hat{\Lambda} w \quad [10.4]$$

where $\beta_{\text{port,c}}$ refers to the portfolio exposure to the carbon measure.

If the forecasted risk neglects to include the contribution due to the $\hat{\sigma}_{fc}^2 \beta_{\text{port,c}}$ term, then forecasted risks will generally be too low relative to realized risks.

To measure the risk impact of an omitted carbon risk factor, each period we simulate two sets of 1,000 "completion" portfolios: those that have exposure to carbon (referred to *carbon reduction portfolios*) and those that do not (*baseline portfolios*). Each sample contains a unique vector of expected returns created in the following way: for each period and simulation sample, we create a random normal vector and residualize that vector to style, industry and country exposures. This score is subsequently re-standardized to have zero mean and a standard deviation of 1. Within each sample, both the carbon reduction portfolios and baseline portfolios have exposure of one to this alpha, which represents the (factor-residual) expected returns of a representative active manager.

We refer to these as completion portfolios because they reflect active portfolio positions that when combined with a market portfolio, result in an active portfolio with full market exposure and unchanged risk factor exposures. As such, they can be viewed as the active component of a portfolio. This framework also enables all returns and risk to be viewed from the active, i.e. net of market perspective, which is commonplace when evaluating investment portfolios.

In addition to the randomized alphas, the value of carbon exposure is also cross-sectionally standardized to have zero mean and unit standard deviation each month. Besides this transformation, we make no other alterations to the raw GHG emission data, which is highly skewed. The carbon exposure is set to -1 z-scores each period for the carbon reduction portfolios. In addition to being dollar neutral, these carbon reduction portfolios have net zero exposure to Barra's GEM3L country, industry and style factors. Since these portfolio restrictions are all equality constrains, the resultant portfolios have closed-form solutions of[9]:

$$w^* = \Lambda^{-1}[F, f_c, \alpha]([F, f_c, \alpha]^T \Lambda^{-1}[F, f_c, \alpha])^{-1}k \qquad [10.5]$$

where k is a constraint vector consisting of a column of zeros (one for each controlled risk factor), followed by a value of -1 for the carbon constraint, and a value of 1 for the alpha constraint: $(0,0,\ldots,0,-1,1)$. Λ is set to the specific risk estimates, sourced from Barra's GEM3L model. All weights can be levered up or down from this baseline to alter the amount of carbon reduction. We plot these security weights for a representative carbon reduction portfolio in Figure 10.2; the portfolio takes large short positions in the most positively skewed GHG emitters, while offsetting those positions with small positive weights in near-zero emitters.

If the economic consequences of carbon emissions were a diversifiable factor, it would not have a systematic impact on portfolio returns. Errors in measured risk in both groups of portfolios – carbon reduction and baseline – should be similar when measured over the same period. Any observed bias would simply reflect the quality of the risk model as measured over this time period. Variations in bias would, of course, be observed as the result of the active management associated with the random expected return model used in building the portfolio – but the resulting distribution should not differ between baseline and carbon-reduced portfolios. In contrast, if carbon exposure is systematically priced as factor, the bias statistics for the portfolio with carbon exposure will be significantly higher. For each set of random portfolios, we follow standard practice by calculating a bias statistic defined as $\text{Bias}_i = stdev(\frac{r_t}{\hat{\sigma}_t})$. For unbiased estimates of risk, the expectation is that this measure equals one; a value of Bias_i less than (greater than) 1 means that realized risks are lower (higher) than forecasted.

9 The general solution for a minimize risk objective ($min\ w`\Omega w$) subject to linear equality constraints ($X`w = k$) is equal to: $w^* = \Omega^{-1}X(X`\Omega^{-1}X)^{-1}k$. Because this optimal solution is closed-form, we do not need to rely on a quadratic optimizer to create our portfolios.

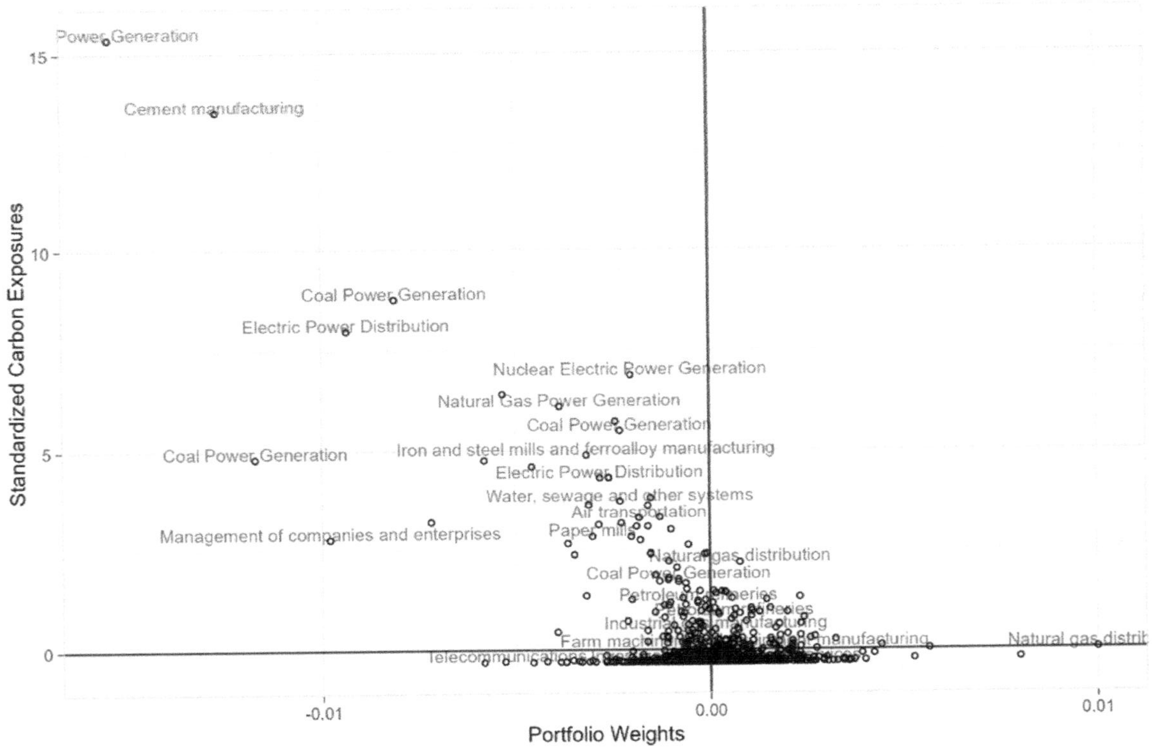

The time span presented is 2015:02; universe is the MSCI World. Scope 1 and Scope 2 emissions sourced from Trucost are divided by firm revenue and standardized (vertical axis). Portfolio weights (horizontal axis) are mean variance optimized to have -1 exposure to carbon factor, neutral to all Barra's factors, and have minimum risk (using specific risk estimates from Barra's GEM3L).

Figure 10.2. *Carbon portfolio weights*

While quantifying the magnitude of bias is useful, there is no guarantee that the level of bias will be invariant over time. We next evaluate the relationship between the returns to the carbon reduction portfolios and various measures of carbon return. The goal of this next analysis is to ascertain the extent to which investors can gain insight from relating return to the constrained factor to portfolio performance. In addition to the specific carbon factor return (i.e. the same targeted in the portfolio construction process), we also estimate the factor return to three other carbon measures, which include security carbon sensitivity. We then relate the returns of the carbon reduction portfolios (the 1,000 series are indexed at *j*) to each of these measures of factor return by estimating the regression relationship between the portfolio return (R_{jt}) and the factor return (f_{c_t}):

$$R_{jt} = a_j + b_j f_{c_t} + \varepsilon_{jt} \qquad [10.6]$$

If the factor return is representative of the return carbon-related returns, we would expect this regression to have a meaningful R^2. In the case where the factor return is the return to the factor used in portfolio construction, the R^2 should be high almost by definition. Any of the reduction in R^2 from unity is a direct result of the random expected return vector used in the portfolio construction process. For other proxies of the return to the carbon-related factor, the regression R^2 will provide insight into the ability of using different measures in performance evaluation in such a portfolio. The size and significance of the intercept in this regression (a_j) also indicates whether such a regression-based factor approach can be used to gain understanding into the performance of these types of portfolios. Specifically, we examine whether the average intercept across all series is significantly different either than zero or the average portfolio return ($\overline{R_{jt}}$).

10.3. Data description

While data on company fundamentals and returns are available for long time periods, the same cannot be said for data on GHG emissions. In this study, we use data from Trucost which is one of the most widely used data vendors. However, because of the limited availability of data prior to 2015, this study covers the time period 2015–2021[10]. Notably, this time interval coincides with the negotiations and increased adoption of the Paris Climate Agreement. Stock return data is from

10 Trucost supplies estimates of carbon data for firms that do not directly disclose their data (Bolton and Kacperczyk, Signaling through Carbon Disclosure 2020). Their Table 7 shows no changes in emissions when firms begin to disclose their emissions data, indicating that the Trucost estimates are unbiased.

Compustat, and the risk model is a fundamental-based risk model[11] that uses country factors, industry factors as well as style risk factors reflecting variables that explain the cross-section of stock returns (such as market capitalization, beta, value, price momentum, quality). The investment universe of interest is the MSCI World Developed Market index. While the time period covered by data is short by the standard of financial econometrics, it does reflect the real-world challenge of managing exposures to factors that do not have a deep history.

Our analysis of performance is based on the carbon footprint at the portfolio level. As there is some debate in the industry on the appropriate measure for carbon footprinting a portfolio, we calculate the GHG emissions using commonly used measures referred to as carbon intensity (CI) and weighted average carbon intensity (WACI). In all calculations, emissions are defined as the sum of issuers' Scope 1 and Scope 2 GHG emissions[12].

A portfolio's CI allows institutional investors to measure the volume of carbon emissions per dollar of sales by the portfolio's companies. CI is calculated by dividing the portfolio's total Carbon Emissions (weighted by the investor's ownership share) by the portfolio's total sales (also weighted by the investors' ownership share)[13]. Although efficiency at the company level is best measured using industry-specific output (e.g. per tons of steel, miles flown), this is not possible in intra-industry comparisons. The intuitive interpretation of this metric is that it presents the amount of carbon reduction per dollar value of revenue. Also, this measure is "portable"; meaning carbon emissions for active portfolio weights are additive to the emissions of the market portfolio. When evaluating portfolios at the market-wide level, the "ownership share" is unity for each security and does not enter into the calculation.

---

11 We source the MSCI Barra GEM3L risk model which has 11 style factors, 34 industry factors and country-specific factors. For further details see *Barra Global Equity Model Empirical Notes* (Morozov et al. MSCI, January 2012.)

12 We focus on Scope 1 and Scope 2 direct emissions because they are less susceptible to estimation error in carbon exposures. For example, Trucost uses an Environmentally Extended Input-Output Model (EEIO), which relies heavily on the BEA Make-and-Use table values to produce estimates of indirect emissions stemming from firms' supply chains. Although Scope 3 emissions, which rely on indirect emissions, are important for key industries (such as automobile manufacturing and oil refining), the advantage of direct emissions are that they are easily tied out across vendors. See the discussion of data sources contained in Bolton and Kacperczyk (2021).

13 Specifically, given any vector of portfolio weights w_p, we define the Carbon Intensity for portfolio i as: $\frac{\sum w_i * C_i}{\sum w_i * R_i}$, where defining the equity market value for firm i as E_i, $C_i = \frac{\text{Carbon}_i^{raw}}{E_i}$ and $R_i = \frac{\text{Revenue}_i^{raw}}{E_i}$.

While building portfolios with lower emissions by this CI measure, we only compare portfolios that are similar in terms of their overall output (i.e. revenues) being the same. Keeping revenues the same and reducing emissions per unit of investment allows us to avoid the trivial solution of investing in smaller companies where the emissions may be smaller. Investing in such companies is obviously not an equilibrium solution because we are computing this measure for the market portfolio. This preferred measure is less used in the industry, but the metric is appropriate for our analysis where we are comparing the performance of distinct portfolios with identical carbon exposures.

The second measure of emissions used to evaluate the carbon footprint is referred to as a portfolio's WACI. This is achieved by calculating CI (Scope 1 + Scope 2 emissions in tons of CO_2/\$M sales) for each company and calculating the average by portfolio weight[14]. We report this measure along with the CI as it is widely used in the industry for carbon footprinting. With this measure portfolio, exposure is simply the weighted average of the individual company scores – making it easy to evaluate the impact of adjusting portfolio weights as well as attribution analysis and portfolio decomposition. By using an alternate measure, we also highlight the challenges even with the simple task of the measuring the carbon benefits of a portfolio, as there does not exist a one-to-one correspondence between the different measures.

Two additional carbon proxies are drawn from the literature[15] which examine whether stock characteristics or risk sensitivities better explain the cross-section of security returns and risk[16]. The benefits of characteristics are that they are timely and easier to interpret. Since sensitivities are estimated on historical data, they are slower to update, and have both estimation error and modeling error. However, the benefit of sensitivities is that they only require returns and a proxy for the underlying factor and can be estimated for securities that have gaps in carbon data coverage (which is typical outside of developed markets and in earlier time periods). Furthermore, because they are market based measures they likely capture the forward-looking risk facing a company – incorporating issues such as supply chain disruptions, crop yields and building codes – which are not necessarily reflected in historical GHG emissions.

---

14 $WACI_i = \sum w_i \frac{C_i}{R_i}$.

15 A series of papers by Daniel and Titman conclude that the Fama-French value and size premia are better captured by firm exposures. On the contrary, MSCI Barra show that industry betas better explain the cross-section of returns than GICS-based industry dummy exposures, which has been rolled out to the latest generation of models.

16 We opt to use the term "carbon sensitivity" throughout this paper instead of "carbon beta", in order to avoid the confusion between "carbon beta" and "market beta".

Carbon sensitivities are estimated through a regression of daily security returns on daily carbon characteristic factor returns, controlling for the relevant market return. Returns to this characteristic factor return are calculated using a daily cross-sectional regression identical to that employed in the earlier modeling of monthly factor returns:

$$r_{daily} = FB + f_c b_c + \epsilon \qquad [10.7]$$

The f_c measures are based either on the CI or WACI measure, which produce the respective sensitivity. To improve the statistical characteristics of the f_c measures, and to facilitate interpretability, they have been standardized to have a cross sectional mean of 0 and standard deviation of 1 each time period.

Using these daily factor returns, we estimate sensitivities at the security level with the following time-series regression, re-estimated each month over the trailing year:

$$\tilde{r}_t^i = \gamma_{t-1}^i f_{c,t-1} + \gamma_t^i f_{c,t} + \gamma_{t+1}^i f_{c,t+1} + \varepsilon_{it} \qquad [10.8]$$

where $\tilde{r}_t^i$ is the security residual returns after having been purged market returns using the following regression:

$$r_t^i = \gamma_{t-1}^{m,i} \mathrm{r}_{t-1}^m + \gamma_t^{m,i} \mathrm{r}_t^m + \gamma_{t+1}^{m,i} \mathrm{r}_{t+1}^m + \tilde{r}_t^i \qquad [10.9]$$

In this context $\gamma_t^{m,i} = \gamma_{t-1}^{m,i} + \gamma_t^{m,i} + \gamma_{t+1}^{m,i}$ is interpreted as the CAPM market beta, estimated over the contemporaneous day and those prior and following to correct for asynchronous global trading. And

$$\hat{\gamma}_t^i = \gamma_{t-1}^i + \gamma_t^i + \gamma_{t+1}^i \qquad [10.10]$$

is the security's sensitivity to the carbon characteristic factor return[17]. In our analysis, the universe of securities and the market return are both based on the MSCI World Developed Markets index. Similar to our carbon exposure X_c, we standardize the carbon sensitivity in the cross-section to have a mean of 0 and standard deviation of 1. As shown in Figure 10.3, the cross-sectional correlations between the underlying carbon exposures and the sensitivity-derived measures are uniformly positive, with means centered around 0.1 and values approximately 0.15 in later periods.

17 The results we present for carbon sensitivity are not sensitive to alternative specifications.

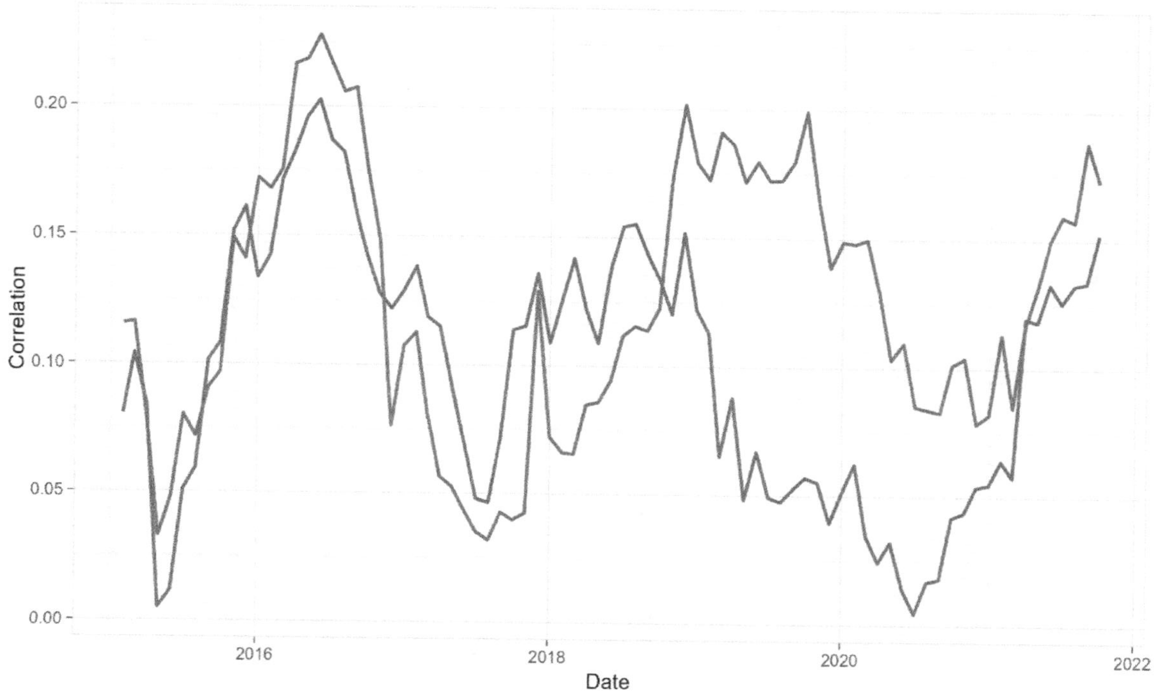

The time span covers 2015:02 to 2021:10; universe is the MSCI World. Scope 1 and Scope 2 emissions are sourced from Trucost. The blue (red) series is the cross-sectional correlation between emissions scaled by revenue (market capitalization) and carbon sensitivities. The methodology behind carbon sensitivity estimation is described in this chapter.

Figure 10.3. *Cross-sectional correlations between carbon exposures and sensitivities*

10.4. Results

We first compare the composition of the two sets of portfolios in Table 10.1. This study encompasses the time period 2015:02 to 2021:10, a period of 81 months, so the statistics shown in the table correspond to 81,000 observations (1,000 portfolios over 81 time periods). By design we expect the portfolios targeting CI to have exposure of minus one standard deviation, and the zero-carbon portfolios to have no exposure to this same factor. Inspecting the raw CI distribution, we note an average portfolio reduces CO_2 emission by more than 1,000 tons per million dollars invested. Using the alterative WACI z-score measure, the carbon reduction portfolios have a significant and consistent reduction in their carbon footprint (on average -0.677 with a standard deviation of 0.105). As intended, the zero-carbon portfolios have no exposure to either the CI measure or the WACI measure. The leverage of the carbon reduction portfolios is slightly higher, which is not surprising given that they have the additional requirement to maintain a significantly lower level of carbon exposure.

In Table 10.2, we show the average return and risk of both sets of portfolios. We first compute summary statistics for each portfolio, and in each table, we report the distribution of the relevant statistic for each group of 1,000 portfolios. Over this time period, the carbon reduction portfolios realize an average information ratio (IR) of 0.559, which is not surprising given the investor interest in this type of exposure. The zero-carbon portfolios have an information ratio of zero, consistent with the notion that these portfolios do not have any exposure to a systematic driver of returns in this time period. The difference in the distribution of IRs of these groups of portfolios is striking – with none of the carbon reduction portfolios having a negative realized information ratio.

For each portfolio, we then compute the realized bias statistic as described earlier. Thus, there is a single bias statistic associated with each portfolio for a total of 1,000 measures for each of the carbon reduction and baseline groups of portfolios. The distribution of these bias statistics is shown in Figure 10.4. The average bias statistic for the zero carbon portfolios is 1.031 that is close to the unbiased expectation of 1. The distribution of the bias statistic is centered on 1. For the carbon reduction portfolio however, the distribution of the bias statistic is significantly different from one, with the calculated bias statistics for all 1,000 portfolios being greater than 1. The average bias statistic for this group of portfolios is 1.212. This suggests that the realized risk of the carbon based portfolio are typically 20% too high relative to forecasts, with some cases as much as 30–40% too high. Importantly, because of the factor neutral construction of these portfolios, this difference cannot be attributed to a poor country, industry or style factor forecast, but are directly attributable to the performance of the omitted carbon risk factor.

Portfolio	Statistic	N	Mean	Standard deviation	Minimum	Pctl_25	Median	Pctl_75	Maximum
Carbon reduction	Carbon intensity	81,000	-1,090	197	-1,643	-1,176	-1,045	-951	-825
Carbon reduction	Carbon intensity(z-score)	81,000	-1	0	-1	-1	-1	-1	-1
Carbon reduction	WACI	81,000	-564	93	-772	-634	-573	-497	-308
Carbon reduction	WACI (z-score)	81,000	-0.677	0.105	-0.91	-0.773	-0.661	-0.596	-0.278
Carbon reduction	Annualized predicted risk	81,000	0.009	0.001	0.006	0.007	0.008	0.01	0.012
Carbon reduction	Leverage	81,000	0.577	0.02	0.505	0.563	0.576	0.59	0.665
Baseline	Carbon intensity	81,000	0	0	0	0	0	0	0
Baseline	Carbon intensity(z-score)	81,000	0	0	0	0	0	0	0
Baseline	WACI	81,000	-0.01	20.9	-118.1	-13.7	-0.08	13.6	133
Baseline	WACI (z-score)	81,000	0	0.025	-0.122	-0.017	0	0.016	0.111
Baseline	Annualized predicted risk	81,000	0.005	0.001	0.004	0.004	0.005	0.005	0.007
Baseline	Leverage	81,000	0.411	0.005	0.391	0.408	0.411	0.414	0.431

The time span covers 2015:02 to 2021:10; universe is the MSCI World. Scope 1 and Scope 2 emissions are sourced from Trucost. Each period, we simulate 1,000 portfolios that have zero exposure to Barra's factors, exposure of 1 to a random normal vector, minimum risk and z-score of either 0 ("baseline") or -1 ("carbon reduction") to carbon intensity.

Table 10.1. *Summary statistics: monthly characteristics*

Portfolio	Statistic	N	Mean	Standard deviation	Minimum	Pctl_25	Median	Pctl_75	Maximum
Carbon reduction	Mean return	1,000	0.006	0.002	0.000	0.005	0.006	0.007	0.012
Carbon reduction	Standard deviation return	1,000	0.011	0.001	0.009	0.010	0.011	0.011	0.012
Carbon reduction	Information ratio	1,000	0.559	0.185	0.001	0.434	0.556	0.674	1.193
Baseline	Mean return	1,000	0.000	0.002	-0.006	-0.001	0.000	0.001	0.006
Baseline	Standard deviation return	1,000	0.005	0.000	0.004	0.005	0.005	0.005	0.007
Baseline	Information ratio	1,000	-0.012	0.400	-1.296	-0.280	-0.010	0.238	1.455

The time span covers 2015:02 to 2021:10; universe is the MSCI World. Scope 1 and Scope 2 emissions are sourced from Trucost. Each period, we simulate 1,000 portfolios that have zero exposure to Barra's factors, exposure of 1 to a random normal vector, minimum risk and z-score of either 0 ("baseline") or -1 ("carbon reduction") to carbon intensity. We then calculate the realized return for each of the 1,000 portfolios, and present descriptive statistics of the average returns, the standard deviation, and information ratio of these portfolios.

Table 10.2. *Summary statistics: annualized return measures*

We now pivot to the time-series attribution analysis to attempt to explain the active returns to the carbon reduction portfolios. In Table 10.3, we present descriptive statistics and correlations of the time-series factor returns based on the four candidate carbon measures. The monthly realized standard deviation of these payoffs ranges from 16 to 41 bps, depending on the carbon proxy. For comparison, this is in the same range as the volatility estimates for the book-to-price, market cap, liquidity, leverage and dividend yield Barra style factors. Somewhat surprisingly, the correlations between CI and WACI values are relatively low, with many correlations near zero. The correlation measures between the exposures and sensitivity-based portfolio returns are higher, in the range of 0.3–0.4. Keeping in mind that these variables attempt to capture the same economic effect, these correlations are much lower than with those observed with financial factors. For example, correlations between the payoffs to different measures of valuation such as earnings to price, cash flow to price, or book to price are typically in excess of 0.6.

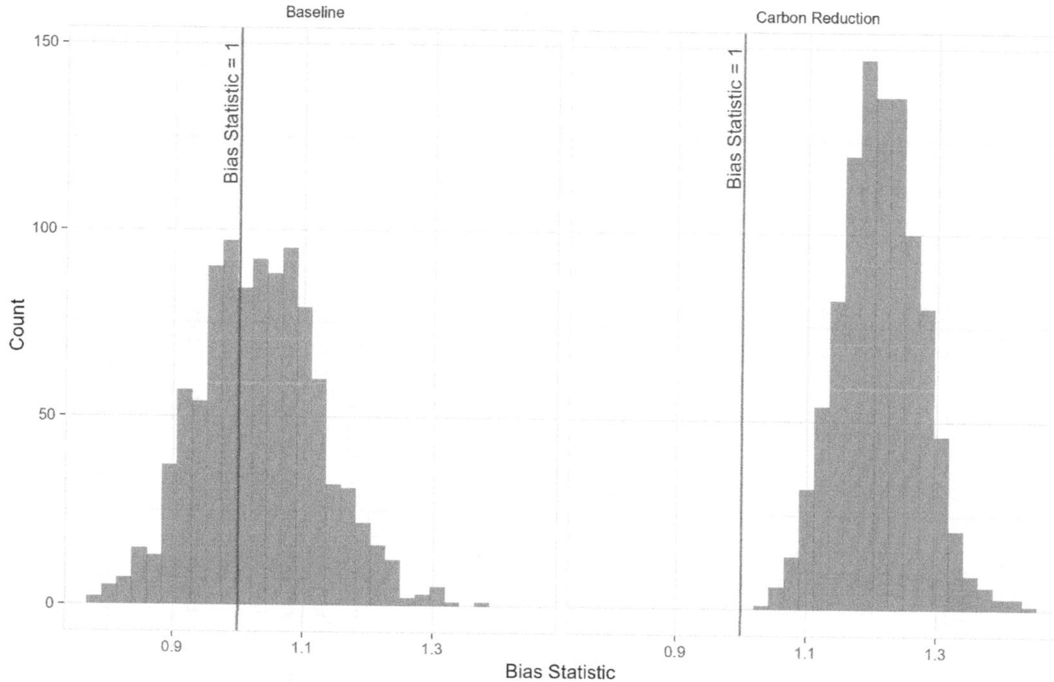

The time span covers 2015:02 to 2021:10; universe is the MSCI World. Scope 1 and Scope 2 emissions are sourced from Trucost. Each period, we simulate 1,000 portfolios which have zero exposure to Barra's factors, exposure of 1 to a random normal vector, minimum risk and either a z-score of zero (left panel) or -1 (right panel) to carbon intensity. The frequency of bias statistics are plotted for each of the 1,000 portfolios, which are defined as the standard deviation (across time) of realized returns divided by forecasted risks. The reference red line is one, reflecting an unbiased measure of risk.

Figure 10.4. *Histograms of bias statistics*

Measure	Carbon intensity	Carbon intensity sensitivity	WACI	WACI sensitivity
Mean	-0.0003	0.0003	-0.0005	-0.0005
Standard deviation	0.0029	0.0041	0.0016	0.0036
Minimum	-0.0053	-0.0079	-0.0049	-0.0113
Maximum	0.0106	0.0117	0.0028	0.0074
Correlations: Pearson\Spearman				
Carbon intensity		0.26	0.12	-0.03
Carbon intensity sensitivity	0.29		-0.04	0.40
WACI	0.11	-0.03		0.47
WACI sensitivity	-0.03	0.44	0.43	

The time span covers 2015:02 to 2021:10; universe is the MSCI World. Scope 1 and Scope 2 emissions are sourced from Trucost. Factor returns are defined as coefficients to the standardized measures below, from a multivariate regression that have Barra's factors as controls and are inverse-specific variance weighted.

Table 10.3. *Summary statistics and correlations: monthly payoffs*

With this range of correlations, we expect the four carbon-based return measures to offer different explanatory power within the time-series risk attribution framework. For each carbon reduction simulated portfolio, we regress that portfolio's realized returns on the corresponding carbon factor returns. We focus on the R^2 and the estimated intercepts from this regression, since both metrics are invariant to the differing mean and standard deviation of the four-candidate factor returns. Recall that the carbon reduction portfolios have exposure to both carbon and a random expected return component. As the returns to the expected return factor are not included in the regression, we would expect the regression to not fully explain the time series of returns for each portfolio. To the extent that the factor portfolio captures the return to the carbon-based exposure, we expect the intercept to be close to zero.

The histograms of the R^2 are presented in Figure 10.5, with a reference red line at zero. The R^2 are comparatively small for all carbon proxies except for the CI measure, which is centered around 0.7 is not surprising given that this is the same measure used in the portfolio construction. The sensitivity specifications all have lower R^2 than their characteristic counterparts.

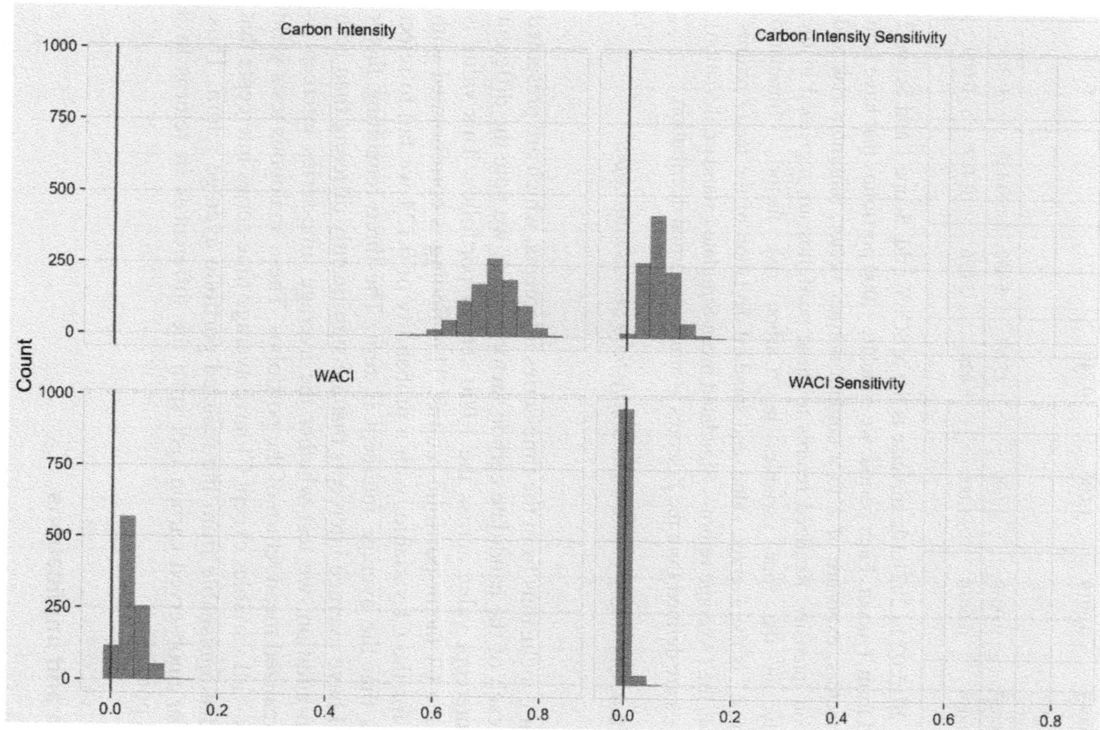

The time span covers 2015:02 to 2021:10; universe is the MSCI World. Scope 1 and Scope 2 emissions are sourced from Trucost. Each period, we simulate 1,000 portfolios that have zero exposure to Barra's factors, exposure of 1 to a random normal vector, minimum risk and z-score -1 to carbon intensity. Realized returns to these portfolios are regressed separately on four variables. The distribution of the R-squared is presented, with the red reference line at zero. Sensitivity calculations are described in this chapter.

Figure 10.5. *Time-series R-squareds*

Carbon risk factor	Simulated mean	Simulated standard deviation	Number of portfolios	Z-score: average return	Z-score: zero return	Pr{Z-Score <AR}	Pr{Z-Score >0}
Carbon intensity	0.0022	0.0019	1,000	1.930	1.155	0.027	0.124
Carbon intensity sensitivity	0.0065	0.0019	1,000	-0.240	3.351	0.595	0.000
WACI	0.0037	0.0020	1,000	1.152	1.892	0.125	0.029
WACI sensitivity	0.0059	0.0019	1,000	0.028	3.096	0.489	0.000

The time span covers 2015:02 to 2021:10; universe is the MSCI World. Scope 1 and Scope 2 emissions are sourced from Trucost. Each period, we simulate 1,000 portfolios that have zero exposure to Barra's factors, exposure of 1 to a random normal vector, minimum risk and z-score of -1 to carbon intensity. Realized returns to these portfolios are regressed in the time-series separately on the four variables in "Carbon risk factor", "Simulated mean/Standard deviation" is the average value, and standard deviation of the factor returns. Z-scores are defined as the (Average return – Simulated mean/Simulated standard deviation). Associated probabilities are defined from the Z-scores assuming a normal distribution.

Table 10.4. *Summary statistics: monthly time series intercepts*

We draw similar conclusions from the time-series intercepts, which are presented in Table 10.4. For each of the candidate carbon factor returns, we note the different average monthly intercept value across the 1,000 sets of portfolios. This value is closest to zero for the CI factor portfolio returns. This finding is corroborated with the statistical tests for the CI variable; with a probability of 0.124 we fail to reject the null hypothesis that the average intercept is zero. The three remaining factor returns variables all have average intercepts that are significantly different than zero at the 5% level. In addition, we test whether the average time-series intercept is different from the realized mean return of the portfolios. These analogous tests yield similar conclusions – all variables except CI have average time-series intercepts that are statistically indistinguishable from the realized portfolio average return. Thus, no other proxies for unobserved carbon risk span the differences in returns in a carbon-constrained portfolio.

10.5. Conclusion and implications

Fundamental risk models are widely used to build and manage equity portfolios with exposure to sustainability or ESG-related exposures. These models are built specifically to capture the risk of fundamental tilts such as those associated with style, sector and county-specific exposures. Users of these types of models should recognize the measurement error associated with using standard fundamental models to target systematic sustainability or ESG factors in an investment portfolio. This is especially true in the case of factors like GHG emissions, which capture an economy-wide non-diversifiable portfolio exposure. When the targeted factor is undiversifiable and missing from the risk model, statistical theory would imply that the active risk of such portfolios will be understated. In the specific case considered here, we find the realized risk to be approximately on average 20% higher than expected, with this excess volatility ratio being as high as 40% in some cases. The magnitude of this bias is significant and must be accounted for when building portfolios with these types of exposure. This risk will also not materialize in any kind of deep historical simulations that are used to evaluate the active risk of such portfolios – for example, using historical US stock return data from 1929 to 2021 – as they often reflect emergent risks not present in historical returns.

Because of historical data limitations, the investor targeting such exposures cannot easily quantify the risk associated with taking on this type of exposure. We find that commonly used attribution techniques, such as using factor returns or the returns to characteristic portfolios, are of limited use in explaining active returns to carbon reduction portfolios. The sole exception to this finding is the instance when the same factor used in portfolio construction is also used in the attribution process. This may be an artifact of the GHG emissions variable considered here – as the correlation between the exposures and payoffs to the other measures is relatively low. Given this lack of high correlation, the inability to explain returns is not surprising, but is likely the biggest challenge for investors looking to attribute risk and return in GHG themed portfolios.

The implication from these findings is that investors should not rely solely on commonly used risk models to accurately assess investment risk in sustainability themed portfolios. Rather, investors should recognize the misspecification error present in the model and heuristically adjust their risk forecasts upwards to reflect the additional risk present in the portfolio. The economic costs associated with sustainability-related issues are real, but investors must realize that existing risk models cannot be relied upon to produce unbiased measures of risks for portfolios with active tilts on such dimensions.

10.6. References

Aswani, J., Raghunandan, A., Rajgopal, S. (2022). Are carbon emissions associated with stock returns? SSRN 3800193, SSRN papers, Columbia Business School [Online]. Available at: https://papers.ssrn.com/sol3/papers.cfm?abstract_id=3800193 [Accessed 2022].

Bloomberg Businessweek (2021). The ESG mirage. *Bloomberg Business Week*, December 9 [Online]. Available at: https://www.magzter.com/stories/Business/Bloomberg-Businessweek/The-ESG-Mirage.

Bolton, P. and Kacperczyk, M. (2020). Signaling through carbon disclosure. SSRN 3755613, SSRN papers [Online]. Available at: https://ssrn.com/abstract=3755613. [Accessed 2020].

Bolton, P. and Kacperczyk. M. (2021). Do investors care about carbon risk. *Journal of Financial Economics*, 142(2), 517–549.

Matos, P. (2020). *ESG and Responsible Institutional Investing Around the World: A Critical Review*. CFA Institute Research Foundation.

MSCI (2018). MSCI global low carbon target indexes methodology [Online]. Available at: https://www.msci.com/eqb/methodology/meth_docs/MSCI_Low_Carbon_Target_Indexes_Methodology_May2018.pdf.

S&P Dow Jones Index Education (2020). Spotlight on Japan: How carbon-efficient indices can shape the ESG landscape [Online]. Available at: https://www.spglobal.com/en/research-insights/articles/spotlight-on-japan-how-carbon-efficient-indices-can-shape-the-esg-landscape.

11

Absolutely Sustainable Investing Across Asset Classes with Paris-aligned Benchmarks: An Application to AP2

11.1. Introduction

Absolutely sustainable investing means reducing a portfolio's greenhouse gas (GHG) emissions in line with the 1.5-degree trajectory of the Intergovernmental Panel on Climate Change (IPCC). This instructs a portfolio to target Net Zero GHG emissions in a disciplined process, which requires the portfolio to be not only relatively better than its benchmark in a given year in time but also to be better than

Chapter written by Claes EKMAN, Andreas G.F. HOEPNER, Peter MANNERBJÖRK, Tomas MORSING and Gabija ZDANCEVICIUTE. Authors are listed alphabetically.
The underlying EU TEG work is based on the excellent and tireless efforts of Claudia Bolli, Manuel Coeslier, Delphine Dirat, Steffen Hoerter, Jean-Christophe Nicaise Chateau, Sebastien Lieblich, Sara Lovisolo, Veronique Menou, Cesare Posti, Chantal Sourlas and Jean-Yves Wilmotte. Professor Hoepner also gratefully acknowledges scientific support on the EU TEG work from Theodor Cojoianu, Saphira Rekker, Fabiola Schneider and Theresa Spandel. This publication has emanated from research supported in part by Science Foundation Ireland (SFI) under Grants 18/CRT/6183 and 19/FIP/AI/7539 as well as Laudes Foundation Grant GR-075812.

itself last year through at least 7% annual reductions in absolute GHG emissions. EU Paris-aligned Benchmarks and EU Climate Transition Benchmarks represent such absolutely sustainable investing (Hoepner et al. 2019a, 2019b). This chapter will introduce the former concept and describe how Andra AP-fonden (AP2 or the Fund, hereafter) is ensuring that its liquid portfolio follows the absolutely sustainable investing concept.

AP2 is one of four buffer funds (AP1-AP4) in the Swedish state pension system. AP2's main objective is to be of maximum benefit to the pension system while managing its funds in an exemplary way through responsible investments and responsible ownership without compromising on the main objective.

Sustainability is a part of AP2 investment beliefs. AP2's sustainability work aims to create and protect value and to contribute to sustainable development based on an economic, environmental and social perspective. By including sustainability aspects in the analysis and investment processes, the Fund gains broader and better information as a basis for decisions. The Fund is also committed to maintaining a high level of trust by taking the lead on sustainability issues, while also contributing to the development of sound practices within asset management.

As of today, striving for sustainable development, in which we meet our needs without compromising the ability of future generations to meet their needs, entails huge challenges in many areas, including climate change, biodiversity, poverty and corruption. Being a responsible pension fund, it is not a question of if, and no longer a question of when, but rather a question of how the Fund can contribute to a more sustainable world.

One of the more, if not the most, urgent and undisputable sustainability issues is climate change. It has been on the agenda for decades – analyzed, discussed, recognized and addressed to a varying extent by policymakers, academia and companies. The urgency and complexity of this matter is not least manifested by the Paris agreement. How to integrate climate change issues into the investment process has not been put into practice by investors in a systematic and scientifically based manner until relatively recently. Even so, there is still a large degree of uncertainty on how to take climate change into account in portfolio construction.

The EU Paris-aligned Benchmark (PAB) framework provides science-based guidance for investors on how to align their portfolios with the goals of the Paris agreement. AP2 has chosen to adopt this framework, thus being at the forefront of the transition to a low-carbon economy. Today, close to 50% of AP2's assets are Paris-aligned, including developed and emerging market equity as well as global corporate bond portfolio, with actively managed Swedish equities expected to implement PAB in 2022.

However, even though there is now a framework in place for how investors can meet the Paris climate goals, many challenges remain. The following sections discuss the hurdles encountered by AP2 in the process of PAB implementation. The chapter briefly outlines the technical considerations relevant to PAB integration; however, a significant discussion is dedicated to the conceptual aspects, such as Paris-aligning in specific asset class investments or the enterprise value including cash (EVIC) and inflation factor adjustments.

First, we provide a brief introduction to the development and the requirements of climate benchmarks. We then dive into the case study of AP2's PAB implementation by outlining the context in which AP2 operates and the development of sustainability within the fund. We further focus on the general considerations, such as the interpretation of regulations from a pension fund perspective, PAB backtesting as well as data-related aspects. Finally, we discuss the implementation of PAB across the asset classes and finish the chapter with a discussion on GHG-intensity definition and EVIC-inflation adjustment.

11.2. The climate benchmarks

According to the EU Technical Expert Group on Sustainable Finance (TEG) report (Hoepner et al. 2019a), climate benchmarks incorporate financial objectives as well as GHG emission reduction targets to transition to a low-carbon economy. The carbon reduction targets should be in line with the IPCC scientific evidence and implemented through the selection and weighting of the benchmark constituents.

In March 2018, the European Commission introduced an action plan on sustainable finance that aims to align the financial flows with climate and sustainable development targets (European Commission Action Plan 2018). The document outlined reorienting of capital flows, managing financial risks arising from climate change and fostering transparency and long-termism of economic activity as the key pillars. The legislative package that followed included four proposals, including the low-carbon benchmarks. The benchmarks were first introduced in the September 2019 final EU TEG report (Hoepner et al. 2019a) with the regulation published in November 2019 (Regulation (EU) 2019/2089) and the delegated regulation, outlining the minimum EU climate benchmark standards, published in July 2020 (Commission Delegated Regulation (EU) 2020/1818).

The regulation (Regulation (EU) 2019/2089) introduced two types of climate benchmarks – a Climate Transition (CTB) and a Paris-aligned (PAB) Benchmark. Both have similar objectives but differ in their level of ambition. Table 11.1 compares the minimum standards applied.

11.2.1. *Minimum benchmark requirements*

The two key differences in the minimum benchmark requirements are the intensity reduction and the activity exclusion criteria. Climate Transition Benchmarks are required to decarbonize at minimum 30% at the date of implementation compared to the investable universe, whereas Paris-aligned Benchmarks must decarbonize at minimum by 50%. Both benchmarks apply the 7% year-on-year decarbonization criteria and Scope 3 emissions phase in over up to four years.

Both benchmarks apply exclusions to companies involved in activities related to controversial weapons, cultivation or production of tobacco as well as companies that violate the United Nations Global Compact (UNGC) principles or the Organisation for Economic Cooperation and Development (OECD) Guidelines for Multinational Enterprises (see footnote 2). However, PABs must additionally exclude companies whose revenues, in specified sectors, exceed the revenue thresholds. Specifically, it is required to exclude companies that derive more than a set percentage of their revenues from coal, oil, gas or fossil fuel power generation, in order to support the transition to renewable energy sources. In contrast, no such exclusion criteria apply to CTBs.

The benchmarks also incorporate weight constraints on high impact sectors to ensure financing towards sustainable solutions. The Delegated Act (Commission Delegated Regulation (EU) 2020/1818) specifies that the climate benchmark exposure to high impact sectors (see footnote 4) should be no less than that of the underlying universe. This constraint is introduced to prevent widespread disengagement with the "heavy" sector companies and to ensure financing for the transition of these businesses. The high impact sector weight criteria apply to equity securities only, while fixed income sector weights remain unrestricted.

Minimum standards	EU CTB	EU PAB
Risk-oriented minimum standards:		
Minimum Scope 1+2(+3)[1] carbon intensity reduction compared to investable universe	30%	50%
Year-on-year decarbonization trajectory	7%	7%
Scope 3 phase-in	By 23rd December 2024	By 23rd December 2024
Baseline exclusions	Yes Controversial Weapons Societal norms violators[2]	Yes Controversial Weapons Societal norms violators
Activity exclusions	No	Coal (1%+ revenues) Oil (10%+ revenues) Natural Gas (50%+ revenues) Electricity producers with carbon intensity of lifecycle GHG emissions higher than $100gCO2_0$/kWh (50%+ revenues)
Significant Harm Exclusions	Violators of DNSH as per EU Taxonomy Article 9 (by 31st December 2022)	Violators of DNSH as per EU Taxonomy Article 9
Weight constraints[3]	"High impact"[4] sector exposure should be no less than that of the underlying investable universe for equity benchmarks	"High impact" sector exposure should be no less than that of the underlying investable universe for equity benchmarks

Table 11.1. *Minimum standards for the EU CTB and EU PAB (Hoepner et al. 2019a, Commission Delegated Regulation (EU) 2020/1818)*

1 Scope 3 being phased in during a four-year time period.

2 Societal norms include UNGC Principles, OECD Guidelines for Multinational Enterprises and the six Environmental Objectives: (1) climate change mitigation; (2) climate change adaptation; (3) sustainable use and protection of water and marine resources; (4) transition to a circular economy, waste prevention and recycling; (5) pollution prevention and control; (6) protection of healthy ecosystems.

3 The sector weight constrains apply to equity securities only.

4 High impact sectors are defined as NACE sections A to H and section L. Namely, Agriculture, Forestry and Fishing; Mining and Quarrying; Manufacturing; Electricity, Gas, Steam and Air Conditioning Supply; Water Supply; Sewerage, Waste Management and Remediation Activities; Construction; Wholesale and Retail Trade; Repair of Motor Vehicles and Motorcycles; Transportation and Storage; Real Estate Activities.

11.2.2. Benchmark decarbonization and inflation adjustment

In order to match the Paris Agreement goals, climate benchmarks incorporate carbon reduction targets in line with the IPCC 1.5 degree scenario (UNFCCC 2015). The alignment is translated to an annual decarbonization trajectory of 7% for both benchmarks. This means that a benchmark should reduce the GHG emissions by at least 7% year-on-year calculated geometrically. Practically, the decarbonization implementation differs slightly for equities and fixed-income benchmarks.

In the case of equities, a benchmark has to reduce its carbon footprint based on carbon intensity measure. GHG intensity is defined as a company's absolute GHG emissions divided by its enterprise value including cash (EVIC) as specified in the Delegated Act (Commission Delegated Regulation (EU) 2020/1818). The *total* index level GHG intensity at time t is thus defined as the weighted sum of individual constituent total GHG emissions divided by the constituent EVIC at time t.

$$Index\ level\ GHG\ intensity(t) = \sum w_i(t) \frac{GHG_i(t)}{EVIC_i(t)} \qquad [11.1]$$

where i is a benchmark constituent. However, if EVIC of the investable universe would increase year-to-year, the GHG intensity would reduce as a result. Therefore, the carbon intensity calculation should be inflation-adjusted by dividing EVIC by the inflation adjustment factor or multiplying the index-level carbon intensity by the adjustment factor.

$$Inflation\ adjustment\ factor\ (t_m) = \frac{\frac{1}{n(t_m)} \sum EVIC_i(t_m)}{\frac{1}{n(t_{m-1})} \sum EVIC_i(t_{m-1})} \qquad [11.2]$$

$$\begin{matrix} Inflation\ adjusted \\ GHG\ intensity(t_m) \end{matrix} = \begin{matrix} Index\ level \\ GHG\ intensity(t_m) \end{matrix} \times \begin{matrix} Inflation\ adjustment \\ factor\ (t_m) \end{matrix} \qquad [11.3]$$

where t_m is the time period in the time range $1, \ldots, m$, and n is the number of entities.

Consequently, the 7% year-on-year reduction target can be expressed as the nominal benchmark intensity ratio multiplied by the inflation adjustment factor.

$$\frac{\begin{matrix} Index\,level \\ GHG\,intensity(t_m) \end{matrix}}{\begin{matrix} Index\,level \\ GHG\,intensity(t_{m-1}) \end{matrix}} \leq 0.93 \qquad [11.4]$$

In the case of fixed income, the delegated climate benchmarks regulation (Commission Delegated Regulation (EU) 2020/1818) specifies two scenarios. If a debt security issuer has equity listings in the public market, the carbon targets can be measured and the 7% reduction applied to either the GHG intensity or the absolute GHG emissions. However, if the fixed income issuer does not have an equity listing in a public market, the 7% reduction must be applied to absolute GHG emissions.

The inflation adjustment factor is a particularly important consideration in ensuring that benchmarks achieve the Net Zero targets. Adjusting the GHG intensity for EVIC increase (or decrease), as a result of inflation (or deflation), is designed to prevent the reduction in GHG intensity as a result of a change in EVIC. If unaccounted for, an increase in portfolio EVIC could result in lower GHG intensity while no reductions in absolute GHG emissions occurred. Therefore, applying the 7% reduction requirement simply to the nominal benchmark intensity ratio can lead to decisions that are not compatible with the Paris Agreement, as asset price inflation can decrease the intensity ratio even though emissions are constant or growing.

11.3. Absolutely sustainable investing

The five points we consider relevant for absolutely sustainable investing are:

1) reduction of GHG emissions versus market benchmark in a given year;

2) at least 7% year-on-year reduction of GHG emissions;

3) Net Zero GHG emissions by 2050;

4) integration of full scope three emissions;

5) use of the precautionary principle in GHG data estimation.

The EU climate benchmarks are tools for absolutely sustainable investing. The literature generally agrees that an absolutely sustainable framework should 1) set absolute targets based on sustainable levels of impact rather than relative targets on internal values (Bjørn et al. 2015, 2020; UNFCCC 2015) and 2) increase the positive environmental impact (Bjørn et al. 2020). Paris-aligned and Climate Transition Benchmarks further extend this notion to a list of five points.

The two principles from the literature are fundamental to the climate benchmark framework. The benchmark methodology was designed to facilitate the capital reallocation to sustainable businesses which in turn is intended to finance the positive environmental solutions. Furthermore, the Net Zero 2050 objective together with the integration of the IPCC 1.5 degree scenario define a clear external target while the mechanics of benchmark decarbonization further impose absolute targets.

The 30% or 50% initial emissions reduction requirement, for CTB and PAB, respectively, is defined in reference to a baseline year. Similarly, the 7% annual decarbonization is applied with respect to previous year's emissions. Both, the year one and the following year targets will result in a reduction in absolute GHG emissions, eventually bringing it to Net Zero, considering the inflation adjustment has been implemented.

Net Zero 2050 emissions target, together with decarbonization against the market benchmark and annual decarbonization target are the core pillars of absolutely sustainable investing. However, the Paris-aligned and Climate Transition Benchmarks further add that an absolutely sustainable investing framework should integrate Scope 3 GHG emissions data and should use the precautionary principle in estimating the missing GHG data. As outlined in the delegated regulation (Commission Delegated Regulation (EU) 2020/1818), both climate benchmarks should include GHG Scope 3 data in their decarbonization targets for every sector. However, the quality of the GHG emissions data is currently not yet at the level that would make it possible to set quantified thresholds for every investable entity. Therefore, the benchmark regulation contains a phase-in period for Scope 3 data in different sectors (see Article 5).

Given the quality of the emissions data, the benchmark administrators will likely have to perform estimations, especially of Scope 3 data. The implementation of the precautionary principle in data estimation is crucial to disincentivize company non-reporting (Hoepner and Rogelj 2021). The precautionary principle errs on the side of the planet by estimating the non-reported Scope 3 emissions, for example, in line with the worst sector performers. The benchmark regulation further encourages the implementation of the precautionary principle by requiring it to disclose as part of estimation methodology where applicable.

11.4. Case study: implementation of PAB at Andra AP-fonden

11.4.1. *The Swedish pension system and the AP-funds*

In the 1990s, Sweden reformed its benefit pension system (ATP), which had been in place since the 1960s. The reform was far-reaching and brought about by the fact that it had become economically and demographically untenable. The reformed system is contribution-defined, sometimes called Notional Defined Contribution (NDC), and consists of two parts: Income pension and Premium pension. AP2 is part of the income pension system, which is a so-called pay-as-you-go system. Individuals contribute to the income pension system with 16% of their salary in pension fees (up until a certain threshold). Each Swedish krona paid in fees gives the individual a pension right, while the actual money is used to finance pensions to current retirees. When retiring, currently between the age of 62 and 68 years, the accumulated pension rights are divided by the conditional life expectancy at the retirement age and paid out monthly as a pension. A later retirement will thus give a higher pension since more pension rights will be accumulated which are subsequently divided by a shorter remaining life expectancy.

The pension rights are indexed to Swedish wage inflation, with an automatic balancing mechanism regulating the indexation in order for the system to be financially stable. If the system becomes underfinanced, the balancing mechanism reduces the indexation until the system stabilizes.

To handle short-term imbalances between pension fees paid to the system and pension payments, four buffer funds are part of the system. When the new Swedish pension system was launched in 2001, SEK 134 billion (EUR 14.5 billion at the time) were allocated to each of the four buffer funds.

AP1-AP4 are government agencies regulated by Swedish law via the Swedish National Pension Insurance Funds Act. According to this Act, the primary objective of the buffer funds is to manage the assets to the greatest possible benefit to the pension system. The goal is to maximize long-term returns in relation to the risk in the investments. Risk and return shall be interpreted in terms of outgoing pensions. To meet the objective of the Act, AP2 has chosen a strategic portfolio that aims to minimize any future negative impact on pensions caused by the automatic balancing mechanism.

In its original version in 2000, the Act governing the AP funds highlighted the importance of ethical and responsible investing. In 2019, a revision of the Act raised responsible investing and ownership to be a secondary objective for the buffer funds. According to the revised act, the AP funds shall be managed in an exemplary manner through responsible investments and ownership. Particular attention shall be paid to how sustainable development can be promoted without compromising the overall goal of the investment activities.

11.4.2. *Development of sustainability integration and benchmarks at AP2*

When AP2 was founded in 2001 the portfolio was basic, consisting of Swedish government bonds, Swedish and foreign equities and Swedish real estate. Sustainability, as we know it today, was not part of the investment strategy, and most of the assets were externally managed. Today, the portfolio is very different. It is more diversified and has expanded across asset classes as well as markets. Most of the assets are internally managed and sustainability is highly integrated into the investment processes.

AP2's use of, and work with, benchmarks in equities and corporate bonds has evolved continuously. Currently, AP2 has equity exposure in the Swedish market, developed markets (DM), and emerging markets (EM) including mainland China (A-Shares). All benchmarks used are customized to some degree, and some are aligned with the PAB framework.

DM equities is the largest single asset class in the Fund's strategic portfolio. Therefore, much of the development and work with benchmarks at AP2 has its origin in this asset class. Equity benchmarks for other regions have drawn on the knowledge acquired in DM equities. This knowledge has also been used when developing benchmarks for corporate bonds.

The first benchmark used for DM equities was market capitalization-weighted, with sustainability mainly incorporated by the exclusion of companies violating international conventions. The allocation to DM equities was modest in the beginning, at merely 3%, but increased rapidly and is now more than 20% of the Fund's portfolio. Gradually over time, fundamentally weighted benchmarks and different types of smart beta indices (such as value-weighted and low risk-weighted) were introduced. As the Fund's focus on responsible investments grew, exclusions of companies were complemented with engagement and voting at annual general meetings, often in collaboration with other investors. In 2009, AP2 published for the first time a measure of the carbon footprint of its portfolio in listed global equities, within the framework of the Carbon Disclosure Project.

In 2005, AP2 decided to form a quantitative investment team, running both benchmark replication and active management on top of the benchmarks in equities. The approach chosen was to do as much as possible in-house, only being dependent on data from external providers. This was the starting point of building an internal body of knowledge within portfolio construction and factor investing, which in many ways has shaped the current use of benchmarks in equities and bonds at AP2. Quantitative strategies are applied to equities in DM and EM (excluding China A-shares), corporate bonds (USD, EUR and GBP), Swedish government bonds, as well EM hard and local currency bonds.

By 2015, AP2 began to divest from companies considered to be at high financial risk due to climate change. In practice, this meant selling specific companies in the energy sector. In 2015, ESG factors were also introduced into active management, giving the fund an opportunity to expand the scope of sustainability and to make more granular sustainability decisions, as compared to the binary decision of whether to include or exclude.

At the time, in addition to market capitalization-weighted benchmarks, AP2 had more than a handful of smart beta indices for equities in developed and emerging markets. There was a discussion of complementing these indices with ESG indices. However, instead of increasing the number of indices, the fund decided to develop two multi-factor benchmarks for global equities, one for DM and one for EM, incorporating all the various properties and characteristics important to AP2, including ESG. This was done internally, many thanks to in-house quantitative expertise that had been accumulated over the years in combination with a well-developed infrastructure for quantitative investing. In early 2018, the old index structure was replaced with two internally developed multi-factor benchmarks. Alongside more traditional risk premia, ESG factors were also used in the benchmark construction, and Scope 1 and Scope 2 GHG intensities were reduced by approximately 25%.

Since 2015, responsible investors have been grappling with how to align their portfolios with the Paris Agreement and how to contribute to the transition to a green economy. In this context, AP2 has chosen to adjust benchmarks and portfolios in line with the PAB regulation. The main challenges for the fund in this project were not of technical nature, given AP2's know-how within index construction. Rather, the challenges lay in interpreting the regulations from a pension fund perspective, putting this into a practical context, and finding relevant and necessary data needed to align with PAB. In November 2020, Paris-aligned indices for global equities (DM and EM) and global corporate bonds were launched.

The development of the benchmark for corporate bonds has followed a path similar to global equities with a lag of a few years. The work on developing an alternatively weighted multi-factor index for corporate bonds had already started when the decision was taken to incorporate the PAB requirements into AP2 indices and portfolios. In 2019, the fund launched an internally developed index for developed markets corporate bonds which is both alternatively weighted, with tilts towards desirable fundamental factors, and in line with the PAB regulation.

The benchmark setup in global equities and corporate bonds is very similar. Twice a year in equities, and monthly in bonds, the investable universe is updated and new constituent weights are calculated internally. These weights are sent to an index provider, who is responsible for the daily updates of constituent weights due to market movements and corporate events, as well as for the calculation of benchmark returns. AP2 owns the construction of the benchmarks but outsources the day-to-day maintenance and performance calculations.

The Swedish equities asset class has been treated differently, mainly because it is not managed quantitatively. It is managed by a team of fundamental portfolio managers and analysts, working with engagement, dialogue and voting to a much higher extent than in the global quantitative mandates. Two benchmarks are used, one internally developed for Swedish large-cap stocks, and one market capitalization-weighted for Swedish mid- and small-cap companies. Work is ongoing to make both benchmarks for Swedish equities compliant with PAB criteria.

11.4.3. *Implementing the EU Paris-aligned Benchmark at AP2*

11.4.3.1. *Why PAB?*

For a long time, AP2 has had an ambitious agenda regarding sustainability, especially focusing on climate change. The fund started measuring emissions as early as 2009 and has actively been reducing its carbon footprint since 2018. In 2016, a decision was made to align AP2's portfolios with the Paris Agreement. However, at the time there was no generally accepted method for investors to align their portfolios with the targets of the agreement. Therefore, when the criteria for Paris-aligned benchmarks were introduced, this framework was a welcome tool for AP2 to put the alignment decision into practice. The outspoken link between PAB and the commitments laid down in the Paris Agreement, or more specifically the IPCC 1.5°C scenario, enabled AP2 to take the next step in formalizing and further sharpening the sustainability work related to climate change. In AP2's view, PAB offers a framework to reach Net Zero GHG emissions by 2050 without any major disadvantages or obvious shortcomings.

AP2 believes that a Paris-aligned Benchmark can be constructed without compromising the risk-return profile. The fund further finds the PAB framework attractive because:

– PAB takes a science-based approach, building on IPCC research and recommendations which hopefully adapts as new research and data become available.

– PAB offers a public and transparent framework developed by a credible third party, which decreases the risk of being accused of greenwashing.

– More practically, PAB offers a definition of what is considered a fossil fuel company (which is not a clear cut) and takes a holistic approach to the financing of the climate transition.

11.4.3.2. *How does AP2 apply PAB?*

The Regulation (EU) 2019/2089 introducing the EU climate benchmarks establishes minimum requirements to be met by index providers in order to label index products as Paris-aligned. One objective of the new regulation is to increase comparability and transparency of climate benchmark methodologies, thus helping investors and other users of climate benchmarks to make informed decisions regarding different index products. Therefore, the Delegated Act includes articles related to disclosure and transparency requirements (such as disclosure of ESG data, methods of estimation of missing GHG emissions data, etc.), giving guidance on how and what to report to meet the PAB labeling.

AP2 is a government agency regulated by Swedish law and not an index provider. Therefore, Regulation (EU) 2019/2089 (and the updated Commission Delegated Regulation (EU) 2020/1818) does not formally apply to an entity such as AP2. Still, AP2 believes that the PAB framework is important and the best practice available, for the reasons outlined above. Consequently, the fund's internally developed benchmarks are constructed to be compliant with the articles in Chapter II[5] and Article 14[6] of the Delegated Act. One exception is the use of EVIC-inflation adjustment. Since the year-on-year decarbonization must not be driven by changes in price (or changes in debt/cash), and instead should change as a result of reduction of absolute GHG emissions, the EVIC-inflation adjustment, as defined in the Delegated Act (Article 7), may need to be analyzed more regarding its sufficiency in certain scenarios. For instance, a re-definition of the population-based inflation adjustment (see equation [11.2]) or aligning the EVIC-inflation factor with individual securities rather than the entire benchmark may prove to be more viable alternatives. Further research is currently ongoing investigating alternative methods for improved EVIC-inflation adjustment.

11.4.3.3. *Implementing PAB*

Early in 2020, AP2 decided to investigate the possibility of developing benchmarks for global equities (DM and EM excluding China A-shares) and global corporate bonds to align with the regulation of a PAB. Beyond PAB, several other aspects also had to be considered. Among the more significant desired properties were:

– to maintain or improve expected risk-adjusted return;

– to achieve exposure to desired characteristics, for example, sustainability;

– feasibility (specifically reasonable turnover).

Development and implementation of a PAB in-house requires relevant competencies and experience, as well as adequate technical solutions in terms of data and infrastructure. Most of this was already in place at AP2, thanks to more than 10 years of quantitative investing, but there were nonetheless obstacles to

5 Minimum standards for the design of the benchmark methodology. Regulates decarbonization, exclusions and sector exposures of the benchmark.

6 Disclosure of the decarbonization trajectory.

overcome. Some were of a more general nature, such as how to backtest a benchmark aligned with PAB, how to handle data issues and how to ensure that not only benchmarks but also (actively managed) portfolios are aligned with the PAB criteria. There were also challenges of a more specific nature depending on asset class or universe, such as the handling of non-listed companies in a corporate bond PAB, and how to deal with the sector constraint in a small national equity universe where all sectors might not be well represented. Some of these aspects will be further discussed in the next sections.

The complexity of creating a PAB makes the task well suited for a quant team with experience in optimization. Data from different sources are used which must be structured and cleaned. Constraints and exclusions at both company and portfolio level are used in the benchmark construction. The benchmark's GHG emissions intensity is decreased and its exposure to desirable fundamental factors is controlled. A quant team with knowledge of backtesting, factor investing, portfolio optimization, risk modeling and trading cost modeling has experience based on problems similar to those entailed in constructing an optimized index. With data and data management processes already in place, adding the required parts of PAB is a manageable task. The real challenge is setting up the framework and processes needed to handle the index construction process itself.

11.4.3.4. *General aspects*

11.4.3.4.1. Backtesting

The choice of a benchmark is an important part of the fund's investment process since it defines the exposure of AP2's strategic portfolio to a particular asset class. Before choosing a benchmark, a thorough analysis ensures that the index meets AP2's objectives regarding risk-adjusted expected return and sustainability, and that its diversification characteristics fit well in the total strategic portfolio. One step of the evaluation process is a backtest where the index characteristics are analyzed. Given the sustainability improvement a PAB would give, it is important to verify that the risk-adjusted return of the PAB is better than, or at least not worse, than its non-PAB counterpart.

Backtesting is one of the most important tools for quantifying the effects of different investment strategies. In this context, the purpose of backtesting is to study how an index would have performed historically. Backtesting a PAB means ensuring that the PAB requirements are fulfilled throughout a historical period and studying the index characteristics over that time. The return of the PAB is compared with the return of the non-PAB counterpart during the same period. During the backtesting period which AP2 used to analyze PAB, there was no evidence found that a PAB-compliant benchmark had a lower expected return than a non-PAB portfolio, after controlling for risk and factor exposures. AP2 therefore concluded that adjusting the portfolio to meet the PAB requirements is in line with the objectives of AP2 and can be done without compromising risk-return characteristics.

The general criticism of backtesting is that historical performance does not guarantee future performance. With a PAB it is an extra challenge to evaluate and draw conclusions from backtesting, as the awareness and treatment of emissions have changed significantly over the last decades. However, given the lower interest in reducing GHG emissions and sustainable investing historically, a reasonable hypothesis is that this backtesting would favor GHG-intensive companies, if anything. This hypothesis was not confirmed in the results. It is also possible to argue that the increasing interest in sustainable investing and the growing focus on the transition to a sustainable economy should result in better returns going forward than backtests show.

Fixed income has additional challenges related to backtesting. Mapping company-level data, like GHG emissions and other reported data, to the relevant backtest dataset in a structured and automatic way is not as straightforward as in equities. There are bonds issued at different levels of a company's structure, and every bond must be linked to a specific company which must be linked to an identifier to access company-level data sources. To verify and backtest the PAB, this exercise also needs to be done for the whole backtesting history. The historical mapping process is further complicated by mergers and other corporate actions, making the PAB construction process more complicated compared to equities.

11.4.3.4.2. Technical needs

Data

It is crucial to have relevant and correct data when constructing PAB. Apart from the ordinary index data, like constituent prices and outstanding amounts, several other data points are required to meet the quantitative targets set out by the EU regulation, including:

– decreasing the GHG intensity by 7% per annum;

– excluding instruments issued by companies with revenues from fossil fuels (including fossil fuel generated energy) exceeding a defined threshold value;

– ensuring exposure to "high climate impact sectors" (not applicable to bonds).

Data to address these requirements have shortcomings.

Companies' reporting on emissions data can be absent, inadequate or delayed. Also, there exist parallel reporting standards and different methods for the estimation of missing data. Taken together, these issues potentially lead to large differences between data vendors, which are discussed further below.

The quality of the revenue data is dependent on the granularity of the companies' reports, i.e. to which extent revenues are broken down into different business activities. Lack of standards and transparency might cause biases in the data leading to suboptimality in the benchmark construction.

The exposure constraint to "high climate impact sectors" is based on the classification system "Statistical classification of economic activities in the European Community" (Nomenclature statistique des Activités économiques dans la Communauté Européenne, NACE). NACE is a classification system seldom used by index providers or investors. Therefore, a mapping from the NACE system to a more commonly used sector classification system (such as BICS, GICS and ICB) is most often required. In general, there is not a unique mapping between different systems, leading to different results depending on which classification system is adopted.

Lack of standards or auditing of reporting on sustainability data is a major challenge for responsible investors. There is an obvious risk that companies take advantage of these shortcomings to appear more environmentally friendly, which can result in continued financing of unsustainable activities .

Optimizer

When constructing a PAB, several constraints must be satisfied. As noted previously, the most prominent relate to GHG reduction, fossil revenues and sector allocation. Fossil revenue data is the basis for exclusions and are straightforwardly implemented. The NACE sector allocation and GHG reduction constraints are more complex since they are measured at the portfolio level. It is not uniquely defined how to adjust index-constituent weights to meet requirements and constraints at a portfolio level. It is possible to develop heuristic or naïve methods to exclude companies or decrease constituent weights through a pre-defined algorithm (or by trial and error), resulting in an index meeting portfolio level targets. However, such methods entail a risk of modifying other index characteristics in an uncontrolled and maybe undesirable way. For example, constructing a PAB of an alternatively weighted factor index using a heuristic method can severely affect exposures to traditional systematic risk factors of high importance.

To handle such shortcomings, AP2 uses optimization algorithms when constructing PAB. It is very natural to manage constraints, such as those in the PAB regulation, in an optimizer. From an implementation point of view, they do not differ from other constraints when solving optimization problems. At AP2, optimizers have been used for years in solving similar problems, for example, in creating internally developed multi-factor indices, and also in the active management of portfolios. The main challenge for AP2, from this perspective, was to understand the constraints and to identify relevant data to be used as input to the optimizer. Creating a PAB, therefore, fitted very well into the existing processes at AP2.

11.4.3.4.3. Estimation of missing emissions data

Decarbonization is central in the EU climate benchmarks methodology and GHG emissions data play a significant role when creating climate benchmarks. Companies' disclosure of GHG emissions has improved over the last decade, but missing emissions data and understatement of emissions are still significant problems to handle. These issues are more pronounced in emerging markets and small-cap companies.

Using the Greenhouse Gas Protocol (WRI and WBCSD) has over the years become the standard for reporting GHG emissions. Most companies reporting GHG emissions are using the "GHG Protocol Corporate Accounting and Reporting Standard" (WRI/WBCSD 2001), giving guidance on reporting direct emissions (Scope 1) and indirect emissions due to the purchase of electricity (Scope 2). However, some companies report according to an alternative standard called "Corporate Value Chain (Scope 3) Standard" (WRI/WBCSD 2011), and therefore also supply a figure of all indirect emissions other than emissions due to the purchase of electricity (Scope 3). Many companies that report according to the first standard, but not the second, still supply Scope 3 figures, but these might not cover all indirect emissions and therefore underestimate the companies' emissions. Only companies using both standards will give a full accounting of all scopes. AP2 has not been able to find data on whether a company reports according to one or both standards. Even though Scope 3 data today is of lower quality than the other scopes and needs special attention when estimated, it is still important to take Scope 3 emissions into account in a low-carbon benchmark. Decarbonization based only on Scope 1 and Scope 2 emissions could lead to counterproductive results, as many industries and services that need to transform to be in line with the Paris Agreement mainly have indirect emissions.

There are several data vendors providing emissions data on all scopes. Usually, they provide data taken from companies' reports, and also specific estimates of the company emissions. Even though AP2 has access to emissions data from several providers, there are still a considerable number of companies with missing data for one or more of the scopes. Therefore, developing internal estimation methods have been part of the index construction process.

A company's emissions depend to a large extent on its type and size of operations. Companies might have several distinct types of operations making the estimation more complicated. In the development of Paris-aligned Benchmarks, AP2 has investigated two different approaches to estimate missing emissions data.

The first method assumes that each company has one type of operation defined by a sector classification system, for example, GICS or ICB, and uses Sales as an approximation for the size of the emitting activities. The method estimates a linear regression model with the logarithm of GHG emissions, for a given scope, as a dependent variable and the logarithm of Sales as an independent variable for those companies with the same type of operations. Expressed in simplified terms as:

$$log(emissions_i) = \mu + \beta log(sales_i) + \varepsilon_i \qquad [11.5]$$

where μ and β is the intercept and slope, respectively, for the sector.

This gives a model to predict the GHG emissions for companies with missing data with the same or similar type of operations. Sales seem to be the best readily available approximation for the size of emitting activities. It would seem natural to group companies with as much granularity as possible according to the sector classification system, giving a more specific description of the operation, but this must be balanced against the number of companies in each group needed to obtain a credible estimate. Pooling of similar subsectors might be necessary to obtain a sufficient number of companies in the estimation of each regression model.

Companies suspected of underestimating their emissions, especially Scope 3 emissions, in the sense of having an unrealistically low carbon intensity compared with their peers, were omitted when fitting the model. Estimating emissions of these companies using the fitted model gives emissions data more in line with expectations. To ensure a more conservative approach, companies not reporting or obviously understating their reporting might be penalized in the estimation, applying United Nation's (1992) precautionary principle: "If in doubt, err on the side of the planet not the side of the company".

This approach of estimating missing GHG emissions data can be generalized using more sophisticated statistical models.

The second method attempts to estimate the emissions intensity of a specific type of operation or activity, for example, coal power generation or cement manufacturing. This method requires data on companies' distinct types of operations and the proportion of the total Sales derived from each type of operation. The next step of the method is to fit a model assuming that the *i*th company's emissions are the sum of the emissions from the company's distinct type of operations, where the size of a specific operation is approximated by $Sales_{ij}$, i.e. the Sales attributed to operation j. In simplified terms, it is expressed as:

$$log(emissions_i) = \sum_j \mu_j + \beta_j log(sales_{ij}) + \varepsilon_i \qquad [11.6]$$

where μ_j and β_j is the intercept and slope, respectively, for operation j. $Sales_{ij}$ is company i's Sales in operation j.

This method works reasonably well in groups of companies where the number of operations is relatively small compared to the number of companies, such as utility companies. For other sectors, as well as the full universe of companies, where the number of different types of operations is large, the estimated model parameters are not stable enough to be used in practice. This might be solved by pooling similar types of operations into new groups, and thereby reducing the number of model parameters. Further research is being conducted to advance this method (see Hoepner and Rogelj (2021), for general guidance on precautionary principle-based estimation).

When estimating missing emissions data for corporate bonds, some companies do not have listed shares, as they are privately owned. Therefore, fundamental data like Sales is less readily available, complicating the use of the first method. One solution to this problem is to start by estimating Sales using the companies' total amount of outstanding bonds, which is data that could be found for all companies in AP2's corporate bond universe. Once missing Sales figures have been estimated, the first method can be used to estimate missing emissions data.

11.4.4. *Specific aspects*

11.4.4.1. *Global equities*

An implementation of the PAB framework is probably most straightforward in a European mid- and large-cap universe of listed equities. There are several reasons for this:

1) In general, European companies have relatively well-developed sustainability reporting and are urged to improve further, not least due to initiatives like the EU Corporate Sustainability Disclosure Directive.

2) A sufficient number of constituents provide degrees of freedom in allocating benchmark weights and reduce the risk of having a few "high emitters" driving the intensity of the benchmark.

3) All major industrial sectors or business areas are well represented, which allows flexibility to meet the exposure constraints.

Still, AP2 decided to implement PAB on a global scale. The PAB framework is in use for both a developed market mid- and large-cap universe as well as for an emerging market counterpart. The construction and implementation of these two indices were both relatively straightforward. However, in emerging markets, the core data quality is generally of poorer quality, such as data on emissions (especially Scope 3 emissions) and data on revenues from fossil fuels. As a result, emerging market benchmarks are more dependent on estimations. Therefore, constituent weights might change when reporting on emissions data becomes more standard in these markets.

11.4.4.2. *Swedish equities*

In early 2021, a project was launched to align AP2's Swedish equity indices to the PAB framework. Up to recently, AP2's allocation to Swedish equities is divided into two non-PAB indices, one internally developed smart beta index consisting of the 60 largest companies and one market-weighted small-cap index. All capital allocated to Swedish equities is managed internally. The Swedish equity team runs an active investment process with a relatively high tracking error against the benchmarks. As part of the investment process, the team has an ongoing dialogue with portfolio companies about financial developments and also regarding business ethics and sustainability.

Adapting the Swedish benchmarks revealed some specific issues that merit discussion. In a global market capitalization-weighted benchmark, exclusions of companies with revenues from fossil fuels are sufficient to achieve the initial reduction of 50% of GHG intensity, which the PAB framework requires. In a small national benchmark, depending on the sector composition, this is not guaranteed.

To adjust the GHG intensities in the Swedish equity benchmarks, AP2 has chosen to change the individual constituent weights proportionally to the company's GHG emissions to Sales intensity until the GHG emissions to EVIC intensity criteria is met. This method was chosen to handle the concern that fluctuations in stock prices would result in large changes in individual benchmark weights. Also, using Sales intensity facilitates a more constructive dialogue with companies. The method is implemented in an optimization where the PAB constraint on climate intensive sectors is maintained.

The issue of missing emissions data is also more pronounced in the Swedish universe due to the larger proportion of small-cap companies. To achieve a better precision of the GHG emissions estimates, the Swedish universe is complemented with the global universe in the modeling process.

For many companies, a shift towards more sustainable technologies and practices requires external financing, where pension funds like AP2 have a role to play. Such companies, with a reliable transition plan, are problematic to handle within the framework of PAB since they are important for the transition but not well handled by the regulation. AP2's understanding of the regulation is that a high-emitting company having a reliable transition plan is not treated differently than a high-emitting company without such a plan.

11.4.4.3. *Corporate bonds*

Traditionally, measuring GHG emissions has been more common for equity than bond portfolios. However, companies might just as well finance their activities by issuing bonds. If this source of financing is not included in the PAB framework, issuing bonds could be a way for companies with non-sustainable activities to continue obtaining financing. Since bonds have a finite life and must be refinanced when they mature, bondholders can have an impact on the financing cost for non-sustainable companies.

When constructing an optimized multi-factor bond benchmark, there are several important differences compared with creating an equity multi-factor benchmark. A broad corporate bond benchmark has a significantly larger number of instruments than its equity counterpart, as several different instruments issued by the same company are usually included in the bond benchmark. The instruments have different characteristics such as duration, callability and seniority features, which make the index less homogeneous than an equity index. Bonds are also frequently issued by different entities in the corporate structure. In addition, corporate bonds and equities differ in their exposure to risk factors. The exposure to common risk factors like currency, interest rate level and spread typically explains a larger part of the variation in corporate bond returns compared to equity returns, where the company-specific component of returns is larger.

Companies without listed equity complicate the construction of an optimized corporate bond benchmark. One issue is the lower data coverage for these companies as compared to listed companies. Also, calculating the GHG intensity is not possible using EVIC since it is not readily available. The PAB regulation offers two options for handling this problem; either to use absolute GHG emissions for the whole universe or to use absolute GHG emissions only for the subset of non-listed companies.

Many PAB requirements are the same for equity and corporate bond benchmarks. One notable difference is the NACE sector constraint which is not applied to corporate bonds. In other words, a corporate bond PAB can be divested from specific sectors. The reason is that bondholders have no voting rights; thus, the possibility to influence companies towards a more sustainable business model by voting is limited. Also, bond investors are more active in the primary market than equity investors are, meaning that a sector constraint for a bond PAB could force investors to continue financing non-sustainable sectors.

For a benchmark to add value to an investment process, it must be replicable. Given the relative illiquidity, higher trading costs and the more manual trading process in the corporate bond market compared to the equity market, the replicability of the optimized benchmark needs to be verified. With no focus on the tradability of a corporate bond benchmark, the optimized benchmark may show desirable backtesting properties but may be expensive or even impossible to replicate and there is a risk that a portfolio tracking the optimized benchmark will not improve returns after trading costs. The evolution of electronic trading could potentially help mitigate replicability issues, at least to some extent.

Because of the difficulties mentioned, creating a backtesting environment for a bond universe to draw reliable conclusions is not straightforward. If the research framework is not designed to handle systematic exposures, like duration or currency exchange rates, results could be clouded by returns coming from factors that were not controlled for in the investigation. If potential backtesting biases are not handled correctly, the research results could be an effect of these biases and erroneous conclusions could be drawn. These issues have a higher risk of influencing the results for bonds compared to equities as the exposure to common risk factors is larger. For example, if the backtesting portfolio is long duration during a period of falling interest rates, it would potentially have a high return not because of the exposure to desired factors but because of the long duration position and decreasing interest rates.

AP2's backtesting, which controls for relevant systematic exposures, shows that implementing the PAB requirements will not decrease the expected return of the corporate bond portfolio.

11.4.5. *Discussion*

11.4.5.1. *GHG intensity definition*

Since the "The Report on Benchmarks" was published in September 2019, criticism has been raised and suggestions for improvements have been presented. One of the most frequent objections is the use of EVIC as the denominator when calculating the GHG intensity. Sales is often proposed as a more natural choice since it is more closely related to emissions and is less volatile, which implies a reduced turnover.

Firstly, AP2 agrees that Sales is a better approximation for the size/volume of a company's emitting activities than EVIC. That is the reason why Sales is used as an explanatory variable in the models that AP2 uses for predicting missing emissions data.

Secondly, EVIC, indeed, has larger volatility than Sales. However, it is important to compare the two quantities in a correct metric considering the given application. A PAB shall be rebalanced annually to meet the decarbonization target, which is a reduction of 7% of the benchmark's carbon intensity as compared to the previous year in real terms (i.e. after adjusting for EVIC inflation). The first thing to note is that it is not the cross-sectional volatility of EVIC or Sales that is of interest, but rather the cross-sectional volatility of the *change* over a year. Furthermore, since the decarbonization target is a *relative* reduction, it is the cross-sectional volatility of *relative* change (percentage change) that is the correct metric to explore, not the *absolute* change. Analysis over the last two decades on global equities (DM and EM) indicates that the volatility of relative change in EVIC is approximately one-third higher compared to that of Sales. Analyzing absolute changes (or absolute values) shows EVIC being almost three times more volatile as compared to Sales. It seems possible that some of the criticism of EVIC as a denominator stems from a misinterpretation of the underlying data where the volatilities of the two quantities are compared with an incorrect metric (i.e. cross-sectional volatility instead of cross-sectional volatility of the *change* over a year).

An attractive feature of EVIC is the natural way of attributing the intensities to stock and bond holders (ignoring the cash part). This can be useful to investors investing in both asset classes. By splitting a company's emissions proportionally between stock and debt holders (i.e. a company's emissions E_i is broken into two parts: $E_{i,equity} = E_i \frac{MV_i}{MV_i+D_i}$ and $E_{i,bond} = E_i \frac{D_i}{MV_i+D_i}$, where MV_i is the market value and D_i is the total debt of company i). Having w_i denoting the benchmark (or portfolio) weight of constituent i, the sum of the equity and the bond-based intensities adds up to the benchmark (portfolio) intensity:

$$\sum w_i \frac{E_i}{EVIC_i} = \sum w_i \frac{E_i}{MV_i + D_i} = \sum w_i \frac{E_{i,equity}}{MV_i} + \sum w_i \frac{E_{i,bond}}{D_i} \qquad [11.7]$$

Such attribution is not possible using Sales as the denominator.

A final comment on the "high turnover" criticism: this concern has its origin in a view that PAB is a yearly exercise in rebalancing weights to meet the 7% reduction of GHG-intensity. AP2's focus is on organic self-decarbonization, instead of rebalancing. The objective of PAB is to have companies reducing their emissions on a consistent basis, which if done correctly, means that the weightings do not actually need to change very much. AP2 strongly believes in this development, and that, inter alia, policy changes and investor engagement will drive companies in this direction.

11.4.5.2. *EVIC-inflation adjustment*

The argument for applying the requirement of a 7% reduction per annum to the EVIC-inflation adjusted ratio is valid and non-controversial. Applying the 7% reduction requirement simply to the nominal benchmark intensity ratio, i.e. to

$$Nominal\ benchmark\ intensity\ ratio = \frac{Benchmark\ Level\ GHG\ Intensity(t_m)}{Benchmark\ Level\ GHG\ Intensity(t_{m-1})} \quad [11.8]$$

can lead to decisions that are not compatible with the Paris Agreement, as asset price inflation can result in a decrease in the intensity ratio even if emissions are constant or growing.

In the definition of the index intensity (see equation [11.1]), EVIC appears in a sum as a reciprocal (i.e. $\frac{1}{EVIC}$). This seemingly innocent observation has consequences on the sufficiency of the inflation adjustment factor (see equation [11.2]). It is not obvious why, however, the observation has support in simulations and partly in mathematical theory (Jensen's inequality).

A deep dive into this subject is beyond the scope of this chapter, nevertheless an important point for further research. To produce meaningful insights, it is essential to truly understand why an adjustment to the nominal intensity ratio is needed. There are at least two, related yet not identical, motivating factors for the need of inflation adjustment:

1) a change in index GHG intensity must be driven by a change in GHG emissions;

2) a change in index GHG intensity must not be driven by a change in EVIC.

There is a subtle difference between these two statements. The first claims that any change in the nominal benchmark intensity ratio due to changes in price, debt, and/or cash shall be compensated for. It is important to note that changes in price will affect the benchmark weights. The second claims that any change in the nominal benchmark intensity ratio due to changes in EVIC must be compensated for, ignoring the benchmark weights.

There are at least two different approaches that can be used studying this problem:

1) using a population-based adjustment factor. That is, the same adjustment is applied to all constituents. The Delegated Act proposed inflation adjustment is an example of a population-based adjustment;

2) using individual adjustments. That is, each constituent is adjusted individually.

Thorough research is needed to better understand the consequences of each choice, not least in relation to the target of reduction of absolute emissions.

11.5. Conclusion

Absolutely sustainable investing means reducing a portfolio's GHG emission in line with the 1.5-degree trajectory of the Intergovernmental Panel on Climate Change (IPCC). This instructs a portfolio to target Net Zero GHG emissions in a disciplined process, which requires the portfolio to be not only relatively better than its benchmark in a given year in time but also to be better than itself last year through at least 7% annual reductions in absolute GHG emissions. EU Paris-aligned Benchmarks and EU Climate Transition Benchmarks represent such absolutely sustainable investing (Hoepner et al. 2019a, 2019b).

For AP2, the alignment of benchmarks with the PAB framework is an important part of creating a sustainable investment portfolio. The climate benchmarks comprehensively integrate the principles of the Paris Agreement providing a framework for absolutely sustainable investing.

Close to 50% of AP2's assets under management have been aligned with the PAB framework. The fund has developed PABs for developed and emerging market equities, corporate bonds and Swedish equities (to be implemented in 2022). As a responsible investor, AP2 finds the science-based approach, using IPCC research and recommendations to formulate the framework for EU climate benchmarks, to be the most credible alternative when decarbonizing benchmarks and investment portfolios. Furthermore, AP2 attaches high value to the PAB regulations being public and transparent.

The construction of a PAB is strongly dependent on company-specific data, and thereby on the companies' willingness to report on this data. In a global context, there are significant differences in the quality of sustainability data, seen across countries and regions, and also across industries. Such differences might potentially cause systematic biases when constructing a PAB in a global universe. It is therefore of importance that sustainability data is of at least the same quality as fundamental data (such as income statement data and balance sheet data). AP2 sees no reasons to why sustainability data should not be as timely, accurate and standardized as fundamental data. Nevertheless, it is not an option to await a general improvement in the quality of sustainability data, given the urgency of the climate crisis.

Apart from company-specific data, we also encourage further research into the quality of corporate emissions reduction targets (e.g. Net Zero by 2050 or -45% absolute emissions by 2030 on a 2010 base year as suggested by COP26). On a particular technical challenge (outside the scope of this chapter), while the argument for EVIC inflation adjustment procedure is valid and non-controversial, its technical complexity may result in scenarios where GHG emissions and average EVIC-inflation adjustment would be constant but subtle shifts occur in the underlying distribution of security EVICs, resulting in inflation factor under- or over-adjusting. However, research on this topic is ongoing.

The PAB framework signals that companies must focus on reducing emissions to attract new investments. In the end, the transition to a Net Zero world means that companies and industries must change, reducing their climate impact. AP2 believes that the framework of EU climate benchmarks is an important concept in guiding this transformation.

11.6. References

Bjørn, A., Diamond, M., Owsianiak, M., Verzat, B., Hauschild, M.Z. (2015). Strengthening the link between life cycle assessment and indicators for absolute sustainability to support development within planetary boundaries. *Environmental Science & Technology*, 49, 6370–6371. DOI: 10.1021/acs.est.5b02106.

Bjørn, A., Chandrakumar, C., Boulay, A.-M., Doka, G., Fang, K., Gondran, N., Hauschild, M.Z., Kerkhof, A., King, H., Margni, M. et al. (2020). Review of life-cycle based methods for absolute environmental sustainability assessment and their applications. *Environmental Research Letters*, 15, 083001.

Commission Delegated Regulation (EU) (2020/1818 of 17 July 2020). Supplementing Regulation (EU) 2016/1011 of the European Parliament and of the Council as regards minimum standards for EU Climate Transition Benchmarks and EU Paris-aligned Benchmarks, OJ L.

Communication from the Commission to The European Parliament, The European Council, The Council, The European Central Bank, The European Economic and Social Committee and the Committee of the Regions (2018). Action plan: Financing sustainable growth. Paper, European Commission.

Hoepner, A.G.F. and Rogelj, J. (2021). Emissions estimations should embed a precautionary principle. *Nature Climate Change*, 11, 638–640.

Hoepner, A.G.F., Masoni, P., Kramer, B., Slevin, D., Hoerter, S., Humphreys, N., Viñes Fiestas, H., Lovisolo, S., Wilmotte, J.-Y., Latini, P. et al. (2019a). TEG final report on climate benchmarks and benchmarks' ESG disclosure. Report, European Commission, Brussels.

Hoepner, A.G.F., Masoni, P., Kramer, B., Slevin, D., Hoerter, S., Humphreys, N., Viñes Fiestas, H., Lovisolo, S., Wilmotte, J.-Y., Latini, P. et al. (2019b). Handbook of climate transition benchmarks, Paris-aligned benchmark and benchmarks' ESG disclosures. Handbook, European Commission, Brussels.

Regulation (EU) (2019/2089 of the European Parliament and of the Council of 27 November 2019). Amending Regulation (EU) 2016/1011 as regards EU Climate Transition Benchmarks, EU Paris-aligned Benchmarks and sustainability-related disclosures for benchmarks, OJ L.

UNFCCC (2015). Adoption of the paris agreement. proposal by the president. Draft Decision/CP.21.

World Business Council for Sustainable Development, World Resources Institute (2001). The Greenhouse Gas Protocol: A Corporate Accounting and Reporting Standard, 1st edition. World Business Council for Sustainable Development.

World Business Council for Sustainable Development, World Resources Institute (2011). Greenhouse gas protocol corporate value chain (Scope 3) accounting and reporting standard.

WRI, WBCSD. Greenhouse Gas Protocol [Online]. Available at: https://ghgprotocol.org/standards.

12

Delegated Philanthropy in Mutual Fund Votes on Climate Change Externalities

This chapter studies the votes of institutional investors on shareholder resolutions instructing corporations to mitigate climate change externalities. Our sample includes 238 US fund families that voted on 14,409 different shareholder resolutions at 2,700 companies over the period from 2013 to 2016. We find that, in line with the delegated philanthropy logic, fund families that have larger proportions of responsible investments display a larger support for resolutions on climate change. This result holds (i) especially for fund families with a large percentage of SRI, (ii) whether ISS favors or not these resolutions, (iii) when these resolutions end up being close call votes and (iv) when we focus only on fund families that have voted more than 50 or 100 resolutions on environmental and social issues.

12.1. Introduction

One of the most important negative externalities produced by corporations is greenhouse gas emissions that constitute one of the main sources of global warming (Intergovernmental Panel on Climate Change 2022): the cost of pollution of such emissions is not fully reflected into market prices and taxes and is thus borne in part

---

Chapter written by Marie BRIÈRE, Sébastien POUGET, Martin SCHMALZ and Loredana URECHE-RANGAU.

Sébastien Pouget gratefully acknowledges the TSE Sustainable Finance Center's financial support in the context of the foundation TSE-Partnership, an academic organization under the aegis of the Toulouse School of Economics, financially supported by various partners, a list of whom is available at: https://www.tse-fr.eu/tsepartnership?lang=en. The list of partners of the TSE Sustainable Finance Center is available at: www. https://www.tse-fr.eu/sustainablefinance?tabs=1. Martin Schmalz acknowledges support from the Deutsche Forschungsgemeinschaft under Germany's Excellence Strategy – EXC 2126/1-390838866.

by society at large. According to a report by Trucost, a leading extra-financial analysis firm, environmental externalities alone represented, in 2008, 7% of revenues for the major 3,000 companies over the world (see Mattison et al. 2011). According to the Stern (2006) review, climate change is expected to trigger a drop in global gross domestic product of around 5% per year, a drop that could even be as large as 20% in some catastrophic scenarios. This may have a dramatic impact on firms' operations and on societies around the world (see, for example, Cruz and Rossi-Hansberg (2021)). It also indicates that climate change constitutes a systematic risk that, if material, could severely impair the value of portfolios.

In this chapter, we study the drivers of institutional investors' votes at general assembly meetings on shareholder resolutions to mitigate corporate impact on climate change. One motivation that could induce institutional investors to take externalities into account when voting at general assembly meetings is related to delegated philanthropy (Benabou and Tirole 2010). When shareholders are also citizens, consumers, workers and taxpayers, absent perfectly competitive and complete markets, they care about corporate policies' impact on their welfare, over and above the cash they receive from the firm (see Grossman and Stiglitz 1977; Hart 1979; Gordon 2003; Morgan and Tumlinson 2019). They may thus want firms to internalize externalities such as their impact on climate change. In this view, firms internalize externalities, even though this is not financially profitable, because this is globally best for their shareholders, given their preferences. To study this issue, we use the proportion of assets under management invested in equity that are part of a Socially Responsible Investment (SRI) fund as a proxy for the preferences of institutional investors' clients towards mitigating negative externalities. We then test whether the proportion of assets under management in SRI is positively related to investors' propensity to vote in favor of negative externality mitigation at general assembly meetings, and in particular, climate change mitigation.

We address this question by studying mutual fund votes at general assembly meetings on environmental and social (E&S) issues.[1] We ask whether fund families' votes on climate change, and more generally on E&S issues, depend on whether these families include more or less responsible investments in their portfolio. We study 238 fund families' votes during the 2013–2016 period on 14,409 different shareholder resolutions at 2,700 companies.

Our main variable of interest is the frequency at which a fund family votes in favor of shareholder resolutions requesting to mitigate climate change and other negative externalities (or to produce more positive ones). We control for several

1 Our focus on corporate engagement at general assembly meetings complements the empirical studies of behind-the-scene engagements on E&S issues offered by Dimson et al. (2015), Barko et al. (2018), and Azar et al. (2019).

variables that may affect a fund family's voting policy, including its age, size, type of investment management, proportion of equity funds and of retail funds.

We find that the proportion of responsible funds in a family's assets under management increases its level of support for shareholder resolutions on climate change. This result also holds when focusing on other topics, clearly related to externalities, such as human rights, discrimination issues and compensation restrictions. It also holds when focusing on more broadly defined environmental and social issues. Moreover, robustness tests show that our results are valid when we split our sample period into two subperiods – 2013–2014 and 2015–2016. We thus complement the literature that studies mutual fund votes on shareholder resolutions at general assembly meetings, including Davis and Kim (2007), Matvos and Ostrovsky (2008), Morgan et al. (2011), He et al. (2021), by focusing on the proportion of responsible investments and on externality-related resolutions.

Our results suggest that the frequency of a fund family votes in favor of shareholder resolutions requesting firms to mitigate climate change increases with the proportion of SRI funds' assets under management within the family. This result is much less pronounced when we focus on shareholder resolutions on other governance and financial issues but holds for other externalities related to human rights and discrimination issues.

We submit our results that fund families with larger proportions of responsible investments support more climate change and other externality-related E&S resolutions to various robustness tests. Our results hold if we restrict our sample to include only the resolutions supported by the proxy advisor ISS. This is remarkable because ISS has been shown to support these resolutions often (Bolton et al. 2020). Our results also hold if we consider only the resolutions that end up being close-call votes. Thus, fund families with large proportions of responsible investments do not fear that the resolutions may end up gathering majority support from shareholders.

Our results have several practical implications. First, our findings might be useful for investors who would like to benchmark or evaluate their voting behavior on climate change and other ESG issues against the universe of mutual fund families. A second implication is for regulators: fund families should be incentivized to know the preferences of their clients regarding climate change and other corporate externalities and to refine their voting policy to make it more in line with these preferences. As indicated by Hart and Zingales (2017), shareholder votes are crucial in this respect.

12.2. Sample, data sources, variables and descriptive statistics

Our analysis is at the fund family level: each mutual fund in our sample is associated with a fund family based on information collected on the Internet and on Bloomberg. We merge data on mutual fund votes, characteristics and holdings, and aggregate these data at the fund family level.

12.2.1. *Mutual fund votes*

We use the ISS database that provides the votes of US mutual funds at general assembly meetings as reported in the SEC N-PX filings. We focus on votes on shareholder resolutions. We aggregate these votes at the fund family level to reflect the fact that a lot of funds use a centralized voting policy, in particular among the largest ones (see Fichtner et al. 2017). There are 329 fund families in our initial sample. After matching with the other databases discussed below, i.e. CRSP and Thomson Reuters, we obtain a dataset including 238 fund families that voted on 14,409 different shareholder resolutions at 2,700 companies.

For each fund family and each calendar year in our sample, i.e. from 2013 to 2016, we compute the proportion of votes in favor of the proposal for climate change mitigation as well as for different broad categories of shareholder resolutions: environmental and social resolutions, together and separately[2]. As a benchmark, we also measure the support for other governance and financial issues.

For comparison purposes, we also compute fund families' support for human rights and discrimination, which are clearly related to corporate externalities. We also include, as a specific topic, compensation restrictions linked to externalities that include resolutions that call for limiting the compensation of executives or for linking this compensation to social targets. Indeed, as shown, for example, by Benabou and Tirole (2016), an excessive reliance on financial metrics to judge the performance of executives might lead to excessive bonuses and to some important aspects of corporate performance being overlooked.

Our sample includes 290 environmental resolutions, 323 social resolutions and 10,963 other governance and financial resolutions. Our classification is indicated in the Appendix to this chapter (section 12.5). Regarding specific topics, we have the following numbers of different resolutions: 83 for climate change, 55 for human rights, 68 for discrimination and 153 for compensation externalities. Various fund families have voted for these resolutions so that our data include a cross-section of

2 We also have data on shareholder resolutions on executive compensation but we do not include this issue in our investigation.

voting behavior that provides us with more statistical power than this relatively low number of resolutions suggests.

Our analysis is based on the premise that environmental (and social) resolutions are bound to reduce negative corporate externalities and promote positive ones. Indeed, most of these resolutions either request corporations to provide information on these issues (improving corporate stakeholders' ability to exercise pressure on corporations on these issues) or request actions to be taken. For example, climate change resolutions include proposals on topics such as "Report on Climate Change", "GHG emissions", "Climate change action" or "Publish two-degree scenario analysis". Examples of proposals include "Report on Methane Emissions Management and Reduction Targets", "Approve Strategic Climate Change Resilience for 2035 and Beyond" and "Adopt Quantitative GHG Goals for Products and Operations". Most of these resolutions are requesting companies to report on their emissions. This is in line with the survey evidence reported by Ilhan, Krueger, Sautner and Starks (2020) showing that institutional investors find climate risk reporting to be as important as financial reporting.

12.2.2. *Mutual fund characteristics*

We use all the share classes of US mutual funds in the CRSP database to compute the following variables at the level of the fund family and at the end of each calendar year. The proportion of socially responsible investment (SRI) funds (*% SRI funds*) is computed across the family's equity funds. A fund is classified as SRI if its name includes the following words or radicals: "SRI", "social", "ethic", "respons", "ESG", "sustain", "impact", "green" and "environ". We will use this proportion as a proxy for the preference of a fund family's clients towards corporate social responsibility.

The total net assets under management are the sum of total net assets of all funds in the family, in millions. The variable *Total assets (ln)* in the tables below represents the natural logarithm of the total net assets. The proportion of equity funds (a variable labeled *% equity funds* in the tables) is computed by dividing the net asset value of the equity funds by the total net asset value of the family.

The proportions of retail and index funds (*% retail funds* and *% index funds*, respectively) are also computed across the family's equity funds. Index funds correspond to categories B, D and E in the CRSP database. The age of a fund family is the number of years since the first fund appearing in CRSP was created. The average expenses ratio of a fund family is computed across funds weighted by their net asset value. The expense ratio of a fund is based on the total amount paid by its clients to cover the fund's operating expenses.

For a given year, the idiosyncratic risk of a fund family is measured as the annualized volatility of the residuals of a regression of the total daily returns of the fund family on the daily excess returns of the US market (all stocks listed on NYSE, Amex and Nasdaq) as given by Kenneth French on his website (we request to have more than 50 observations for a given fund family; fund families with fewer than 50 daily returns for a given year are thus not part of our sample). The total daily return of the fund family is equal to the average daily return of the various funds within the family (for simplicity, we use simple averages, but our results hold if we use weighted averages).

12.2.3. *Mutual fund holdings*

Mutual funds' quarterly holdings are obtained through Thomson Reuters database. This database provides security holding information for all registered mutual funds that report their holdings with the SEC, plus 3,000 global funds. We aggregate quarterly holdings at the level of the fund family by doing the sum of the holdings of all the funds belonging to this family.

For each family, as a measure of investment horizon, we compute the investor turnover of Derrien et al. (2013). Investor turnover is computed as follows. For each fund family j, each quarter t and each stock i, we compute the proportion of stock i held by fund family j at t-12 (i.e. three years before) and sold between t-12 and t. If the weight of a stock has increased over the three-year period, the turnover is set to zero for this stock. The turnover of the fund family at quarter t is computed as the sum of the turnover of the stocks held in portfolio, weighted by the proportion of stock i in fund family j's portfolio at quarter t-12. A fund family's investor turnover for a given year is the average of its turnover at each quarter of the year. Investor turnover is between 0 and 100%. This measure has been used by Derrien et al. (2013) and also, for example, by Harford et al. (2018).

12.2.4. *Descriptive statistics*

Table 12.1 reports summary statistics for our sample that includes 238 fund families over four years, 2013–2016.

	Mean	Std. Dev.	Min	Max	Obs
Panel A: Support to shareholder resolutions					
Climate	**39%**	36%	0%	100%	989
Human rights	**21%**	31%	0%	100%	870
Discrimination	**24%**	30%	0%	100%	1,071
Compensation restrictions	**40%**	36%	0%	100%	1,141
ES	**32%**	30%	0%	100%	1,308
E	**31%**	32%	0%	100%	1,170
S	**34%**	33%	0%	100%	1,268
G executive compensation	**37%**	34%	0%	100%	1,210
Other G and financial	**44%**	27%	0%	100%	1,346
Panel B: Asset manager characteristics					
% SRI funds	**1.18%**	8.81%	0.00%	100.00%	1,373
Turnover	**55%**	22%	5%	100%	1,373
Number of stocks	**756**	1,020	1	4,820	1,373
Total assets	**94,352**	437,491	3	7,211,405	1,373
% equity funds	**78%**	28%	0%	100%	1,373
% retail funds	**52%**	37%	0%	100%	1,373
% index funds	**11%**	27%	0%	100%	1,373
Age of asset manager	**31.4**	22.6	0.1	96.5	1,373
Average expense ratio	**0.95%**	0.51%	0.00%	4.08%	1,373
Idiosyncratic volatility	**4.58%**	3.23%	0.63%	42.75%	1,373

Table 12.1. *Summary statistics for all fund families*

Panel A of Table 12.1 shows that the average support for shareholder resolutions by fund families in our sample ranges from 31% to 44% depending on the type of resolutions, environmental, social, and other governance and financial issues. Support for the specific topics is more dispersed, from 21% for human rights to 40% for compensation restrictions.

Panel B of Table 12.1 describes the characteristics of fund families in our sample. The average fund family in our sample holds 1.18% of assets under management in socially responsible investment (SRI) funds, 78% in equity funds, 52% in retail funds and 11% in index funds. The level of idiosyncratic risk is relatively low and equal to 4.58%. The average number of stocks in portfolio and assets under management are 756 and USD 94.1 billion, respectively. In our regressions, we use the natural logarithm of these variables to mitigate heteroscedasticity issues. The turnover appears relatively high, at 55% on average.

12.3. Empirical analysis

Our empirical analysis is based on regressions of the proportion of fund family support for shareholder resolutions onto the type of fund family and fund family characteristics. Our focus is on how the percentage of SRI assets under management in a fund family affects its votes on climate resolutions and more generally environmental and social resolutions. We also study resolutions on human rights, discrimination and executive compensation restrictions because we are interested in understanding whether the percentage of SRI funds induces fund families to fight against negative corporate externalities, in line with the delegated philanthropy logic. We include in our regression year fixed effects, and report robust standard errors for our estimated coefficients.

12.3.1. *Impact of the percentage of SRI on the support for climate resolutions*

Table 12.2 displays the results of our main regressions of support on shareholder resolutions onto various fund family variables. Column (1) focuses on our entire time period, 2013–2016, column (2) on the early time period, 2013–2014, and column (3) on the later time period, 2015–2016. The dependent variable is the support to climate resolutions.

	Dependent variable: support for climate resolutions		
	(1)	(2)	(3)
	2013-2016	2013-2014	2015-2016
% SRI funds	0.688***	0.763***	0.652***
Turnover	0.190*	0.171	0.190*
Number of stocks (ln)	0.039**	0.050	0.035
Total assets (ln)	-0.011	-0.008	-0.010
% equity funds	0.002	-0.000	0.009
% retail funds	-0.092	-0.076	-0.093
% index funds	0.070	0.197	0.032
Age of oldest fund	-0.002*	-0.002	-0.002
Average expense ratio	-5.497	15.550	-11.940**
Idiosyncratic volatility	1.163	0.023	1.349*
Constant	0.223	-0.036	0.287
Observations	445	131	314
R-squared	8%	10%	9%

Table 12.2. *Support for climate resolutions. Note: p-values computed with robust standard errors. *** $p<0.01$, ** $p<0.05$, * $p<0.10$*

Our main finding is that the percentage of SRI within the fund family is positively associated with the support to climate resolutions. This association is very strong from a statistical point of view. It holds both for the earlier and the later period in our sample. It is also economically significant: a one standard deviation increase in the percentage of SRI (9%) induces an increase in the support of 6.2% (9% x 0.688).

12.3.2. *Resolutions on other corporate externalities*

Table 12.3 shows that the percentage of SRI also affects the support to shareholder resolutions on other corporate externalities. Indeed, this variable is associated with a coefficient that is statistically significant for all the other externalities we study, namely, human rights, discrimination and compensation restrictions, at 0.465, 0.350 and 0.494, respectively. However, the magnitude of the effect is smaller than for climate resolutions. A statistical test of difference between coefficients shows that these differences are statistically significant at $p<0.05$ and $p<0.01$ for human rights and discriminations, respectively.

Together, these results indicate that fund families with a higher percentage of SRI support more shareholder resolutions bound to mitigate corporate negative externalities, and particularly climate change externalities.

This result is also apparent in Table 12.4 that shows that support for broader environmental and social resolutions is also stronger the higher the percentage of SRI is. Table 12.4 also shows that the percentage of SRI is also positively associated with more support for other governance and financial resolutions. However, this positive association appears weaker than for climate change resolutions; a statistical test shows that the difference in coefficients between the climate change resolutions (0.688 in Table 12.2) and the other governance and financial resolutions (0.214 in Table 12.4) is significant with $p<0.01$. This result suggests that the stronger support displayed for climate change resolutions is not simply due to the general tendency of SRI to induce more support for shareholder resolutions.

	Dependent variable: support for shareholder resolutions		
	(1)	(2)	(3)
	Human rights	**Discrimination**	**Compensation restrictions**
% SRI funds	0.465***	0.350***	0.494***
Turnover	0.226**	0.125**	0.051
Number of stocks (ln)	0.035**	0.026**	0.029*
Total assets (ln)	-0.022**	-0.018***	-0.009
% equity funds	-0.089	-0.012	-0.046
% retail funds	-0.025	-0.020	-0.093*
% index funds	0.054	0.087**	0.043
Age of oldest fund	-0.000	0.000	-0.001*
Average expense ratio	-4.196	-0.404	9.785**
Idiosyncratic volatility	-0.486	0.164	0.126
Constant	0.199	0.092	0.319**
Observations	353	494	582
R-squared	11%	18%	9%

Table 12.3 displays the results of regressions of support on shareholder resolutions onto various fund family variables. Column (1) focuses on human rights, column (2) on discrimination and column (3) on compensation restrictions. These three columns deal with other issues than climate change that are clearly related to externalities.

Table 12.3. *Support for other identified externalities. Note: p-values computed with robust standard errors.* *** $p<0.01$, ** $p<0.05$, * $p<0.10$

	Dependent variable: support for shareholder resolutions			
	(1)	(2)	(3)	(4)
	Environmental	**Social**	**Environmental and social**	**Other gov and financial**
% SRI funds	0.611***	0.574***	0.619***	0.214***
Turnover	0.221***	0.201***	0.206***	0.212***
Number of stocks (ln)	0.039***	0.045***	0.041***	0.041***
Total assets (ln)	-0.013	-0.030***	-0.022***	-0.009
% equity funds	-0.003	-0.083	-0.056	-0.063
% retail funds	-0.063	-0.121***	-0.083**	-0.065**
% index funds	0.069	0.097*	0.102**	0.055
Age of oldest fund	-0.001	-0.000	-0.000	-0.000
Average expense ratio	-4.333	-1.682	-2.813	-2.661
Idiosyncratic volatility	0.740	0.747	0.816	0.498
Constant	0.123	0.320***	0.210**	0.245**
Observations	574	647	673	702
R-squared	10%	12%	10%	8%

Table 12.4 displays the results of regressions of support on shareholder resolutions onto various fund family variables. Column (1) focuses on environmental resolutions (E), column (2) on social (S) and column (3) on environmental and social (ES) resolutions. These three columns deal with issues that are related to externalities. Column (4) offers, as a benchmark, the support for shareholder resolutions on other governance and financial issues.

Table 12.4. *Support for broader categories of resolutions. Note: p-values computed with robust standard errors. *** p<0.01, ** p<0.05, * p<0.10*

12.3.3. *Drivers of support for climate change resolutions*

To better understand what drives the higher support for climate change induced by the percentage of SRI in the fund families, we create various dummy variables. A first dummy takes the value 1 if the fund family has SRI funds and 0 otherwise. We then decompose this dummy into two indicator variables: LowSRI is a dummy that takes the value 1 if the fund family holds a percentage of SRI that is below the median percentage of SRI detentions by all fund families (which is equal to 0.50%); HighSRI is a dummy that takes the value 1 if the fund families hold a percentage of SRI that is below the median percentage of SRI detentions by all fund families.

Panel A in Table 12.5 shows that the simple fact of having some SRI within the fund family is not sufficient per se to display a larger support for climate resolutions. It is slightly associated with larger support for environmental resolutions and other governance and financial resolutions but not for climate change and social resolutions.

Likewise, Panel B in Table 12.5 indicates that fund families with a relatively low percentage of SRI do not support climate change resolutions more than otherwise identical fund families that do not have SRI. On the contrary, it shows that fund families which have a percentage of SRI larger than the median have a larger level of support: their support is larger by 20.7 percentage points. This result suggests that holding some SRI is not enough to influence the voting policy of a fund family. Only when the fund family holds a relatively large amount of SRI does it support more climate change resolutions (and more generally, environmental and social resolutions). This result suggests that SRI investors in fund families that have relatively little SRI funds might not see their preferences reflected in the voting policy of their fund family.

We now study what influences the link between the percentage of SRI fund and the support for climate change resolutions. Table 12.6 shows that such a link is stronger for fund families with a larger proportion of index funds, lower average fees and a larger size (in terms of total assets under management). The stronger link with indexing strategies might be a sign that these funds that cannot vote with their feet tend to exert more voice to change corporate behavior. The impact of lower average fees might indicate that funds that care more about their customers' welfare also vote more in favor of climate resolutions. The impact of a larger size might be related to the fact that larger funds have more resources to develop an active voting policy.

	Dependent variable: support for shareholder resolutions				
	(1)	(2)	(3)	(4)	(5)
PANEL A	Climate	Environmental	Social	Environmental and social	Other gov and financial
Fund family has SRI fund	0.069	0.094**	0.037	0.068*	0.079***
Turnover	0.211**	0.240***	0.214***	0.221***	0.223***
Number of stocks (ln)	0.034*	0.034**	0.042***	0.037***	0.038***
Total assets (ln)	-0.013	-0.016**	-0.031***	-0.024***	-0.012*
% equity funds	-0.012	-0.019	-0.092*	-0.069	-0.073*
% retail funds	-0.076	-0.053	-0.112***	-0.074**	-0.063*
% index funds	0.051	0.042	0.088	0.083*	0.032
Age of oldest fund	-0.002*	-0.001	-0.000	-0.000	-0.000
Average expense ratio	-6.104	-4.620	-1.952	-3.075	-2.746
Idiosyncratic volatility	1.122	0.688	0.711	0.781	0.487
Constant	0.272	0.179	0.350***	0.254**	0.282***
Observations	445	574	647	673	702
R-squared	5%	8%	10%	7%	9%

Robust standard errors in parentheses

*** $p<0.01$, ** $p<0.05$, * $p<0.1$

	Dependent variable: support for shareholder resolutions				
	(1)	(2)	(3)	(4)	(5)
PANEL B	Climate	Environmental	Social	Environmental and social	Other gov and financial
LowSRI	-0.093	-0.048	-0.082*	-0.065	0.078**
HighSRI	0.207***	0.201***	0.131***	0.172***	0.0787***
Turnover	0.191**	0.225***	0.203***	0.210***	0.223***
Number of stocks (ln)	0.033*	0.033**	0.041***	0.036***	0.038***
Total assets (ln)	-0.010	-0.013*	-0.029***	-0.022***	-0.012*
% equity funds	-0.011	-0.022	-0.093*	-0.071	-0.073*
% retail funds	-0.087	-0.060	-0.115***	-0.078**	-0.063*
% index funds	0.054	0.045	0.091*	0.087*	0.032
Age of oldest fund	-0.002	-0.001	0.000	-0.000	-0.000
Average expense ratio	-5.449	-4.231	-1.691	-2.807	-2.746
Idiosyncratic volatility	1.296*	0.850	0.837	0.917	0.487
Constant	0.251	0.161	0.335***	0.239**	0.282***
Observations	445	574	647	673	702
R-squared	7%	10%	11%	9%	9%

Robust standard errors in parentheses
*** $p<0.01$, ** $p<0.05$, * $p<0.1$

Table 12.5 displays the results of regressions of support on shareholder resolutions onto various fund family variables. The percentage of SRI funds is replaced by a dummy variable that accounts for, in Panel A, the presence of at least one SRI fund, while, in Panel B, for a percentage of SRI funds lower (LowSRI) or higher (HighSRI) than the median within the fund family. Column (1) focuses on climate-related resolutions, column (2) on environmental (E), column (3) on social (S) and column (4) on environmental and social (ES) resolutions. These four columns deal with issues that are related to externalities. Column (5) offers, as a benchmark, the support for shareholder resolutions on other governance and financial issues.

Table 12.5. *Regressions with SRI dummies*

	Dependent variable: support for climate resolutions			
	(1)	(2)	(3)	(4)
% SRI funds*% retail funds	-4.038			
% SRI funds*% index funds		2.327**		
% SRI funds*expense ratio			-169.300*	
% SRI funds*total assets (ln)				0.132*
% SRI funds	3.639	0.547***	2.521**	-0.389
Turnover	0.191**	0.192**	0.192**	0.192**
Number of stocks (ln)	0.038**	0.039**	0.038**	0.038**
Total assets (ln)	-0.011	-0.011	-0.011	-0.011
% equity funds	-0.003	0.004	0.003	0.003
% retail funds	-0.085	-0.096	-0.094	-0.092
% index funds	0.072	0.059	0.065	0.070
Age of oldest fund	-0.002*	-0.002*	-0.002*	-0.002*
Average expense ratio	-5.375	-5.441	-5.298	-5.556
Idiosyncratic volatility	1.198	1.204	1.212	1.183
Constant	0.224	0.221	0.220	0.224
Observations	445	445	445	445
R-squared	8%	8%	8%	8%

Table 12.6 displays the results of regressions of support for climate resolutions onto various fund family variables. Column (1) introduces an interaction variable between the percentage of SRI funds and the percentage of retail funds within a fund family, column (2) between the percentage of SRI funds and the percentage of index funds within a fund family, column (3) between the percentage of SRI funds and the expense ratio and column (4) between the percentage of SRI funds and the total assets.

Table 12.6. *Regressions with percentage of SRI interacted. Note: p-values computed with robust standard errors.* *** $p<0.01$, ** $p<0.05$, * $p<0.10$

12.3.4. *Robustness*

This section discusses the results of various robustness analyses.

Table 12.7 focuses on shareholder resolutions for which the proxy advisor ISS recommends support. Table 12.8 focuses on resolutions to which ISS is opposed. Our results hold in these two cases. The association between the percentage of SRI in the fund family and support for climate resolutions appears even more pronounced when ISS is advising to vote against these resolutions. This indicates that fund families with SRI are less influenced by ISS than other fund families.

We then investigate voting behavior of institutional investors for shareholder resolutions which turn out to be close calls, i.e. for which vote results are around the majority threshold. The idea is to test whether support for climate resolutions is not turned down by the high likelihood for these resolutions to gather majority support.

Table 12.9 shows that fund families with a higher percentage of SRI display a larger support for climate resolutions than otherwise similar families, even for close votes. However, this additional support appears weaker than for other resolutions (compare the coefficients 0.487 in Table 12.9 and 0.688 in Table 12.2). This additional support on close votes on climate issues is not different from the additional support on other governance and financial resolutions (compare the coefficients 0.487 and 0.455 in Table 12.9).

Our analysis is based on behavior at the fund family level. Some families might have voted for only a few resolutions, and this might affect our results. We offer two robustness tests to control for such a potential effect. Panel A in Table 12.10 displays the result of a regression on a sample of fund families that have voted for more than 100 resolutions on environmental and social resolutions, and Panel B in Table 12.10 displays the results for more than 50 such resolutions. Table 12.11 displays the results of weighted least square regressions that weight each observation, corresponding to a given fund family, by the number of resolutions voted. The results in Tables 12.10 and 12.11 show that our findings are robust.

	Dependent variable: support for shareholder resolutions				
	(1)	(2)	(3)	(4)	(5)
	Climate	**Environmental**	**Social**	**Environmental and social**	**Other gov and financial**
% SRI funds	0.698***	0.689***	0.658***	0.664***	0.410***
Turnover	0.227**	0.182*	0.224**	0.251***	0.294***
Number of stocks (ln)	0.051**	0.057***	0.053***	0.060***	0.073***
Total assets (ln)	-0.011	-0.015	-0.034***	-0.026***	-0.007
% equity funds	-0.025	-0.025	-0.103	-0.079	-0.024
% retail funds	-0.108	-0.166***	-0.185***	-0.156***	-0.176***
% index funds	0.094	0.089	0.130*	0.106	0.024
Age of oldest fund	-0.002**	-0.002*	-0.001	-0.001	-0.001*
Average expense ratio	-5.389	2.529	2.670	0.882	0.481
Idiosyncratic volatility	1.169	0.659	0.465	1.013	0.618
Constant	0.238	0.220	0.440***	0.293*	0.236*
Observations	419	521	634	652	670
R-squared	10%	9%	10%	10%	12%

Table 12.7 displays the results of regressions of support on shareholder resolutions onto various fund family variables. The focus is on resolutions that were supported by the proxy advisor ISS. Column (1) focuses on climate-related resolutions, column (2) on environmental (E), column (3) on social (S) and column (4) on environmental and social (ES) resolutions. These four columns deal with issues that are related to externalities. Column (5) offers, as a benchmark, the support for shareholder resolutions on other governance and financial issues.

Table 12.7. *Robustness regressions on resolutions supported by ISS.*
*Note: p-values computed with robust standard errors. *** $p<0.01$, ** $p<0.05$, * $p<0.10$*

	Dependent variable: support for shareholder resolutions				
	(1)	(2)	(3)	(4)	(5)
	Climate	**Environmental**	**Social**	**Environmental and social**	**Other gov and financial**
% SRI funds	0.850***	0.572***	0.390***	0.583***	0.0573
Turnover	-0.031	0.073*	0.042	0.043	0.097***
Number of stocks (ln)	0.003	0.015*	0.009	0.011	0.007
Total assets (ln)	-0.011	-0.009*	-0.006	-0.009*	-0.003
% equity funds	-0.032	0.033	0.001	0.009	0.004
% retail funds	-0.024	0.026	0.046**	0.041**	0.065***
% index funds	-0.047	0.011	-0.011	0.012	-0.009
Age of oldest fund	0.001	0.001*	0.000	0.001	-0.001
Average expense ratio	-2.462	-4.206*	-2.246	-2.360	-3.774**
Idiosyncratic volatility	-0.048	-0.373	-0.053	-0.246	0.027
Constant	0.198	-0.009	0.022	0.031	0.041
Observations	279	484	554	625	644
R-squared	16%	12%	6%	11%	5%

Table 12.8 displays the results of regressions of support on shareholder resolutions onto various fund family variables. The focus is on resolutions that were NOT supported by the proxy advisor ISS. Column (1) focuses on climate-related resolutions, column (2) on environmental (E), column (3) on social (S) and column (4) on environmental and social (ES) resolutions. These four columns deal with issues that are related to externalities. Column (5) offers, as a benchmark, the support for shareholder resolutions on other governance and financial issues.

Table 12.8. *Robustness regressions on resolutions not supported by ISS.*
*Note: p-values computed with robust standard errors. *** $p<0.01$, ** $p<0.05$, * $p<0.10$*

	Dependent variable: support for shareholder resolutions				
	(1)	(2)	(3)	(4)	(5)
	Climate	Environmental	Social	Environmental and social	Other gov and financial
% SRI funds	0.487***	0.565***	0.576***	0.564***	0.455***
Turnover	0.202	0.267	0.195	0.253**	0.324***
Number of stocks (ln)	0.014	0.041	-0.001	0.007	0.069***
Total assets (ln)	-0.029	-0.037	-0.051***	-0.042***	-0.028***
% equity funds	-0.308**	-0.125	-0.179**	-0.185**	-0.104*
% retail funds	-0.172	-0.252**	-0.174**	-0.192***	-0.159***
% index funds	0.199	0.142	0.172**	0.159**	0.050
Age of oldest fund	-0.003	-0.003	0.000	-0.001	-0.000
Average expense ratio	-15.580*	-12.300*	-10.960	-12.840**	-4.541
Idiosyncratic volatility	1.370	-0.246	-0.360	0.189	0.325
Constant	1.135***	0.932**	1.181***	1.086***	0.486***
Observations	112	131	359	391	611
R-squared	18%	17%	13%	12%	10%

Table 12.9 displays the results of regressions of support on shareholder resolutions onto various fund family variables. The focus is on resolutions that ended up with a close-call vote (between -10% and +10% around the majority threshold). Column (1) focuses on climate-related resolutions, column (2) on environmental (E), column (3) on social (S) and column (4) on environmental and social (ES) resolutions. These four columns deal with issues that are related to externalities. Column (5) offers, as a benchmark, the support for shareholder resolutions on other governance and financial issues.

Table 12.9. *Robustness regressions on close-call votes. Note: p-values computed with robust standard errors. *** p<0.01, ** p<0.05, * p<0.10*

	Dependent variable: support for shareholder resolutions				
	(1)	(2)	(3)	(4)	(5)
PANEL A	Climate	Environmental	Social	Environmental and social	Other gov and financial
% SRI funds	0.796***	0.648***	0.762***	0.733***	0.226***
Turnover	-0.060	-0.029	-0.017	-0.024	0.208**
Number of stocks (ln)	0.012	0.014	0.019	0.017	0.043***
Total assets (ln)	-0.020	-0.027**	-0.040***	-0.035***	-0.021**
% equity funds	-0.302**	-0.261***	-0.123	-0.173**	0.013
% retail funds	-0.336***	-0.259***	-0.352***	-0.294***	-0.213***
% index funds	0.214**	0.237***	0.202***	0.207***	0.108**
Age of oldest fund	-0.001	-0.000	0.002	0.001	0.001
Average expense ratio	17.890**	15.350**	16.200**	14.730**	8.073
Idiosyncratic volatility	-0.400	-1.273	-0.643	-0.735	-1.345
Constant	0.692**	0.641**	0.557**	0.578***	0.272*
Observations	135	135	135	135	135
R-squared	28%	32%	32%	31%	23%

	Dependent variable: support for shareholder resolutions				
	(1)	(2)	(3)	(4)	(5)
PANEL B	**Climate**	**Environmental**	**Social**	**Environmental and social**	**Other gov and financial**
% SRI funds	0.728***	0.667***	0.643***	0.644***	0.164***
Turnover	0.238*	0.283**	0.241**	0.250**	0.232***
Number of stocks (ln)	0.055**	0.056***	0.048***	0.050***	0.040***
Total assets (ln)	-0.031**	-0.033***	-0.046***	-0.040***	-0.027***
% equity funds	-0.125	-0.084	0.023	-0.020	-0.047
% retail funds	-0.292***	-0.228***	-0.282***	-0.244***	-0.091*
% index funds	0.189**	0.180***	0.156***	0.161***	0.127***
Age of oldest fund	0.000	0.000	0.002*	0.001	0.001
Average expense ratio	4.120	1.831	2.965	1.940	4.034
Idiosyncratic volatility	1.094	0.299	-0.158	0.003	-0.898
Constant	0.304	0.224	0.308*	0.273	0.385***
Observations	210	210	210	210	210
R-squared	21%	24%	26%	25%	16%

Table 12.10 displays the results of regressions of support on shareholder resolutions onto various fund family variables. Panel A reports the results for fund families that voted for at least 100 resolutions while Panel B reports the results for fund families that voted for at least 50 resolutions. Column (1) focuses on climate-related resolutions, column (2) on environmental (E), column (3) on social (S) and column (4) on environmental and social (ES) resolutions. These four columns deal with issues that are related to externalities. Column (5) offers, as a benchmark, the support for shareholder resolutions on other governance and financial issues.

Table 12.10. *Robustness regressions with number of voted resolutions.*
*Note: p-values computed with robust standard errors. *** $p<0.01$, ** $p<0.05$, * $p<0.10$*

	Dependent variable: support for shareholder resolutions				
	(1)	(2)	(3)	(4)	(5)
	Climate	**Environmental**	**Social**	**Environmental and social**	**Other gov and financial**
% SRI funds	0.740***	0.634***	0.665***	0.663***	0.174***
Turnover	0.093	0.125	0.117	0.120	0.261***
Number of stocks (ln)	0.024	0.026	0.023	0.024	0.038***
Total assets (ln)	-0.030**	-0.032***	-0.046***	-0.039***	-0.021***
% equity funds	-0.192**	-0.148**	-0.091	-0.111*	-0.020
% retail funds	-0.247***	-0.178***	-0.249***	-0.210***	-0.106***
% index funds	0.162**	0.166***	0.146***	0.150***	0.094***
Age of oldest fund	-0.000	0.000	0.002*	0.001	0.000
Average expense ratio	6.316	5.277	6.117	5.104	1.573
Idiosyncratic volatility	0.290	-0.252	-0.138	-0.181	-0.956
Constant	0.612***	0.494***	0.570***	0.518***	0.346***
Observations	445	574	647	673	661
R-squared	19%	20%	25%	24%	15%

Table 12.11 displays the results of regressions of support on shareholder resolutions onto various fund family variables estimated with Weighted-Least-Squares (WLS). Column (1) focuses on climate-related resolutions, column (2) on environmental (E), column (3) on social (S) and column (4) on environmental and social (ES) resolutions. These four columns deal with issues that are related to externalities. Column (5) offers, as a benchmark, the support for shareholder resolutions on other governance and financial issues.

Table 12.11. *Robustness regressions with WLS. Note: p-values computed with robust standard errors. *** $p<0.01$, ** $p<0.05$, * $p<0.10$*

12.4. Conclusion

Corporations emit various types of externalities when they operate. Environmental externalities include greenhouse gas emissions that participate in climate change; social externalities include discriminative practices and breaches to human rights. Corporations' choice of governance may also impose negative externalities on other companies, for example, when setting high bonuses to attract and incentivize the most talented CEOs: Benabou and Tirole (2016) show that competition for talents may trigger a bonus culture and excessive pay that is detrimental to overall long-term productivity and welfare.

In this context, fund families that represent the interests of their clients might have an incentive to instruct corporations to limit their negative externalities and increase their positive ones (see, for example, Morgan and Tumlinson (2019), for a theoretical analysis).

This chapter tests whether the percentage of Socially Responsible Investment (SRI), a measure of the importance of externalities for fund families' clients, is associated with a larger support for shareholder resolutions instructing corporations to reduce or communicate on the negative externalities they produce. We offer a special focus on climate change resolutions because that constitutes one of the major societal challenges of our time and because corporations are responsible for a large fraction of greenhouse gas emissions triggering global warming.

To do so, we use ISS data on voting behavior of fund managers, aggregated at the fund family level to reflect the fact that fund families often have a centralized voting policy. We distinguish between different types of resolutions related to externalities: climate change and also human rights, discrimination and compensation restrictions based on environmental and social issues. We also study broader categories corresponding to environmental and social resolutions. We use as a benchmark resolutions on other governance and financial issues.

We merge this data with CRSP data on fund families' characteristics, including the number of stocks held in portfolio, turnover and the percentage of SRI assets under management. Our sample period ranges from 2013 to 2016 and includes 238 fund families that voted on 14,409 different shareholder resolutions at 2,700 companies.

Our analyses are based on regressions of the average level of support for a given year and a given fund family onto the percentage of SRI assets under management and various other control variables. We find that the percentage of SRI is associated with a larger support for resolutions related to the mitigation of climate change and other externalities, controlling for other fund families' characteristics, as well as for resolutions on broader environmental and social issues. We also find that the percentage of SRI is also positively associated with more support for other governance and financial issues, but this association is much lower than the one for climate resolutions. Overall, these findings suggest that fund families engage in delegated philanthropy as defined by Benabou and Tirole (2010): institutional investors adopt the preferences of their clients when engaging corporations to affect their strategy. Absent competitive and complete markets, fund families' clients as citizens, consumers, workers, and tax payers, care about corporate policies' impact on their welfare, over and above their financial impact (see Grossman and Stiglitz 1977; Hart 1979; Gordon 2003; Morgan and Tumlinson 2019). The percentage of SRI is used as a measure of the intensity of clients' willingness to request firms in which they invest to internalize externalities such as their impact on climate change.

Robustness tests further show that our results hold when the shareholder service firm ISS is for and against the resolutions, when votes on resolutions end up being close calls, and when controlling for the number of resolutions voted by fund families.

Our results have several practical implications. First, they provide guidance for institutional investors regarding the important ingredients that enter the determinants of voting policies of a wide variety of asset management companies. This might be useful for investors who would like to benchmark or evaluate their own voting behavior on ESG issues.

Second, our findings suggest that fund families with larger percentage of SRI would like corporations to develop more ambitious policies to limit negative externalities, for example, to mitigate climate change. This suggests that it might be useful to extend the notion of fiduciary duty to include other aspects than the narrowly defined shareholder value, as implied, for example, by the analysis of Morgan and Tumlinson (2019).

Third, another implication for regulators is that fund families should be incentivized to know the preferences of their clients and to make their voting policy more in line with these preferences. As indicated by Hart and Zingales (2017), shareholder votes are crucial to communicate to management how to set up corporate strategy as far as the tradeoff between profits and the common good is concerned. This issue is left for future research.

12.5. Appendix: Classification of shareholder resolutions

This table displays our classification of shareholder resolutions across different categories: environmental, social and executive compensation issues. All resolution topics not listed in this table are classified in the category "Other Governance and Financial".

Environmental	Social	Executive Compensation
Toxic Emissions	Board Diversity	Establish a Compensation Committee
Nuclear Power - Related	MacBride Principles	Establish Director Stock Ownership Requirement
Report on Environmental Policies	Human Rights Risk Assessment	Require Directors Fees to be Paid in Stock
Community -Environmental Impact	Improve Human Rights Standards or Policies	Amend Director/Officer Indemnification/Liability Provisions
Genetically Modified Organisms (GMO)	Plant Closures and Outsourcing	Eliminate or Restrict Severance Agreements (Change-in-Control)
Product Toxicity and Safety	Operations in High Risk Countries	Submit Severance Agreement (Change-in-Control) to Shareholder Vote
Environmental - Related (Japan)	Data Security, Privacy, and Internet Issues	Stock Retention/Holding Period
Operations in Protected Areas	Report on Pay Disparity	Limit/Prohibit Executive Stock-Based Awards
Report on Climate Change	Fair Lending	Death Benefits / Golden Coffins
GHG Emissions	End Production of Tobacco Products	Increase Disclosure of Executive Compensation
Hydraulic Fracturing	Prepare Tobacco-Related Report	Limit Executive Compensation
Climate Change Action	Facility Safety	Submit SERP to Shareholder Vote
Sustainability Activities and Action	Weapons - Related	Link Executive Pay to Social Criteria
Report on Sustainability	Review Foreign Military Sales	Company-Specific--Compensation-Related
Wood Procurement	Review Drug Pricing or Distribution	Performance-Based and/or Time-Based Equity Awards
Renewable Energy	Sever Links with Tobacco Industry	Put Repricing of Stock Options to Shareholder Vote
Energy Efficiency	Reduce Tobacco Harm to Health	Non-Employee Director Compensation
Recycling	Review Tobacco Marketing	Claw-back Compensation in Specified Circumstances
Publish Two Degree Scenario Analysis	Prepare Report on Health Care Reform	Advisory Vote to Ratify Named Executive Officers' Compensation
Animal Welfare	Charitable Contributions	Establish SERP Policy
Animal Testing	Political Contributions Disclosure	Pay for Superior Performance
Animal Slaughter Methods	Political Lobbying Disclosure	Adopt Policy on 10b5-1 Trading Plans
	Political Activities and Action	Adopt Anti Gross-up Policy
	Adopt Sexual Orientation Anti-Bias Policy	Employment Contract
	Report on EEO	Limit/Prohibit Accelerated Vesting of Awards
	Labor Issues – Discrimination and Miscellaneous	Adopt Policy on Bonus Banking
	Holy Land Principles	Adjust Executive Compensation Metrics for Share Buybacks
	Gender Pay Gap	Use GAAP for Executive Compensation Metrics
	Income Inequality	

12.6. References

Anton, M., Ederer, F., Giné, M., Schmaltz, M. (2018). Common ownership, competition, and top management incentives. Working paper.

Appel, I., Gormley, T., Keim, D. (2016). Passive investors, not passive owners. *Journal of Financial Economics*, 121(1), 111–141.

Azar, J. (2017). Portfolio diversification, market power, and the theory of the Firm. Working paper.

Azar, J., Schmaltz, M., Tecu, I. (2018). Anti-competitive effects of common ownership. *Journal of Finance*, 73(4), 1513–1565.

Azar, J., Duro, M., Kadach, I., Ormazabal, G. (2019a). The big three and corporate carbon emissions around the world. Working paper.

Azar, J., Raina, S., Schmaltz, M. (2019b). Ultimate ownership and bank competition. Working paper.

Barko, T., Cremers, M., Renneboog, L. (2018). Shareholder engagement on environmental, social, and governance performance. Working paper.

Bebchuk, L. and Hirst, S. (2019). Index funds and the future of corporate governance: Theory, evidence, and policy. *Columbia Law Review*, 119(8), 2029–2146.

Bebchuk, L., Cohen, A., Hirst, S. (2017). The agency problems of institutional investors. *Journal of Economic Perspectives*, 31(3), 89–112.

Benabou, R. and Tirole, J. (2010). Individual and corporate social responsibility. *Economica*, 77(305), 1–19.

Benabou, R. and Tirole, J. (2016). Incentives and prosocial behavior. *American Economic Review*, 96(5), 1652–1678.

Bolton, P., Li, T., Ravina, E., Rosenthal, H. (2020). Investor ideology. *Journal of Financial Economics*, 137(2), 320–352.

Brav, A., Jiang, W., Li, T. (2019). Picking friends before picking (Proxy) fights: How mutual fund voting shapes proxy contests. Working paper.

Cespa, G. and Cestone, G. (2007). Corporate social responsibility and managerial entrenchment. *Journal of Economics & Management Strategy*, 16(3), 741–771.

Christoffersen, S., Geczy, C., Musto, D., Reed, A. (2007). Vote trading and information aggregation. *Journal of Finance*, 62(6), 2897–2929.

Davis, G. and Kim, E.H. (2007). Business ties and proxy voting by mutual funds. *Journal of Financial Economics*, 85(2), 552–570.

Derrien, F., Kecskés, A., Thesmar, D. (2013). Investor horizons and corporate policies. *Journal of Financial and Quantitative Analysis*, 48(6), 1755–1780.

Di Giuli, A. and Kostovetsky, L. (2014). Are red or blue companies more likely to go green? Politics and corporate social responsibility. *Journal of Financial Economics*, 111(1), 158–180.

Dimson, E., Karakaş, O., Li, X. (2015). Active ownership. *Review of Financial Studies*, 28(12), 3225–3268.

Edmans, A., Levit, D., Reilly, D. (2019). Governance under common ownership. *Review of Financial Studies*, 32(7), 2673–2719.

Fichtner, J., Heemskerk, E., Garcia-Bernardo, J. (2017). Hidden power of the big three? Passive index funds, re-concentration of corporate ownership, and new financial risk. *Business and Politics*, 19(2), 298–326.

Ginglinger, E. and Moreau, Q. (2019). Climate risk and capital structure. Working paper.

Gordon, R. (2003). Do publicly traded corporations act in the public interest? *B.E. Journal of Economic Analysis & Policy*, 3(1), 1–20.

Grossman, S. and Stiglitz, J. (1977). On value maximization and alternative objectives of the firm. *Journal of Finance*, 32(2), 389–402.

Gutiérrez, G. and Philippon, T. (2018). Ownership, concentration, and investment. *AEA Papers and Proceedings*, 108, 432–437.

Hansen, R. and Lott, J. (1996). Externalities and corporate objectives in a world with diversified shareholder/consumers. *Journal of Financial and Quantitative Analysis*, 31(1), 43–68.

Harford, J., Kecskés, A., Mansi, S. (2018). Do long-term investors improve corporate decision making? *Journal of Corporate Finance*, 50(C), 424–452.

Hart, O. (1979). On shareholder unanimity in large stock market economies. *Econometrica*, 47(5), 1057–1084.

Hart, O. and Zingales, L. (2017). Companies should maximize shareholder welfare not market value. *Journal of Law, Finance, and Accounting*, 2(2), 247–274.

He, J. and Huang, J. (2017). Product market competition in a world of cross-ownership: Evidence from institutional blockholdings. *Review of Financial Studies*, 30, 2674–2718.

He, Y., Kahraman, B., Lowry, M. (2018). Mutual fund voting on environmental and social proposals. Working paper.

He, J., Huang, J., Zhao, S. (2019). Internalizing governance externalities: The role of institutional cross-ownership. *Journal of Financial Economics*, 134(2), 400–418.

Hong, H. and Kostovetsky, L. (2012). Red and blue investing: Values and finance. *Journal of Financial Economics*, 103(1), 1–19.

Ilhan, E., Sautner, Z., Vilkov, G. (2019). Carbon tail risk. Working paper.

Ilhan E., Krueger, P., Sautner, Z., Starks, L. (2020). Institutional investors' views and preferences on climate risk disclosure. Working paper.

Intergovernmental Panel on Climate Change (2018). Global warming of 1.5°C. IPCC, Switzerland.

Kraus, A. and Rubin, A. (2010). Reducing managers' incentives to cannibalize: Managerial stock options when shareholders are diversified. *Journal of Financial Intermediation*, 19(4), 439–460.

Krueger, P., Sautner, Z., Starks, L. (2020). The importance of climate risk for institutional investors. *Review of Financial Studies*, 33(3), 1067–1111.

Lindsey, L. (2008). Blurring firm boundaries: The role of venture capital in strategic alliances. *Journal of Finance*, 63(3), 1137–1168.

Massari, M., Gianfrate, G., Zanetti, L. (eds) (2016). Carbon risk and corporate value. In *Corporate Valuation: Measuring the Value of Companies in Turbulent Times*. Wiley, New York.

Mattison, R., Trevitt, M., van Ast, L., Gifford, J., Mnatsakanian, N., Watson, O., Zimmerman, C., Piani, V., Hoepner, A. (2011). Universal Ownership and Environmental Externalities. United Nations Environmental Programme, New York.

Matvos, G. and Ostrovsky, M. (2008). Cross-ownership, returns, and voting in mergers. *Journal of Financial Economics*, 89, 391–403.

Morgan, J. and Tumlinson, J. (2019). Corporate provision of public goods. *Management Science*, 65(10), 4489–4504.

Morgan, A., Poulsen, A., Wolf, J., Yang, T. (2011). Mutual funds as monitors: Evidence from mutual fund voting. *Journal of Corporate Finance*, 17(4), 914–928.

Newham, M., Seldeslachts, J., Banal-Estañol, A. (2018). Common ownership and market entry: Evidence from pharmaceutical industry. Working paper.

Rotemberg, J. (1984). Financial transaction costs and industrial performance. Working paper.

Rubin, A. (2010). Diversification and corporate decisions. *Corporate Ownership and Control*, 3(3), 209–212.

Schmaltz, M. (2018). Common-ownership concentration and corporate conduct. *Annual Review of Financial Economics*, 10, 413–448.

Stern, N. (2006). *The Economics of Climate Change: The Stern Review*. Cambridge University Press, Cambridge.

13

Creditworthiness and Buildings' Energy Efficiency in the Mortgage Market

13.1. Introduction

Energy efficiency represents one of the key planned actions aiming to reduce greenhouse emissions and the consumption of fossil fuel in order to mitigate the impact of climate change. According to the European Commission[1], buildings represent 40% of the total energy consumption and are responsible for 36% of the total CO_2 emissions in Europe. Currently, only 25% of the stock of European buildings

Chapter written by Monica BILLIO, Michele COSTOLA, Loriana PELIZZON, Francesco PORTIOLI, Max RIEDEL and Daniele VERGARI.
The authors acknowledge EU Horizon 2020 project Energy Efficient Mortgages Action Plan (EeDaPP) funded by the European Commission. They thank Luca Bertalot, Jennifer Johnson, Lusyne Keszisjan, Iva Hristova, Gianluca Natalini, Stella Fumarola and the other consortium partners for their support during the project. Ca' Foscari University and Leibniz SAFE units (M. Billio, M. Costola, M. Riedel and L. Pelizzon) contributed to the formalization of the research questions and the implementation of the econometrics methodology to analyze the relationship between building's energy efficiency and credit risk. The CRIF unit (F. Portioli and D. Vergari) contributed to the mortgage loan data collection from pilot banks and the data pool to be used in the analysis. Additional analyses produced during the EeDaPP project can be found at cordis.europa.eu. Finally, they also thank Deborah Leboullenger, Numa Bosc, Charalampos Kouratzoglou, Isidoro Tapia and other members of ABI (Italian Banking Association), EBA (European Banking Authority) and EIB (European Investment Bank) for fruitful discussions. Loriana Pelizzon also acknowledges the Leibniz Institute for Financial Research SAFE for financially sponsoring this research.
For a color version of all figures in this chapter, see www.iste.co.uk/jurczenko/climate.zip.

1 For more details, see https://ec.europa.eu/info/news/new-rules-greener-and-smarter-buildings-will-increase-quality-life-all-europeans-2019-apr-15_en.

is considered as energy efficient. In the 2030 climate and energy framework, the European Commission has set as a minimum target for the improvement in energy efficiency by 32.5%. The process of making buildings energy efficient reveals several advantages in terms of economic stimuli (e.g. the construction of buildings is a driving force component of the GDP) and in terms of investment opportunities, since it provides an increase in the property value. Furthermore, energy efficiency seems to be an attractive market segment also for lenders in the mortgage market. Recent literature has started to investigate whether energy efficiency is related to a lower solvency risk for owners, and consequently, whether it reduces the lending risk for banks and other financial institutions. Most of these studies conclude that there is a negative association between buildings' energy efficiency and owners' default. For instance, Kaza et al. (2014) provide evidence that the US energy-efficient residential buildings – measured by the ENERGY STAR certificate – have a lower probability of default. Guin and Korhonen (2020) focus on the UK residential mortgages and find that energy-efficient buildings show less payment arrears than the inefficient ones. An and Pivo (2020) investigate the commercial mortgage-backed securities market and find that the green buildings have a lower default risk thanks to a lower contemporaneous loan-to-value ratio with respect to the non-green counterpart. Billio et al. (2021) investigate the Dutch residential market and find that mortgages on energy-efficient buildings have a lower probability of default. In addition to the borrower's individual factors, the authors show that there exists an economic channel (e.g. savings coming from reduced energy costs) that mitigates even further the probability of default of borrowers in lower-income groups. Current regulatory credit risk scoring models, such as the standardized approach produced by the Basel Committee on Banking Supervision, do not include energy efficiency features. If energy efficiency represents a significant predictor of mortgage default, it should be included in credit-scoring models of financial intermediaries along with the usual determinants such as an applicant's demographic (e.g. age) and financial information (e.g. income level) and loan-specific characteristics (e.g. loan-to-value ratio). However, most of the studies that have attempted to analyze the European market have encountered several data availability issues. For instance, the non-compulsory disclosure of energy-efficient (EE) investments, the heterogeneity among the energy labels in the union and the General Data Protection Regulation (GDPR) requirements. Therefore, further empirical investigations for the European Union are certainly welcome. In this chapter, we investigate the relationship between energy efficiency and the borrower's solvency risk in the Italian market. Specifically, we analyze a residential mortgage portfolio of four financial institutions which includes about 70,000 loans matched with the energy performance certificate of the associated buildings.[2] In this analysis, we adopt the logit model and the Cox model to test for the relationship between energy efficiency and the

2 The dataset has been made available to Ca' Foscari University of Venice by CRIF S.p.a and is part of the Energy efficiency Data Protocol and Portal (EeDaPP) project funded by the European Commission on the H2020 Programme (H2020-EE-2017-CSA-PPI).

likelihood of default of a mortgage. The logit regression is suitable to model binary outcomes (the dependent variable takes the value of one in the presence of default and zero otherwise), while the Cox regression allows distinguishing between "healthy" and "non-healthy" mortgages (i.e. non-defaulted vs. defaulted) over time. Our findings show that there is a negative relationship between a building's energy efficiency and the owner's probability of default. Findings survive after we account for dwelling, household, mortgage, market control variables, and regional and year fixed effect. In addition, a receiver-operating characteristic (ROC) analysis shows that there is an improvement in the estimation of the mortgage default probability when the energy efficiency characteristic is included as a risk predictor in the model. Finally, results indicate that the degree of energy efficiency also has an impact and that the higher the energy efficiency, the lower is the risk of default. The remainder of the chapter is structured as follows. Section 13.2 introduces the Italian residential mortgage data and the associated energy performance certificates, and then section 13.3 presents the implemented methodologies. Section 13.4 illustrates the empirical analysis and our major results. Finally, section 13.5 concludes the chapter.

13.2. Portfolio analysis

In the following, we employ Italian residential mortgage data that was provided by CRIF s.p.a as part of the Energy Efficient Mortgages Initiative, for the project "Energy efficiency Data Protocol and Portal" (cf. EeDaPP 2020). The original sample was narrowed based on the following criteria. Complete information on the borrower's creditworthiness must be available for each loan. The time period of the sample is restricted from 2012 through to 2019, during which mortgage loans were originated. The loan-to-value (LTV) ratio is capped at 1.1 to eliminate outliers. Borrower type is "individual", with each borrower holding a single mortgage. Either "apartment" or "house" must be selected as the property type. The status of the property falls into one of the following three categories: new/retrofit, used or to be renovated. The years of construction of the buildings range from 1900 to 2019, and each borrower is required to be associated with exactly one building/apartment and vice versa. Our final dataset contains 70,666 individual mortgage loans after applying the above selection criteria.

13.2.1. *Energy efficiency*

We use Energy Performance Certificates (EPC) of buildings to classify them into different energy efficiency categories. Figure 13.1 depicts the EPC distribution within 10-year buckets of building construction years. From the bar charts, we can observe that energy efficiency has improved significantly over time, with the most EE buildings built after 2010.

The distribution of EPCs by year of loan origination is shown in Figure 13.2. Panel A shows the total number of mortgages granted, while Panel B shows the percentage of each EPC category by year of loan origination. According to the latter, between 12 and 15% of loans were originated for buildings with an EPC rating of A or B. Panels C and D deal exclusively with loans in arrears. The first time a borrower is more than 90 days delinquent, the loan is considered in default. As shown in Panel C, the absolute number of delinquent loans decreases with the year of loan origination. Panel D reports the percentage share of each rating within the year of origination.

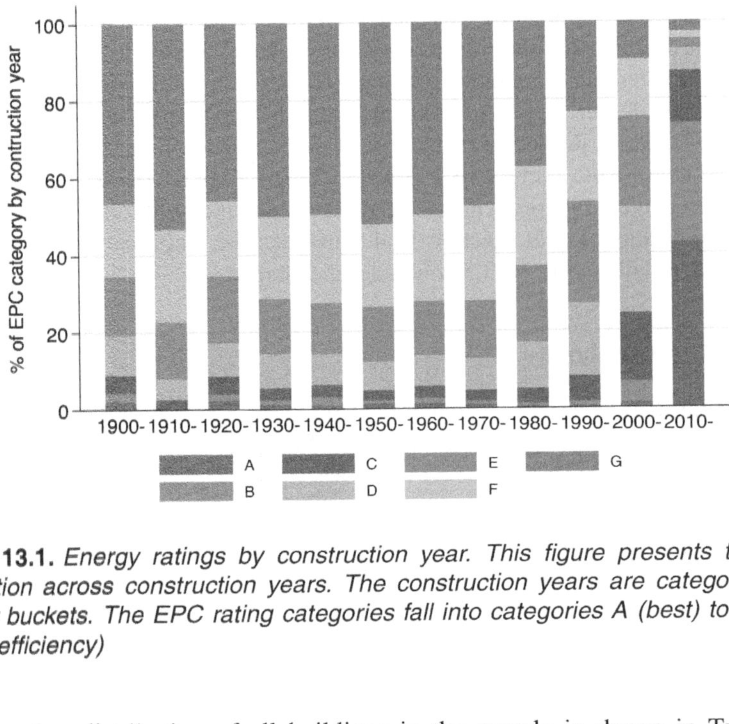

Figure 13.1. *Energy ratings by construction year. This figure presents the rating distribution across construction years. The construction years are categorized into 10-year buckets. The EPC rating categories fall into categories A (best) to G (worst energy efficiency)*

The rating distribution of all buildings in the sample is shown in Table 13.1, while the building distribution across Italian provinces is presented in Table 13.2. Less efficient buildings, especially G-rated buildings, represent the largest share in the sample, whereas A- and B-rated buildings account for roughly 13.1%. The percentage of defaulted mortgages in each rating class is shown in column 3 of Table 13.1. In this regard, it is worth noting that the highest percentage of defaults is linked to the poorest energy efficiency rating G, whereas the lowest share of default is associated with the EPC rating A. Defaulted mortgages account for 1.31% of all mortgages.

Table 13.2 shows that mortgages are not evenly distributed throughout Italian regions, with Lombardy (46.20 %) and Emilia Romagna (29.04%) accounting for the biggest shares. The A-rated buildings represent between 2.4% and 24.2% of all buildings within individual regions. In all regions, the share of defaulted EE mortgages is lower than the share of defaulted non-EE mortgages.

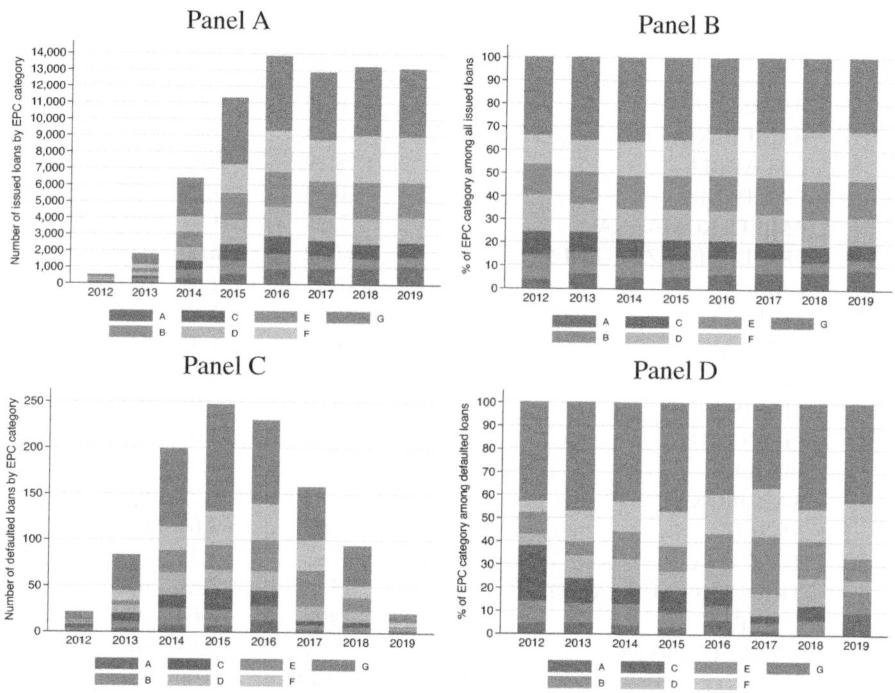

Figure 13.2. *Energy Rating (EPC) distribution of all (Panels A and B) defaulted (Panels C and D) mortgages by year of mortgage origination. The right (left) panels present the percentage share (absolute number) of each rating category given the origination years ranging from 2012 to 2019*

Rating category	All	Defaulted
A	7.01	0.71
B	6.06	1.33
C	7.44	1.20
D	12.37	1.05
E	15.80	1.19
F	18.77	1.12
G	32.55	1.73
Total	100	1.31

Table 13.1. *Rating distribution for all and defaulted mortgages. Column 2 presents the % share of each rating category given the total sample of mortgages. Column 3 indicates the share of defaulted loans within each rating category. The number of unique mortgages is 70,666*

	All		Defaulted	
Property region	By region	EE within region	non-EE	EE
ABRUZZO	0.45	18.50	1.57	0.00
BASILICATA	0.16	18.75	1.79	0.00
CALABRIA	0.22	18.59	0.64	0.00
CAMPANIA	0.33	10.59	1.27	0.00
EMILIA ROMAGNA	29.04	7.60	1.09	0.07
FRIULI VENEZIA GIULIA	0.23	11.32	0.00	0.00
LAZIO	0.82	10.21	2.08	0.00
LIGURIA	0.84	2.18	2.18	0.00
LOMBARDIA	46.20	6.56	1.31	0.05
MARCHE	0.58	5.10	2.67	0.00
MOLISE	0.05	24.24	0.00	0.00
PIEMONTE	9.05	4.44	1.06	0.00
PUGLIA	0.31	15.60	1.38	0.00
SARDEGNA	0.73	12.16	0.77	0.00
SICILIA	4.82	2.67	2.05	0.06
TOSCANA	2.33	3.82	1.58	0.06
TRENTINO ALTO ADIGE	0.74	15.49	0.38	0.00
UMBRIA	0.06	18.60	0.00	0.00
VALLE D'AOSTA	0.35	2.44	2.03	0.41
VENETO	2.68	19.40	0.90	0.11
Total	100	6.93	1.39	0.05

Table 13.2. *Loans distribution (all and defaulted) conforming to the NUTS 2 classification for Italy. Column 2 shows the % share of each region within the total. Column 3 indicates the share of EE buildings (defined as A-rated buildings) within each region. Column 4 (column 5) presents the percentage share of defaulted non-EE mortgages with a region. The total number of mortgages is 70,666*

In the analysis, we define the energy efficiency characteristic by creating a binary variable EE_i that equals 1 if a building i is A-rated (e.g. the top-notch in the EPC classification) and zero, otherwise.

13.2.2. *Descriptive statistics*

The covariates for the correlation analysis are chosen such that mortgage-, dwelling- and borrower-specific effects are accounted for. We use total loan amount, loan-to-value ratio and mortgage term at the time of loan origination as mortgage controls. The total number of monthly installments divided by 12 gives the mortgage term, measured in years.[3] Dwelling-specific controls are property type and status, as well as age of the dwelling at loan origination. The majority of mortgages in the sample is issued for apartment purchases (97.8%), while only a small fraction was used to acquire a house (2.2%). The status of the dwelling is one of three types: new/renovated (32.8%), used (65.6%) or to-be-renovated (1.6%). Borrower-specific controls are the borrower's age at the time of loan origination and their credit score rating. We use the quarterly Italian unemployment rate (at the NUTS1 macroregional level), the monthly inflation rate (year-on-year change in the consumer price index) and the quarterly growth in the house price index (year-on-year change, at the NUTS1 macroregional level) to account for overall macroeconomic conditions. The variables were obtained from Istat, the Italian National Statistical Office.[4] Summary statistics on key borrower, property and mortgage characteristics are reported in Table 13.3. The table distinguishes between non-defaulted (Panel A) and defaulted (Panel B) mortgages. Furthermore, we also distinguish between energy-efficient (EE = 1) and energy-inefficient (EE = 0) dwellings in both panels. We stress again that the energy efficiency indicator EE is set to one if a dwelling's EPC rating is A, and it is zero otherwise. In terms of borrower characteristics, there does not appear to be a significant difference in age at origination between EE and non-EE mortgages. However, defaulted loans exhibit a slightly higher average age than non-defaulted loans in the sample. In terms of creditworthiness, the less creditworthy borrowers are more likely to default. Defaulted non-EE mortgages have the highest average loan-to-value ratio. Borrowers appear to default more often on mortgages with a higher loan amount, higher property values and an earlier year of construction.

The distribution of mortgages by construction year (Panel A), granted loan amount (Panel B) and mortgage origination year (Panel C) is shown in Figure 13.3. Starting from the 1950s, the dataset is well distributed in terms of construction year, but the earlier years are sparsely represented. The average loan amount is EUR 118,032, with only 1% of issued loans exceeding a value of EUR 400,000. The loan sample is relatively new, with 0.68% of loans originated as recently as 2012.

In terms of macroeconomic indicators, the average quarterly total unemployment rate for adults aged 15 and older was 11.18% between Q1 2012 and Q4 2019. During the same period, the average inflation rate was 1.01%, while the house price index decreased on average by 2.3% across regions.

3 The total number of monthly installments is calculated by transforming all reported periodicities (i.e. monthly, quarterly, semi-annually and annually) into monthly.

4 For more information, visit www.istat.it.

Panel A: Non-defaulted

	EE	Mean	Median	Std.	Minimum	Maximum	N
Borrower age at	0	40.4	39.0	10.1	18.0	87.0	64,823
origination	1	40.0	39.0	9.9	18.0	79.0	4,916
Credit score	0	516.7	521.0	42.0	167.0	598.0	64,823
	1	520.2	524.0	38.1	204.0	598.0	4,916
Granted loan amount	0	109,755	98,000	68,420	30,009	3,000,000	64,823
	1	158,486	140,000	121,434	32,000	4,100,000	4,916
Loan-to-value	0	0.65	0.69	0.19	0.04	1.09	64,823
	1	0.60	0.63	0.20	0.06	1.09	4,916
Mortgage term	0	20.69	20.08	6.58	3.00	40.33	64,823
(in years)	1	21.61	20.08	6.55	4.00	40.08	4,916
Construction year	0	1977	1974	25	1900	2019	64,823
	1	2009	2015	19	1900	2019	4,916
Property value	0	181,319	152,000	126,179	32,000	5,310,000	64,823
	1	279,919	240,000	216,434	44,000	6,028,000	4,916

Panel B: Defaulted

	EE	Mean	Median	Std.	Minimum	Maximum	N
Borrower age at	0	41.8	41.0	10.8	20.0	78.0	892
origination	1	43.1	41.0	12.1	27.0	79.0	35
Credit score	0	456.2	485.0	89.3	172.0	579.0	892
	1	433.2	469.0	102.9	179.0	558.0	35
Default since origination	0	26.3	23.0	16.4	5.0	95.0	892
(in months)	1	25.3	23.0	14.1	5.0	62.0	35
Granted loan amount	0	109,994	94,023	118,985	30,722	3,000,000	892
	1	358,305	155,000	1,158,114	38,000	7,000,000	35
Loan-to-value	0	0.66	0.72	0.20	0.07	1.09	892
	1	0.61	0.64	0.21	0.18	0.99	35
Mortgage term	0	21.84	20.63	6.69	3.08	40.00	892
(in years)	1	23.72	25.08	6.10	10.50	30.58	35
Construction year	0	1976	1972	24	1900	2018	892
	1	2007	2013	22	1900	2016	35
Property value	0	184,735	145,000	199,071	38,000.00	4,054,000.00	892
	1	577,571	254,000	1,651,266	111,000	10,000,000	35

Table 13.3. *Summary statistics for loan and borrower characteristics. Panel A (Panel B) provides the statistics for non-defaulted (defaulted). Column 2 discriminates between energy-efficient (EE=1) and energy-inefficient (EE=0) buildings*

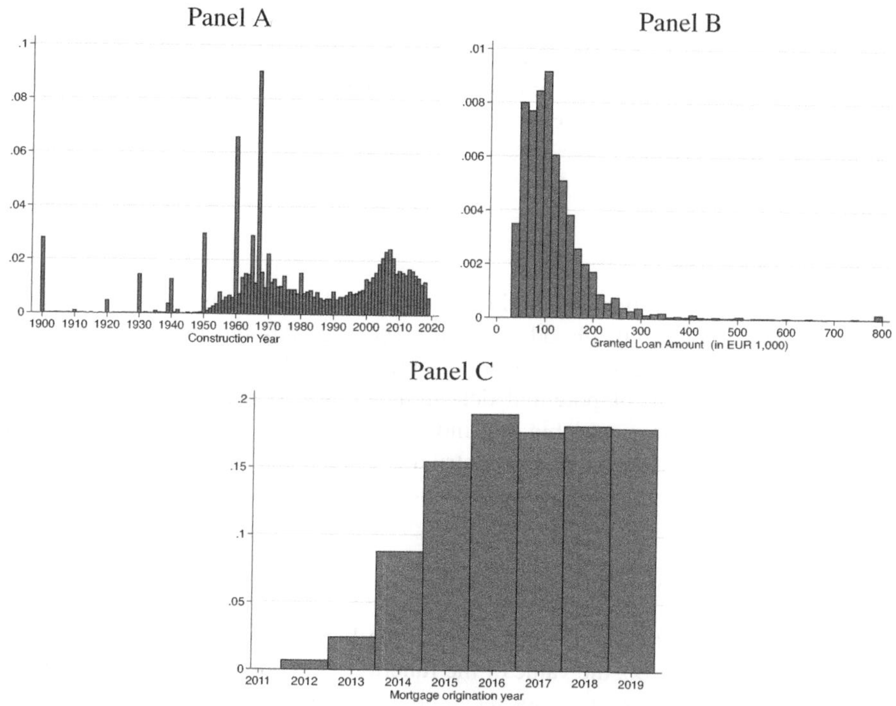

Figure 13.3. *Distribution by construction year and original balance. Panel A indicates the relative frequency of construction year for the considered buildings. Panel B shows the relative frequency of total mortgage original balance. Panel C provides the frequency of the earliest mortgage origination year*

In summary, the portfolio analysis of Italian mortgages indicates that the dataset contains a sizeable amount of defaulted loans, although defaulted EE mortgages represent only a small part of the sample. This might be attributed to the fact that the mortgage sample is relatively recent and defaults typically occur after 10 or more years from the date of origination. However, the sample size is large enough to run a comprehensive analysis and the presented control variables for borrower, mortgage and dwelling characteristics will be used to isolate the relationship between EE and the probability of mortgage default.

13.3. Methodology

Logit regression and survival analysis are two statistical approaches that are commonly used for credit risk evaluation, and we choose to use both for robustness considerations. Our decision is based on the fact that logit regressions are commonly used for cross-sectional datasets, but survival models account for the time dimension, in which the hazard of an event occurring (i.e. default) increases with time. As a result, survival analysis appears to be a good complement to the more static Logit regression. In addition, the Cox model takes into account truncation and censoring in the data. The two subsections that follow give further information on the approaches used and their unique characteristics.

13.3.1. *Logit regression*

Logistic regression is a typical method for examining the link between borrower-level loan information and the likelihood of default (see, e.g., Altman and Saunders 1997; Qi et al. 2020). This regression makes it possible to model a binary outcome variable associated with a group of potential determinants. The dependent variable in our example is a dichotomous variable that indicates whether a borrower has defaulted or not. The simplicity of such a model contributes to its popularity and is derived from the function $f(z)$, which takes a range of values between 0 and 1, and is defined as:

$$f(z) = \frac{1}{1+\exp(-z)}. \quad [13.1]$$

Figure 13.4 shows that as z tends to $-\infty$, the logistic function $f(z)$ approaches zero, and as z goes to $+\infty$, the value of the function $f(z)$ approaches one. Therefore, the logistic function represents an ideal candidate to model (default) probabilities that lie between 0 and 1.

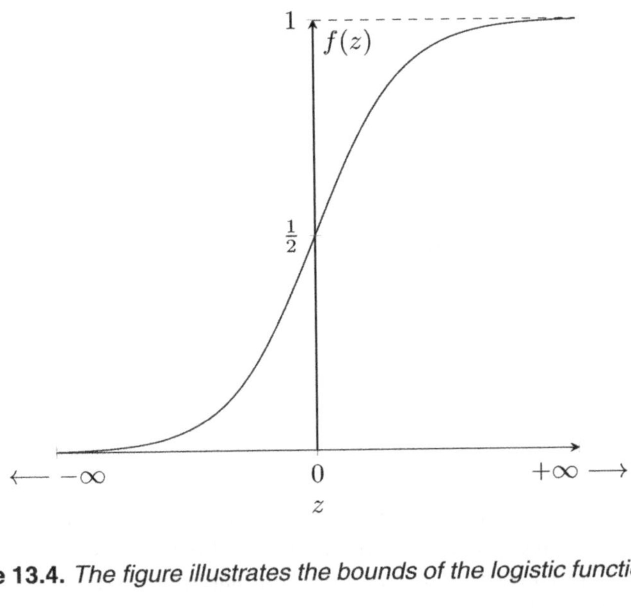

Figure 13.4. *The figure illustrates the bounds of the logistic function $f(z)$*

By defining z as the sum of a linear combination of p covariates x, i.e. $z = \alpha + \beta_1 x_1 + \beta_2 x_2 + \cdots + \beta_p x_p$, the logistic regression model is obtained by substituting z into $f(z)$:

$$\mathbb{P}(Y^i = 1 | x_1, x_2, \ldots, x_p) = \frac{1}{1 + \exp(-(\alpha + \sum_{l=1}^{p} \beta_l x_l))} = G(\beta^i \mathbf{X}^i), \quad [13.2]$$

where α and β_l represent the parameters to be estimated. Therefore, the function $f(z)$ represents the conditional probability of the binary outcome variable Y of a subject i given a set of observed factors $x_1, x_2, \ldots, x_p$. The log-likelihood function is expressed as

$$\log L = \sum_{i=1}^{n} [Y^i \log G(\beta^i \mathbf{X}^i) + (1 - Y^i) \log(1 - G(\beta^i \mathbf{X}^i))]. \quad [13.3]$$

The likelihood is maximized by applying numerical methods such as the Newton–Raphson method to the first-order condition in equation [13.3]. Estimates from logistic and the logit regressions are equivalent since both are retrieved using the maximum likelihood estimator and the logistic function is the inverse of the logit function:

$$\text{logit}^{-1}(\beta^i \mathbf{X}^i) = \text{logistic}(\beta^i \mathbf{X}^i) \quad [13.4]$$

For easier reading, we consider the logit model instead of the logistic since in the latter, the estimates are reported as odds ratios.

13.3.2. *Cox proportional hazards model*

The Cox proportional hazards model (Cox 1972) is one of the most commonly used survival models because it allows for the inclusion of a number of variables and scales them with a baseline hazard rate. The model is defined as follows:

$$h(t, X) = h_0(t) \exp\left(\sum_{l=1}^{p} \beta_l x_l\right), \quad [13.5]$$

where $h_0(t)$ is the baseline hazard function, p is the number of independent variables X, and β_l is the parameter that has to be estimated for lth covariate.

The baseline hazard is a function of time t and does not depend on the set of covariates, which is a key property of the Cox model. In contrast, the exponential function, which does not use the time dimension t, involves only the set of covariates X. This means that the standard formulation of the Cox model reduces to $h_0(t)$ when all covariates x are zero.

The hazard ratio (HR) is defined as the ratio of hazards for two subjects that are included in a study. Assume $t_k^i (t_k^j)$ refers to observation time of subject i (subject j), then we can write the hazard ratio as the estimate of $h(t_k^i, X^i)$ divided by the estimate of $h(t_k^j, X^j)$:

$$\widehat{HR} = \frac{\hat{h}(t_k^i, X^i)}{\hat{h}(t_k^j, X^j)}, \qquad [13.6]$$

where X^i and X^j are the subjects' respective covariate values. From equation [13.6], we can observe that it is possible to estimate the parameters β even though the baseline hazard rate is not specified. Namely, HR can be rewritten as:

$$\widehat{HR} = \frac{\hat{h_0}(t_k^i) \exp(\sum_{l=1}^{p} \hat{\beta}_l x_l^i)}{\hat{h_0}(t_k^j) \exp(\sum_{l=1}^{p} \hat{\beta}_l x_l^j)} = \exp\left(\sum_{l=1}^{p} \hat{\beta}_l (x_l^i - x_l^j)\right) = \theta, \qquad [13.7]$$

where θ is a time-independent constant. The proportional hazard assumption states that for the risk of one person to be proportional to that of another, the HR formulation must remain constant over time. This means that the final expression of the hazard ratio does not include the time variable and that once the values of X^i and X^j are determined, the value of the exponential function becomes time-invariant, as shown in the above equation. This is the formal expression of the proportional hazards assumption. The relation between two subjects can, consequently, be written as $\hat{h}(t_k^i, X^i) = \theta \hat{h}(t_k^j, X^j)$.

A visualization of empirical survival functions can shed light on whether the proportional hazards assumption is valid. The empirical survival function is usually shown using the approach proposed in Kaplan and Meier (1958). The assumption of this non-parametric methodology is that censoring time does not depend on the individual's behavior. The function is defined as:

$$\hat{S}_{t_m} = \prod_{i=1}^{m} \mathbb{P}(T > t_i | T \geq t_i) = \hat{S}(t_{m-1}) \mathbb{P}(T > t_i | T \geq t_i), \qquad [13.8]$$

where t_m represents ordered event times, and the probabilities are approximated by the frequency distribution in the dataset.

Since the observation period in survival analyses is often limited to a certain period of time, it is crucial to distinguish between (i) the date when a subject first takes a risk, (ii) the period when a subject is under observation and (iii) the date when a subject experiences failure. If the date on which an individual first takes a risk is before the start of the observation period, we speak of left truncation. In the context of mortgage analysis, left truncation applies to mortgages that were originated before the first observation date. In general, left-censored subjects can be included in the analysis, but it is critical to consider the subject's time of risk exposure when they enter the observation (i.e. the age of the credit at the start of the study). If the time of default is not observable due to the subject's early exit from the study or early termination of the study, the subject is said to be right-censored. For example, it is not possible to consider the future default of a loan that is still ongoing at the end of the period considered in the survival study. The censoring date for a loan that was honored during the period under investigation, on the contrary, is the date of the last payment, since (theoretically) a default might happen at any time in the future if only the loan term was long enough. The use of a dummy variable for censored observations is a common approach to correct for censoring.

13.4. Results

13.4.1. *Estimates from the logit regression*

As discussed in the previous section, the logit regression model is suitable for modeling binary outcomes like default, where the dependent variable is equal to one if a default occurs and zero otherwise. Table 13.4 shows the regression estimates. Model (1) provides the findings by only including the EE indicator and mortgage-related parameters in the model such as the borrower's credit score, loan-to-value at mortgage origination and loan term. The regression coefficient of −0.5712 for the EE dummy indicates that energy efficiency is significant and negatively associated with the likelihood of mortgage default. Given that this finding could be masked by building or household characteristics, we provide further specifications and include additional control variables in models (2) and (3). Model (2) includes dwelling age as a proxy for a building's general condition. Since more recent buildings are likely to require less renovation costs, the age might influence the borrower's ability to repay the debt. In addition, model (3) takes into account the borrower's age at the time of origination to assess the borrower's attitude towards debt and ability to repay a loan. Models (4) and (5) involve region fixed effects at NUTS 1 level[5] and origination year. Fixed effects of origination year account for the fact that newer loans are less likely to default than older ones as highlighted in the descriptive statistics. Region fixed effects account for regional heterogeneity such as cultural attitude and economic strengths. Finally, model (6) controls for the macroeconomic condition of the economy at the date of loan origination and includes the inflation rate, the unemployment rate and the house price index growth rate. The latter two variables are available at the NUTS 2 regional level. It is important to note that the EE is significant and negatively related to the borrower's risk of default in all the considered specifications.

5 The Nomenclature of Territorial Units for Statistics (NUTS) represents the standard references for geographical coordinates for the classification of countries for statistical purposes. Three NUTS levels are established by Eurostat. The NUTS 1 codes refer to the least granular region specification. In the case of Italy, the NUTS 1 regions are: (i) North-East, (ii) North-West, (iii) Center, (iv) South and (v) Islands.

Dependent variable: default dummy

	(1)	(2)	(3)	(4)	(5)	(6)
EE (A rating)	-0.5712***	-0.5589***	-0.5711***	-0.3988**	-0.3700**	-0.3609**
	[0.1724]	[0.1771]	[0.1769]	[0.1755]	[0.1764]	[0.1763]
Credit score	-0.0159***	-0.0159***	-0.0156***	-0.0151***	-0.0151***	-0.0152***
	[0.0004]	[0.0004]	[0.0004]	[0.0004]	[0.0004]	[0.0004]
Loan-to-value	0.3284*	0.3280*	0.3798*	0.9951***	0.9773***	0.9709***
	[0.1994]	[0.1993]	[0.1995]	[0.2127]	[0.2146]	[0.2147]
Loan term	0.0346***	0.0346***	0.0413***	0.0426***	0.0408***	0.0410***
	[0.0058]	[0.0058]	[0.0060]	[0.0060]	[0.0061]	[0.0061]
Building age		0.0004	0.0001	0.0028**	0.0025*	0.0024*
		[0.0013]	[0.0013]	[0.0013]	[0.0013]	[0.0013]
Borrower age			0.0126***	0.0140***	0.0135***	0.0134***
			[0.0037]	[0.0037]	[0.0037]	[0.0037]
Inflation						13.3975
						[11.0935]
Unemployment						4.2248
						[2.6423]
HPI growth						-2.8562
						[3.8312]
Observations	70,666	70,666	70,666	70,666	70,666	70,666
Dwelling controls	No	Yes	Yes	Yes	Yes	Yes
Household controls	No	No	Yes	Yes	Yes	Yes
Market controls	No	No	No	No	No	Yes
Mortgage controls	Yes	Yes	Yes	Yes	Yes	Yes
Region FE	No	No	No	No	Yes	Yes
Origination year FE	No	No	No	Yes	Yes	Yes
SE	Rob.	Rob.	Rob.	Rob.	Rob.	Rob.
Pseudo R-squared	0.114	0.115	0.116	0.159	0.162	0.163

Table 13.4. *Logit regression results to assess the relationship between buildings' energy efficiency and the default risk of the borrowers. The dependent variable is binary indicating whether a mortgage is in default (e.g. at least three months in arrears) or not. The dummy variable is EE equal to one if a building's energy efficiency rating is A-rated and zero otherwise. Mortgage controls are borrower's credit score, loan-to-value and loan term (in years). Dwelling control is building age at loan origination. Borrower control is borrower's age at loan origination. Market controls are the monthly Italian inflation rate (change in the consumer price index to previous year's value in the same month), quarterly unemployment rate at the regional level and quarterly house price index growth at the regional level. Origination year and NUTS1-region fixed effects (FE) are included where indicated. Standard errors (reported in square brackets) are robust. Statistical significance is denoted by ***, ** and * at the 1%, 5% and 10% level, respectively*

To investigate whether the inclusion of a building's energy efficiency information improves the model's prediction accuracy, we continue with model (6) of Table 13.4 as the baseline specification and perform a receiver-operating characteristic (ROC) analysis. The ROC analysis is a useful tool for evaluating the accuracy of a statistical model that classifies subjects into one of two categories (Metz 1978; Zweig and Campbell 1993). In our case, the Logit model classifies the loans into the categories defaulted and non-defaulted. To measure if this classification improves in precision as the explanatory variable EE is included in the model, we compute the area under the ROC curve (AUC) for both cases, with and without EE. The higher the AUC, the better the model is at predicting defaulted loans as defaulted and the non-defaulted loans as non-defaulted. For model (6), the AUC yields a value of 0.8173. The exclusion of the EE variable results in an AUC equal to 0.8162 (see Panels A and B in Figure 13.5). These results indicate that EE improves the model's prediction accuracy.

As discussed in section 13.2.1, the presented results involve the tightest definition of the variable EE which includes only A-rated buildings. In the Appendix, we relax the previous definition and consider also B-rated buildings as energy efficient (see, Billio et al. 2021). The results show that, even though the EE variable is still negatively associated with mortgage default risk, the statistical significance does not survive in all the specifications that include additional control variables. This highlights that only A-rated buildings have a lower probability of default.

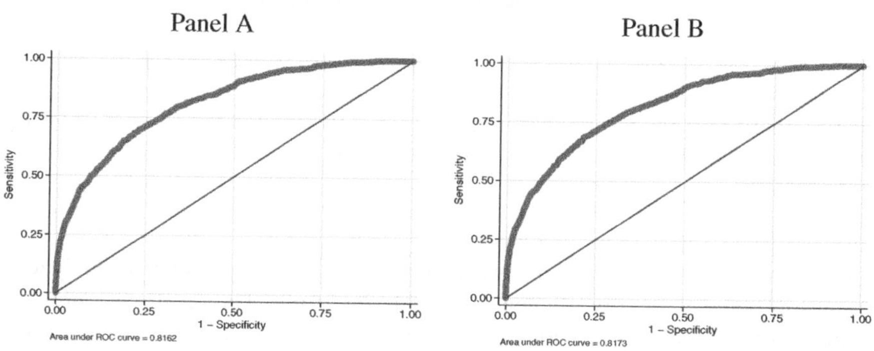

Figure 13.5. *Panel A (B) depicts the ROC excluding (including) the EE, using the model specification (6)*

13.4.2. *Estimates from the Cox regression*

As discussed in the previous section, the Cox model is typically applied in survival studies. For correct application of the model, it is necessary to test if the proportional hazards assumption holds. Figure 13.6 shows the empirical survivor functions for EE and non-EE mortgages, respectively. The plot reveals that the two curves do not intersect or diverge excessively, implying that the proportionality assumption is valid.

According to this finding, the hazard ratio for any two loans may be expected to be constant across time. Furthermore, the survival curves show that EE mortgages, on average, survive longer than non-efficient counterparts, as seen by the growing difference between the two curves. This also underlines the fact that mortgages on EE buildings are less likely to fail.

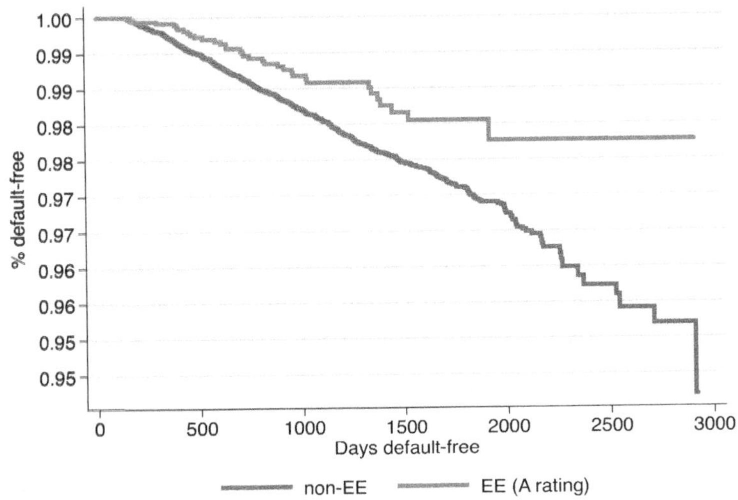

Figure 13.6. *Survivor functions. This figure shows the Kaplan–Meier survival curves for two mortgage groups: mortgages on EE (A-rated) and non-EE buildings. The Log-rank statistical test assesses the equality of survivor functions, and in this case, it yields a p-value of 0.0019. Therefore, the null hypothesis of equality of the two survivor function is not accepted*

Similar to the logit regression, we estimate the Cox regression to analyze the relationship between energy efficiency and survival time. The results are presented in Table 13.5. Model (1) provides the findings by just including mortgage-related characteristics in the model, such as the borrower's credit score, loan-to-value at mortgage origination and loan length. The EE indicator's regression coefficient is negative and significant (-0.4016), corroborating the results of the Logit regression. Accounting for household, dwelling and market control variables does not affect the primary finding on EE, as shown in model specifications (2)–(6).

Analogously to the Logit regression, if we extend the EE definition and also include B-rated, the EE variable is still negatively associated with mortgage default risk, but again the statistical significance fails after considering additional control variables. Also, in this case, A-rated buildings are associated with a lower default risk.

Dependent variable: default dummy						
	(1)	(2)	(3)	(4)	(5)	(6)
EE (A rating)	-0.4016**	-0.3124*	-0.3237*	-0.3358*	-0.3076*	-0.3009*
	[0.1724]	[0.1762]	[0.1758]	[0.1766]	[0.1773]	[0.1773]
Credit score	-0.0143***	-0.0143***	-0.0140***	-0.0140***	-0.0140***	-0.0140***
	[0.0004]	[0.0004]	[0.0004]	[0.0004]	[0.0004]	[0.0004]
Loan-to-value	0.8688***	0.8725***	0.9276***	0.8850***	0.8563***	0.8473***
	[0.2058]	[0.2053]	[0.2046]	[0.2067]	[0.2087]	[0.2087]
Loan term	0.0319***	0.0326***	0.0398***	0.0398***	0.0378***	0.0382***
	[0.0055]	[0.0055]	[0.0058]	[0.0058]	[0.0059]	[0.0059]
Building age		0.0030**	0.0026**	0.0025**	0.0022*	0.0022*
		[0.0012]	[0.0012]	[0.0012]	[0.0013]	[0.0013]
Borrower age			0.0137***	0.0136***	0.0130***	0.0128***
			[0.0036]	[0.0036]	[0.0036]	[0.0036]
Inflation						7.6548
						[11.0186]
Unemployment						3.5620
						[2.5977]
HPI growth						2.0283
						[3.8225]
Observations	70,642	70,642	70,642	70,642	70,642	70,642
Dwelling controls	No	Yes	Yes	Yes	Yes	Yes
Household controls	No	No	Yes	Yes	Yes	Yes
Market controls	No	No	No	No	No	Yes
Mortgage controls	Yes	Yes	Yes	Yes	Yes	Yes
Region FE	No	No	No	No	Yes	Yes
Year FE	No	No	No	Yes	Yes	Yes
SE	Rob.	Rob.	Rob.	Rob.	Rob.	Rob.
Pseudo R-squared	0.0560	0.0563	0.0571	0.0575	0.0587	0.0588

Table 13.5. *Cox regression results to assess the relationship between buildings' energy efficiency and the default risk of the borrowers. The dependent variable is binary indicating whether a mortgage is in default (e.g. at least three months in arrears) or not. The dummy variable EE equals to one if a building's energy efficiency rating is A-rated and zero otherwise. Mortgage controls are borrower's credit score, loan-to-value and loan term (in years). Dwelling control is building age at loan origination. Borrower control is borrower's age at loan origination. Market controls are the monthly Italian inflation rate (change in the consumer price index to previous year's value in the same month), quarterly unemployment rate at the regional level and quarterly house price index growth at the regional level. Origination year and NUTS1-region fixed effects (FE) are included where indicated. Standard errors (reported in square brackets) are robust. Statistical significance is denoted by ***, ** and * at the 1%, 5% and 10% level, respectively*

Dependent variable: default dummy	Logit model			Cox model		
	(1)	(2)	(3)	(4)	(5)	(6)
A/B rating	-0.4013***	-0.3772***	-0.3804***	-0.3919***	-0.3166**	-0.3203**
	[0.1148]	[0.1316]	[0.1316]	[0.1179]	[0.1297]	[0.1295]
C/D rating	-0.3349***	-0.3336***	-0.3503***	-0.3405***	-0.2803***	-0.2952***
	[0.0953]	[0.1055]	[0.1056]	[0.0966]	[0.1037]	[0.1036]
E/F rating	-0.3736***	-0.2346***	-0.2416***	-0.2292***	-0.2077**	-0.2154***
	[0.0808]	[0.0852]	[0.0852]	[0.0805]	[0.0834]	[0.0835]
Credit score	-0.0161***	-0.0151***	-0.0151***	-0.0142***	-0.0139***	-0.0140***
	[0.0004]	[0.0004]	[0.0005]	[0.0004]	[0.0004]	[0.0004]
Loan-to-value	0.3857*	0.9739***	0.9649***	0.8568***	0.8494***	0.8381***
	[0.1980]	[0.2157]	[0.2157]	[0.2076]	[0.2100]	[0.2100]
Loan term	0.0299***	0.0414***	0.0417***	0.0330***	0.0384***	0.0388***
	[0.0056]	[0.0061]	[0.0061]	[0.0055]	[0.0059]	[0.0059]
Building age		0.0008	0.0006		0.0008	0.0006
		[0.0014]	[0.0015]		[0.0014]	[0.0014]
Borrower age		0.0135***	0.0133***		0.0130***	0.0128***
		[0.0037]	[0.0037]		[0.0036]	[0.0036]
Inflation			13.3941			7.7655
			[11.1310]			[11.0023]
Unemployment			5.4038**			4.4800*
			[2.7096]			[2.6457]
HPI growth			-2.4965			2.3931
			[3.8339]			[3.8120]
Observations	71,011	70,666	70,666	70,642	70,642	70,642
Dwelling controls	No	Yes	Yes	No	Yes	Yes
Household controls	No	Yes	Yes	No	Yes	Yes
Market controls	No	No	Yes	No	No	No
Mortgage controls	Yes	Yes	Yes	Yes	Yes	Yes
Region FE	No	Yes	Yes	No	Yes	Yes
Year FE	No	Yes	Yes	No	Yes	Yes
SE	Rob.	Rob.	Rob.	Rob.	Rob.	Rob.
Pseudo R-squared	0.124	0.163	0.164	0.0569	0.0592	0.0594

Table 13.6. *Degree of Energy Efficiency estimates to assess the likelihood of default on mortgages backed by buildings with different degrees of energy efficiency. Logit regression estimates are included from columns 1 to 3, and Cox regression from columns 4 to 6, respectively. The dependent variable is binary indicating whether a mortgage is in default (e.g. at least three months in arrears) or not. The EE dummy variables are built on the four energy efficiency categories: (i) one if the building's rating is A or B, (ii) one if the rating is C or D, (iii) one if the rating is E or F and (iv) one if G and zero otherwise. The latter represents the reference case, i.e. the omitted category in the regressions. Mortgage controls are borrower's credit score, loan-to-value and loan term (in years). Dwelling control is building age at loan origination. Borrower control is borrower's age at loan origination. Market controls are monthly Italian inflation rate (change in the consumer price index to previous year's value in the same month), quarterly unemployment rate at the regional level, quarterly house price index growth at the regional level. Origination year and NUTS1-region fixed effects (FE) are included where indicated. Standard errors (reported in square brackets) are robust. Statistical significance is denoted by ***, ** and * at the 1%, 5% and 10% level, respectively*

13.4.3. *Additional findings*

In the presented analyses, we have focused on the existence of the relationship between a building's energy efficiency and the probability of mortgage default. Given the above findings, we extend the analysis to have a more granular representation of energy efficiency. As in Kaza et al. (2014), we investigate the relationship between a degree of energy efficiency and the likelihood of default by assuming that the higher the energy efficiency of a building, the lower is risk of default. To carry out the new analysis, we build a set of new indicator variables by aggregating the EE ratings into four efficiency classes. Hence, efficiency class 1 involves ratings A and B, class 2 includes ratings C and D, class 3 is assigned to ratings E and F, and class 4 is identified by G-rated buildings. All the previously defined control variables remain unchanged. Table 13.6 includes the regression results for Logit (models (1)–(3)) and Cox regressions (models (4)–(6)). There are two major implications from the regression estimates for efficiency classes 1–3. Firstly, all three regression coefficients are negative and significant, implying that G-rated buildings have the highest risk of mortgage default. Secondly, across all model settings, the regression coefficients show a diminishing trajectory with the degree of energy efficiency. This indicates that more EE buildings have stronger mitigation in the likelihood of default, implying that the degree of energy efficiency matters. As a result, the risk of default is reduced when a building renovation improves the EPC rating by one or two notches (from E to C). These findings remain unchanged by the inclusion of additional control variables. Based on these results, we may deduce that mortgages guaranteed by EE residential buildings have a reduced risk of default. In addition, the results indicate that degree of energy efficiency matters and is correlated with a reduced probability of default.

13.5. Conclusion

Studies on energy efficiency and residential mortgages are crucial in establishing future energy regulations (Billio et al. 2021). These findings also provide intriguing implications in risk management since a reduced risk of insolvency for EE buildings should induce alternative pricing for the borrower, such as lower interest rates. In this chapter, we contribute to this growing body of literature by analyzing the relationship between building with regard to energy efficiency and the risk of default on an Italian portfolio of residential mortgages. Our results indicate a negative and significant association between buildings' energy efficiency and the probability of mortgage default. A receiver-operating characteristic (ROC) analysis confirms that the inclusion of energy efficiency information improves the measurement of the borrower's default risk. Finally, we show that the degree of energy efficiency also matters, i.e. buildings with higher energy efficiency ratings are associated with a relatively lower risk of default. These findings allow us to better understand if and to what extent energy efficiency is desirable in the European mortgage market.

Dependent variable: default dummy

	(1)	(2)	(3)	(4)	(5)	(6)
EE (A/B rating)	-0.2492**	-0.2318*	-0.2374**	-0.1880	-0.1713	-0.1664
	[0.1129]	[0.1191]	[0.1189]	[0.1181]	[0.1184]	[0.1183]
Credit score	-0.0159***	-0.0159***	-0.0157***	-0.0151***	-0.0151***	-0.0152***
	[0.0004]	[0.0004]	[0.0004]	[0.0004]	[0.0004]	[0.0004]
Loan-to-value	0.3378*	0.3384*	0.3897*	0.9995***	0.9810***	0.9746***
	[0.1997]	[0.1995]	[0.1997]	[0.2129]	[0.2148]	[0.2149]
Loan term	0.0344***	0.0344***	0.0410***	0.0425***	0.0407***	0.0409***
	[0.0058]	[0.0058]	[0.0060]	[0.0060]	[0.0061]	[0.0061]
Building age		0.0006	0.0003	0.0027**	0.0025*	0.0024*
		[0.0013]	[0.0013]	[0.0013]	[0.0013]	[0.0013]
Borrower age			0.0124***	0.0139***	0.0134***	0.0133***
			[0.0037]	[0.0037]	[0.0037]	[0.0037]
Inflation						13.5078
						[11.1017]
Unemployment						4.2891
						[2.6441]
HPI growth						-2.8819
						[3.8299]
Observations	70,666	70,666	70,666	70,666	70,666	70,666
Dwelling controls	No	Yes	Yes	Yes	Yes	Yes
Household controls	No	No	Yes	Yes	Yes	Yes
Market controls	No	No	No	No	No	Yes
Mortgage controls	Yes	Yes	Yes	Yes	Yes	Yes
Region FE	No	No	No	No	Yes	Yes
Year FE	No	No	No	Yes	Yes	Yes
SE	Rob.	Rob.	Rob.	Rob.	Rob.	Rob.
Pseudo R-squared	0.114	0.114	0.115	0.159	0.162	0.162

Table 13.7. *Logit regression results to assess the relationship between buildings' energy efficiency and the default risk of the borrowers. The dependent variable is binary indicating whether a mortgage is in default (e.g. at least three months in arrears) or not. The dummy variable EE equals one if the energy efficiency of a building is A- or B-rated, and zero otherwise. Mortgage controls are borrower's credit score, loan-to-value and loan term (in years). Dwelling control is building age at loan origination. Borrower control is borrower's age at loan origination. Market controls are monthly Italian inflation rate (change in the consumer price index to previous year's value in the same month), quarterly unemployment rate at regional level, quarterly house price index growth at regional level. Origination year and NUTS1-region fixed effects (FE) are included where indicated. Standard errors (reported in square brackets) are robust. Statistical significance is denoted by ***, ** and * at the 1%, 5% and 10% level, respectively*

Dependent variable: default dummy

	(1)	(2)	(3)	(4)	(5)	(6)
EE (A/B rating)	-0.2399**	-0.1502	-0.1534	-0.1565	-0.1392	-0.1353
	[0.1109]	[0.1162]	[0.1158]	[0.1160]	[0.1164]	[0.1163]
Credit score	-0.0143***	-0.0143***	-0.0140***	-0.0140***	-0.0140***	-0.0141***
	[0.0004]	[0.0004]	[0.0004]	[0.0004]	[0.0004]	[0.0004]
Loan-to-value	0.8612***	0.8717***	0.9268***	0.8864***	0.8573***	0.8481***
	[0.2064]	[0.2059]	[0.2051]	[0.2072]	[0.2092]	[0.2092]
Loan term	0.0323***	0.0327***	0.0398***	0.0399***	0.0378***	0.0381***
	[0.0055]	[0.0055]	[0.0058]	[0.0058]	[0.0059]	[0.0059]
Building age		0.0029**	0.0026**	0.0025*	0.0022*	0.0022*
		[0.0013]	[0.0013]	[0.0013]	[0.0013]	[0.0013]
Borrower age			0.0136***	0.0135***	0.0129***	0.0128***
			[0.0036]	[0.0036]	[0.0036]	[0.0036]
Inflation						7.7273
						[11.0224]
Unemployment						3.6253
						[2.5980]
HPI growth						2.0206
						[3.8231]
Observations	70,642	70,642	70,642	70,642	70,642	70,642
Dwelling controls	No	Yes	Yes	Yes	Yes	Yes
Household controls	No	No	Yes	Yes	Yes	Yes
Market controls	No	No	No	No	No	Yes
Mortgage controls	Yes	Yes	Yes	Yes	Yes	Yes
Region FE	No	No	No	No	Yes	Yes
Year FE	No	No	No	Yes	Yes	Yes
SE	Rob.	Rob.	Rob.	Rob.	Rob.	Rob.
Pseudo R-squared	0.0560	0.0562	0.0570	0.0574	0.0586	0.0587

Table 13.8. *Cox regression results to assess the relationship between buildings' energy efficiency and the default risk of the borrowers. The dependent variable is binary indicating whether a mortgage is in default (e.g. at least three months in arrears) or not. The dummy variable EE equals one if the energy efficiency of a building is A- or B-rated, and zero otherwise. Mortgage controls are borrower's credit score, loan-to-value and loan term (in years). Dwelling control is building age at loan origination. Borrower control is borrower's age at loan origination. Market controls are monthly Italian inflation rate (change in the consumer price index to previous year's value in the same month), quarterly unemployment rate at the regional level and quarterly house price index growth at the regional level. Origination year and NUTS1-region fixed effects (FE) are included where indicated. Standard errors (reported in square brackets) are robust. Statistical significance is denoted by ***, ** and * at the 1%, 5% and 10% level, respectively*

13.6. Appendix

In this section, we follow Billio et al. (2021) and extend the concept of energy efficiency to B-rated buildings. The model specifications are the same as in the main analysis and the estimation results for the logit and Cox regressions are reported in Tables 13.7 and 13.8, respectively. As can be seen, EE is still negatively related to the probability of default, but the results are weaker and the model coefficient is not significantly different from zero for several specifications. There are several reasons for this. First, the vast majority of loans for A-/B-rated properties were approved recently, so the observation period may not be sufficient to identify many defaults. Model (4) in Table 13.7 shows that the presence of fixed effects for the year of loan origination reduces the statistical significance of the EE variable in the logit estimates. Moreover, credit score is a good predictor of default in all model constellations. This suggests that the control variables used may be missing some important household characteristics that could help separate the EE effect from other household-specific aspects. For instance, environmentally conscious households with higher incomes are more likely to purchase or build an EE dwelling because they can both afford it and benefit from it. Analyses of this type are reserved for future research, as data granularity and availability remain the major obstacles.

13.7. References

Altman, E.I. and Saunders, A. (1997). Credit risk measurement: Developments over the last 20 years. *Journal of Banking & Finance*, 21(11), 1721–1742.

An, X. and Pivo, G. (2020). Green buildings in commercial mortgage-backed securities: The effects of LEED and energy star certification on default risk and loan terms. *Real Estate Economics*, 48(1), 7–42.

Billio, M., Costola, M., Pelizzon, L., Riedel, M. (2021). Buildings' energy efficiency and the probability of mortgage default: The Dutch case. *The Journal of Real Estate Finance and Economics*, 65, 419–450. DOI:10.1007/s11146-021-09838-0.

Cox, D.R. (1972). Regression models and life-tables. *Journal of the Royal Statistical Society. Series B (Methodological)*, 34(2), 187–220.

EeDaPP (2020). Energy efficiency data protocol and portal.

Guin, B. and Korhonen, P. (2020). Does energy efficiency predict mortgage performance? Bank of England working papers 852, Bank of England.

Kaplan, E.L. and Meier, P. (1958). Nonparametric estimation from incomplete observations. *Journal of the American Statistical Association*, 53(282), 457–481.

Kaza, N., Quercia, R.G., Tian, C.Y. (2014). Home energy efficiency and mortgage risks. *Cityscape*, 16(1), 279–298.

Metz, C.E. (1978). Basic principles of ROC analysis. *Seminars in Nuclear Medicine*, 8(4), 283–298.

Qi, M., Scheule, H., Zhang, Y. (2020). Positive payment shocks, liquidity and refinance constraints and default risk of home equity lines of credit at end of draw. *The Journal of Real Estate Finance and Economics*, 1–32.

Zweig, M.H. and Campbell, G. (1993). Receiver-operating characteristic (ROC) plots: A fundamental evaluation tool in clinical medicine. *Clinical Chemistry*, 39(4), 561–577.

14

The Thesis for Green Investing and Other ESG through the Looking Glass of China and the US

14.1. Introduction

Green investing, which falls under "E" in conventional ESG investing, has become dominant in ESG research. Green investing seeks to tilt firm behaviors towards greater care for our (E)nvironment. Where the invisible hand of the market stops, the visible hand of ESG investors moves in to impact the behavior of corporate executives and fund managers. When enough investors reduce the allocation to firms that pollute, the share prices of firms with a bad "E" score fall and their cost of equity increases. Declining share prices also hurts executives whose wealth is concentrated in their companies' equities. The threat of "divesture" by "Green" investors incentivizes these executives to care about environmental issues, when otherwise they might focus exclusively on firm profit maximization.

Advocates of "Green investing", while most of them are willing to sacrifice returns for a better environment, do understand that if "return augmentation" can be assured, the Green movement would be meaningfully more successful. In this chapter, we will specifically explore whether there is a premium to Green investing; if that premium is negative, who should bear the cost? However, firstly, we discuss whether society at large can agree upon the measurement of "Greenness" and the prioritization of "Green solutions". We illustrate with examples from the United States and China to show where there are meaningful disagreements between

Chapter written by Brad CORNELL and Jason C. HSU.

For a color version of all figures in this chapter, see www.iste.co.uk/jurczenko/climate.zip.

different constituency groups on measurement and priorities. It is important to ask: how should we as a society think about Green investing? Whose values, environmental impact assumptions and technological understanding should be reflected? Insofar that there are costs to be borne by fund investors to benefit select firms, entrepreneurs and research institutions as we push to advance environmental protection, how should we think about the rights and responsibilities of Green advocacy?

14.2. Who and what does Green investing impact?

Green investors should consider several important dimensions before investing. Firstly, they must understand what a company's composite "E" score reflects. ESG investing reduces the overall investment opportunity set. Theoretically, it could reduce a portfolio's expected return. It would be counterproductive to incur costs only to impact the wrong issues, behaviors and firms.

Much of the debate around Green investing starts with the premise that we can differentiate between "good" and "bad" companies on issues surrounding the environment; this is often not the case. Unlike profitability and returns, where there are accepted measures for both, what is environmentally responsible or has a positive Green impact are often in the eyes of the beholder. It should come as no surprise that a ranking of companies by Greenpeace bears little resemblance to a ranking by a group focused on renewable energy. In fact, among different alternative energy advocacy groups, there is wild disagreement on the "Greenness" of various technology adaptations.

The problem is further compounded by the observation that the supposedly unbiased experts, the ESG scoring agencies, often disagree on whether a company is "Green" or not. There is a substantial amount of literature documenting the divergence of ESG ratings for the same firms (see Dortfleitner et al. 2015; Semenova and Hassel 2015; Chatterji et al. 2016; Berg et al. 2019; Li and Polychronopoulos 2020). Hsu et al. (2018) and Dimson et al. (2020) examine ESG ratings for companies from top providers such as FTSE Russell, Sustainalytics, Refinitiv and MSCI to find whether (1) correlations between measures are low across the services and (2) whether they disagree on high-profile companies. Facebook, for instance, is ranked at the 1st percentile (among the worst) by Sustainalytics on environment and at the 96th percentile (among the best) by MSCI. The rating organizations differ not only in how to measure the various "E" criteria, but also with respect to which criteria are deemed worthy of measurement.

14.3. Who should set the Green investing agenda?

Our focus is not on whether and how Green can be measured. Other than noting the difficulties associated with measurement, we nonetheless assume that such measurement is possible. A more interesting question is which people, industries, governments or NGOs get to set Green priorities and agendas.

Every society, no matter its form of governance, faces the same fundamental questions in establishing their environmental policies: (1) *who* should set the rules? and (2) *how* should the rules be set? In this chapter, we examine (1) *what* influence should corporations have in directly or indirectly setting environmental policy, and (2) *which* rights should remain with government and public organizations? An important contribution from this chapter is our investigation of how answers to these questions differ between China and the United States.

14.3.1. *Should Green Initiatives be determined by elected civil servants or by rating services, investment funds and corporate CEOs?*

It is difficult to imagine two more strange bedfellows than famed American free-market economist Milton Friedman, who champions "greed is good" and Communist China's President Xi Jinping, who champions greater state influence to ensure "common prosperity". Nonetheless, when it comes to deciding the role of private companies in determining environmental policies, the two are in remarkable agreement. Xi, in our current discussion on Green investing, is of particular interest because of his recent commitment to lead China, a historically high carbon emission country, to achieve carbon neutrality by 2060.

14.3.2. *The Milton Friedman take on who should drive ESG*

In his classic New York Times article, Milton Friedman (1970) anticipated three fundamental issues that lie at the heart of the ESG movement. The three issues, all of which are interrelated, are *standing, taxation, and information*. In this context, standing and taxation are two sides of the same coin. We consider each in turn.

Starting with standing, Friedman puts in this way:

> What does it mean to say that the corporate executive has a "social responsibility" in his capacity as businessman? If this statement is not pure rhetoric, it must mean that he is to act in some way that is not in the interest of his employers. In each of these cases, the corporate executive would be spending someone else's money for a general

> social interest. Insofar as his actions in accord with his "social responsibility" reduce returns to stockholders, he is spending their money.

As the first Friedman quote indicates, standing is closely related to taxation because in pursuing "true" ESG the corporate executives effectively tax shareholders to provide benefits to society at large. In this vein, Freidman goes on to say:

> This process raises political questions on two levels: principle and consequences. On the level of political principle, the imposition of taxes and the expenditure of tax proceeds are governmental functions. Here the businessman – self-selected or appointed directly or indirectly by stockholders – is to be simultaneously legislator, executive and jurist… He becomes in effect a public employee, a civil servant, even though he remains in name an employee of private enterprise. On grounds of political principle, it is intolerable that such civil servants – insofar as their actions in the name of social responsibility are real and not just window-dressing – should be selected as they are now. If they are to be civil servants, then they must be selected through a political process. If they are to impose taxes and make expenditures to foster "social" objectives, then political machinery must be set up to guide the assessment of taxes and to determine through a political process the objectives to be served.

Friedman has identified the central issue related to ESG. As Friedman sagely observed, if ESG is to be more than mere rhetoric, it must differ from the traditional objective – maximization of shareholder (either shareholders in corporations or in funds which invest in companies) wealth. Stakeholder capitalism, for example, is where corporate or investment fund managers pursue environmental goals to deliver "value" to stakeholders other than shareholders.

Let us first focus on stakeholder capitalism as applied to corporations. In its famous statement on the purpose of the corporation, the Business Roundtable (2019) put it this way: "*Each of our stakeholders is essential. We commit to deliver value to all of them, for the future success of our companies, our communities and our country.*" The statement is discordant to an economist. Economics is the science that studies the allocation of scarce resources to alternative uses. A company can always deliver more to one group of stakeholders, say customers, by charging lower prices at the expense of another, stockholders. If everyone is essential, how can it be determined how much one group will be disadvantaged to benefit another? One way to deliver value, to say Apple's customers, is to reduce the price of Apple's products. Doing so, however, would reduce the value of the company and harm shareholders. The tradeoff is obvious, and there are similar tradeoffs involving all other stakeholders. Ignoring such tradeoffs, the Business Roundtable doubles down on this idea by saying, "*we share a fundamental commitment to all of our stakeholders*."

Let us now examine stakeholder capitalism in fund management. Larry Fink, Chairman of Blackrock, in 2021, sent a letter to the CEOs of companies in which BlackRock funds invest, calling on their companies to become more cognizant of ESG. Focusing on the E part of the acronym, Fink stated, "*Climate change has become a defining factor in companies' long-term prospects*." Echoing the Business Roundtable, Fink goes on to say:

> The importance of serving stakeholders and embracing purpose is becoming increasingly central to the way that companies understand their role in society. As I have written in past letters, a company cannot achieve long-term profits without embracing purpose and considering the needs of a broad range of stakeholders.

If all this is to be more than mere rhetoric, it must signal a change in emphasis, a broader view of the mission of the corporation and the investment funds which invest in them. The stakeholder approach, in conjunction with ESG, raises the question of exactly how this "new" approach deviates from traditional shareholder wealth maximization and what should be done when a deviation arises. Even a firm committed to shareholder value maximization will pay close attention to environmental issues. A company that obviously polluted its environment would quickly arouse the ire of regulators and the public and, in so doing, destroy shareholder wealth.

If what Green proponents are advocating is that mangers should take account of environmental regulators and the brand value from being ostensibly Green, then the approach reduces to what can be called "enlightened" value maximization. Green and other ESG investing is then not a repudiation of the traditional approach; it simply reminds managers to apply it wisely. It is only when executives think they should take action that destroys shareholder wealth to promote the interests of non-investor stakeholders, no matter how noble the cause, that a real issue arises.

In this context, the real problem is that Green investing and other ESG approaches provide no clear method for determining how broad that mission should be or how much shareholders will be asked to pay to support it. In addition, there is a practical problem represented by the conflict of interest between large shareholders with the ability to influence management and the small shareholders who do not. BlackRock, as the largest fund company and often one of the top shareholders in most listed firms in the world, has significant influence. If BlackRock instructs CEOs to pursue Green initiatives, which meaningfully deviate from firm value maximization, thus harming other investors, what might be the legal ramification?

14.3.3. *American ESG in conflict with American democracy?*

It is also worth noting that corporate ESG, as Friedman notes, runs afoul of democracy. The democratic ideal is that key public issues, which include environmental policy, must be determined by people elected by citizens based on a one-person, one-vote rule. Transferring policy decisions to corporate executives or investment funds not only removes them from public control, it also violates the one-person, one-vote rule.

Ultimately, corporate executives are appointed by shareholders, and shareholder voting is wealth weighted. Rich shareholders have power far beyond their numbers. This fact has been further complicated by the rise of huge fund managers such as Blackrock and Vanguard, who generally cater to wealthy investors. At this juncture, it remains unclear of how such institutions will exercise their massive voting power. But whatever the ultimate outcome, it is clearly not consistent with a one-person, one-vote rule. American democracy and ESG seem to be at loggerheads on the issue of how much power should be transferred to corporations and fund houses. Friedman suggests that ideally, the democratically elected government should set the appropriate regulations subject to careful consideration of the science and of the cost--benefit tradeoff that its citizens would accept. Corporations and fund houses should then go about their business of maximizing shareholder value subject to those rules. If and when advocacy groups disagree with the environmental policy in place, the democratic approach is to lobby elected representatives to legislate new regulations.

14.3.4. *Who drives environmental protection policy and other ESG issues in China?*

Xi Jinping could not be in more agreement with Milton Friedman. Although China does not have the same democratic process as the United States, the country still must address many of the same environmental issues that exist in the United States and elsewhere in the world. However, the Chinese Communist Party has made it clear that environmental issues fall under the purview of the government, not the country's corporations or investment managers. When asked about ESG investing in China, the Chinese Securities Regulatory Commission (CSRC is China's financial markets regulatory authority) made it very clear that addressing social and environmental issues are responsibilities and agenda for other government agencies and not that of the financial markets or its regulator.

On the environmental front, President Xi Jinping and his government have stated that reaching carbon neutrality is a key objective. Beijing has committed to achieving carbon neutrality by 2060. In this regard, Beijing has mandated disclosures from companies on carbon emission and other Green and sustainability measures with great vigor. With respect to the environment, the government is setting the rules and expecting companies to follow.

The Chinese government has provided a carrot as well as a stick. In recent years, Chinese central and regional governments have made substantial financial subsidies and policy tailwind available for firms in the Green energy space. This has included (1) low-cost equity and debt capital for "Green tech firms" provided by various government-led industry funds, (2) tax subsidy for EV manufacturers and price subsidy for EV buyers, (3) accelerated vehicle licensing approval (it could take years to obtain a license for a traditional Brown energy vehicle in major cities), (4) license fee reduction, etc. Similarly, there is a trend towards increased regulatory headwind for firms in the Brown energy space. Being Green and working to achieve carbon neutrality pays for firms in China, but firms are not empowered to choose their own preferred interpretation of Green.

There is a subtle but important implication in Milton Friedman and Xi Jinping's views on environmental policy. Both agree that the right entity to pressure for stricter environmental regulations is the government. Both do not support domestic corporate executives and fund managers to additionally "tax" their shareholders to implement additional Green agenda. What is not clear is whether either might endorse foreign investors to impact domestic environmental policies through Green investing. For example, what constitutes a proper environmental policy necessarily differs between the United States and China. In the West, a great deal of emphasis has been placed on achieving net zero carbon emissions as quickly as possible. Pressure has been put on companies to announce the steps they will take to reduce emissions and report on their progress. China's policies are more nuanced for economic reasons. GDP per capita in China was only \$10,400 in 2020 compared to \$63,000 for the United States. Consequently, the Chinese government places more emphasis on economic development, even at the expense of added emissions. For example, coal currently accounts for almost 60% of electricity generation. Welch and Cornell (2022) report that China is building, and planning to build, new coal plants with generating capacity equal to one-fourth of the entire electric generation of the United States. The Chinese government has concluded that the economic benefits of the new plants offset the added emissions that come from burning coal.

If Green advocacy groups find China's environmental policy unacceptable relative to the West but have been ineffective at lobbying the Chinese Communist Party for a stricter policy, should global investors then seek to impact environmental practices of listed corporations inside China through Green investing? This is a fascinating exploration that challenges Friedman's critique on the "undemocratic nature" of ESG investing. ESG investing gives unfair influence to wealthy individuals by impacting domestic environmental policies beyond the standard political process. Does the same objection apply when it comes to wealthy foreign advocacy groups wishing to influence domestic environmental policy through Green investing?

14.3.5. *Good intentions but bad skills?*

The final issue is information. Even though it were to be agreed that control over certain environmental issues should be transferred to corporate executives, it is far from clear whether they have the proper information to deal with such issues. Do we expect corporate executives to understand various competing Green technologies and choose the best path forward? Or are we happy with any effort to invest in "Green" and behave "Green" – in effect, is good intention good enough? Plenty of firms and funds today go through a box-checking exercise to signal their commitment to Green without deep research into the actual consequences of their actions.

Consider, for example, solar geo-engineering that involves injecting particles into the upper atmosphere in order to increase reflection of solar radiation, thereby halting global warming. Aldy and Zeckhauser (2020) argue that this is the most economical way to combat climate change. Of course, this argument is controversial. Our point is not to accept or reject the authors' assertion, but to note that this is the type of question that needs to be addressed and studied by a variety of experts in conjunction with the Department of Energy. It is not the sort of issue on which busy executives of say, athletic apparel companies, are able to evaluate. Should the US Department of Energy take a lead role in defining the "E" in ESG? Or should that be left to a combination of for-profit ESG rating agencies, mutual fund managers and corporate CEOs? The Chinese government clearly favors assigning the responsibility of determining the measurements of Green compliance to specialized agencies with domain expertise; instead of letting fund managers and corporate CEOs determine what would best achieve carbon neutrality.

Finally, without the government taking a more central role in providing clarity on the path towards carbon neutrality, honest efforts are likely met with sub-optimal outcomes. In his latest letter to CEOs, BlackRock chairman Larry Fink (2022) has stated:

> Given how central the energy transition will be to every company's growth prospects, we are asking companies to disclose a plan for how their business model will be compatible with a net zero economy – that is, one where global warming is limited to well below 2°C, consistent with a global aspiration of net zero greenhouse gas emissions by 2050. We are asking you to disclose how this plan is incorporated into your long-term strategy and reviewed by your board of directors.

The problem is that any plan for dealing with a transition to a net zero economy depends critically on the regulatory policies the government will put in place to achieve those objectives. However, the US government and its experts are still researching the path forward. Thus, companies are in effect being asked to project future government policy. Again, given that most companies lack the expertise in understanding the science, it is far from clear whether there is a net social benefit from bearing meaningful costs for research and adopting Green policies that must ultimately be passed on to employees and customers.

14.4. Earning a Green alpha?!

One thing many Western companies have suggested, perhaps to placate shareholders, is that pursuing Green objectives is not only virtuous but will also increase both corporate value and investors' returns. There is an element of self-interest here. In fact, there is political pressure for public pension funds, mandated to implement ESG more generally and Green investing more specifically, to pursue these noble causes in a way that also enhances portfolio outcome. To do otherwise has fiduciary implications, leading to an ugly impasse between politicians who represent the values of citizens, and pension investment officers who represent the financial interests of future retirees. Additionally, for investment product providers, ESG funds are difficult to market without the potential of risk-return improvement.

The promise that doing good leads to financial reward can be seen in Mr. Fink's letter. He specifically says that following more stakeholder-oriented goals is the only way to achieve long-term profits. This seems to imply that Green efforts will produce greater value and higher expected returns. Unfortunately, the reality is more complicated and not so sanguine. To begin with, if investors prefer Green companies and are therefore more willing to pay a premium for them, how can buying Green firms lead to higher investor returns?

In equilibrium, it cannot. Fama and French (2007) show that if investors prefer to invest in Green companies, the expected return from investing in those companies will be lower. In effect, the market is equilibrating not only pecuniary risk-adjusted returns, but also total risk-adjusted returns, including the non-pecuniary return from investing in Green companies. As a result, the pecuniary risk-adjusted return on bad companies will be higher than that on good companies. As an illustration of this effect, Hong and Kacperczyk (2009) and Dimson, Marsh and Staunton (2015, 2020) report that returns on "sin" stocks, i.e. companies involved in businesses such as producing alcohol, tobacco and gaming, have higher average returns than otherwise comparable stocks. Similarly, Barber, Morse and Yasuda (2021) find that dual-objective venture capital (VC) funds (those with explicit ESG objectives) earn 4.7 percentage points lower ex-post internal rates of return than traditional VC funds. More recently, a detailed study by Pastor, Stambaugh and Taylor (2021) concludes: "Green assets delivered high returns in recent years. This performance reflects unexpectedly strong increases in environmental concerns, not high expected returns."

There is an easy way to see what is going on here. The market can be interpreted as equilibrating total returns as opposed to financial returns, where the financial return is the traditional measure, and the total return includes the non-pecuniary utility associated with holding a stock. In this context, the non-pecuniary return from holding good ESG stocks would be positive and the non-pecuniary return from holding bad ESG stocks would be negative. Consequently, the financial return for investing in bad ESG stocks would have to exceed that for good ESG stocks in order for the market to be in equilibrium.

More formally, it is important to note that proponents of ESG investing believe that Green energy firms have a lower risk than Brown energy firms. Thus, the return premium, if there is any, from Green investing must be through the alpha channel instead of the risk compensation channel. As is true for any alpha opportunities, for Green signals to generate excess returns, the market has to be inefficient. Specifically, the market has to undervalue Green firms or overvalue Brown firms. This statement, obvious as it may seem, is not trivial from an asset pricing perspective. For Green attributes to be undervalued, they have to, first of all, impact future cash flows. When we examine the hundreds of Green variables defined by various ESG scoring vendors, for many variables, a direct correlation with future cash flow is less than obvious.

For the investors who accurately predicted this shift in Beijing's attitude towards Green ahead of others and who then correctly identified EV makers such as NIO, Xpeng, BYD and battery makers such as CATL as obvious "Green" firms, which will benefit from Beijing's new Green priority, Green investing has paid off handsomely. Similarly, screening out coal miners and firms in the petro supply chain – the obvious "Brown" companies – has led to equity portfolio outperformance. However, now that "Green-energy" companies trade at stratospherically valuation and “Brown-energy” companies trade at deep discounts, would Green investing continue to generate superior return or are future Green investors in China too late to the scene? Has the market already been priced in the policy shift and the changing consumer preference? Perhaps “Green energy” has even become overvalued despite all of the said advantages and is now “glitz and glamor”, while “Brown energy” has become a good value investment, despite facing significant policy headwind and brand value decline. Indeed, even financial market regulators in China have expressed concern regarding the retail investors’ seeming love affair with “Green” energy stocks. The CSRC has slowed approval for “Green” themed ETFs and mutual funds because many asset managers are rushing to the market with a “Green” offering based on the out-worldly historical “Green” stock performance. Based on historical evidence on abnormal Chinese mutual fund flow into hot thematic funds and subsequent fund returns, the likelihood of poor performance for Green investing is high. This would contradict the claim made by ESG investors championing the alpha from Green investing based on casual analysis of recent Green stock performance.

One reason for the confusion regarding the performance of highly rated Green companies in both China and the United States is that ex-post returns realized during a transition period are often conflated with expected returns in market equilibrium. A simple hypothetical example illustrates this issue. Assume that before Green became a popular issue, the required return on both “good” and “bad” Green companies was 7%. As a result of a growing non-pecuniary preference for highly rated Green stocks, over the next 12 years, the required return on good companies drops from 7% to 5% in equal annual increments, while the required return on bad companies remains 7%[1]. Figure 14.1 calculates the theoretical returns on both good and bad companies over the 12-year transition period and beyond. During the transition, the observed returns on the good companies exceed the equilibrium returns because of the drop in the discount rate. Once the transition is over, however, the expected returns for the good companies are 5% compared to an expected return for the bad companies of 7%. As Cornell (2020a, 2020b) observes, confusion arises because media coverage has tended to convey the notion that impressive performance of highly rated Green companies during the transition is indicative of future outperformance; however, the opposite is in fact true.

It should be noted that lower risk-adjusted pecuniary returns for good stocks should be socially beneficial. The lower expected return translates into a lower cost of equity for good firms. This will make it value enhancing for them to take on projects, say switching to renewable energy, which would be rejected if the cost of equity remained at 7%. The lower expected return also means that good firms will

1 It is assumed that each year’s drop in the required return is unexpected.

be more highly valued, but it is precisely this higher valuation that is associated with lower future investor returns in equilibrium.

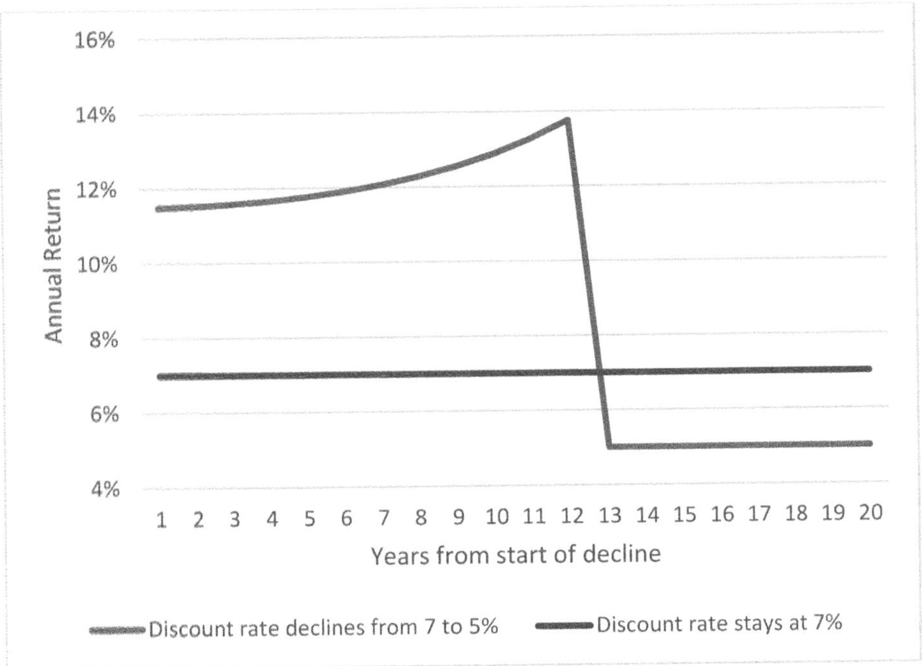

Figure 14.1. *Impact of declining discount rate for good stocks*

Larry Fink (2022) fails to make this distinction clear in his most recent letter to CEOs. Mr. Fink observes:

> Over the course of 2020, we have seen how purposeful companies, with better environmental, social, and governance (ESG) profiles, have outperformed their peers. During 2020, 81% of a globally-representative selection of sustainable indexes outperformed their parent benchmarks. This outperformance was even more pronounced during the first quarter downturn, another instance of sustainable funds' resilience that we have seen in prior downturns. But the story goes deeper. It's not just that broad-market ESG indexes are outperforming counterparts. It's that within industries – from automobiles to banks to oil and gas companies – we are seeing another divergence: companies with better ESG profiles are performing better than their peers, enjoying a "sustainability premium".

It is true that sustainable indexes outperformed during the period Mr. Fink referred to, but that is because it was during the height of the transition period. It does not mean that there will be a sustainability premium going forward.

Investor returns can also experience a short-term boost and valuations will rise if becoming a more highly rated Green firm increases corporate cash flows. However, in this case, there is no need for some new Green purpose of the corporation. Firms attempting to maximize shareholder value will do all they can to improve cash flow in any event.

Another question is whether there are unique Green factor risks affecting expected returns. For example, Engle et al. (2020) developed an algorithm for constructing what they call climate change hedge portfolios, which are equivalent to factor mimicking portfolios that reflect climate change risk. They find that the hedge portfolio has a negative risk premium. They conclude that "*This lower expected return corresponds to the insurance premium paid for the climate hedge portfolio.*" This conclusion is not surprising, assuming that the market is reasonably efficient and factor risks are properly priced, alpha will be unrelated to ESG rating. Good companies may have lower climate-related risks, but that reduced risk translates into lower expected returns.

14.5. Market efficiency and ESG

So far, we have implicitly assumed that the markets in the United States and China are efficient. This may not be true particularly in the short run and particularly in China. Even in the highly competitive US market, there is evidence of inefficiency in the ESG space. One example is the pricing of electric vehicle stocks. EV stock pricing is a prime example of what Cornell and Damodaran (2020) call the "the big market delusion". As Arnott et al. (2021) observe, the hallmark of a big market delusion is when all the firms in an evolving industry rise together even when they are direct competitors. Investors become so enthusiastic that each firm is priced as if it will be a major winner despite the fallacy of composition; the sum of the parts cannot be greater than the whole. In the case of EV stocks, while Tesla was rising 10-fold to over $1 trillion in market capitalization, virtually all of its competitors' stocks were also appreciating even though the global vehicle market was not growing.

The likelihood of inefficiency is even greater in China. We would suspect low price efficiency in a stock market like China, where 85% of the trade volume is contributed by retail speculators known for generating 4000% annual turnover. Given the massive run up in valuation multiples for Chinese Green stocks, when it comes to generating superior return from Green investing, the train has perhaps left the station. The excess return has been earned by the early Green investors at a time when the cost of capital to Green companies was high – when most investors did not embrace Green and did not forecast the dramatic shift in Beijing's attitude towards the environment. As Hsu et al. (2018) argue: the alpha is not driven by "Green", it is driven by the lack of support and capital for "Green" despite the potential upside from a perhaps predictable shift in Beijing attitude towards "Green" as the country emerge. Today, Green, as an initiative, has gone from hard to finance to being flooded with irrationally cheap capital. Thus, the highly inefficient Chinese A-shares market cuts both ways for Green investors. On the one hand, in previous years, the Green theme was underresearched and undervalued. This led to tremendous opportunities for Green investors to generate alpha while also achieving value alignment and impact. However, as undervaluation translated to outperformance over time as the political and consumer attitude shifted, the strong past return has caused unsophisticated retail investors to pile into Green stock and Green funds at very expensive valuation levels. The forward-looking Green alpha is likely to be very negative.

In summary, taking account of possible inefficiency does not lead to the conclusion that highly rated Green firms are likely to be good investments in either China or the United States. If anything, the reverse might be true. Investor enthusiasm has led to events such as the 10-fold increase in the price of Tesla that are difficult to rationalize on the basis of fundamentals. As is always the case, such high prices are indicative of lower future expected returns, not higher.

14.6. Conclusion

Every society must determine its environmental policies. As such, they must also decide how much authority regarding those policies should be transferred to private corporate executives and fund managers. In the United States, Milton Friedman teed up that question in his classic New York Times article in 1970. His conclusion was that rules regarding environmental issues should be set by democratically elected officials, and that companies should then attempt to maximize shareholder value in the context of those rules. Many American proponents of the new stakeholder approach rejecting Friedman's argument. They argue that companies must expand their objective by providing value for all stakeholders. This involves managers setting and pursuing environmental goals that may be contrary to maximizing shareholder value. To this date, the debate remains unresolved in the United States.

There is no such uncertainty in China. Although environmental issues are as front and center in China as they are in the United States, there is no debate on who has the power to determine those policies – the Chinese government. Corporations and investment fund managers are then expected to operate as efficiently as possible given the rules put into place by the government.

With respect to investing, American proponents of ESG have argued that "Green" companies can be expected to offer greater investment returns. While investors in Green companies have generally reaped a one-time windfall during the boom in sustainable investing, long-run equilibrium risk-adjusted returns for Green companies should be no higher than those for Brown companies and may well be less for two reasons. Firstly, if investors receive a non-pecuniary benefit from holding Green companies, then the financial return is likely to be depressed compared to that for Brown companies even in an efficient market. Secondly, to the extent that there are market inefficiencies, the prices of Green companies appear to have been driven to frothy levels in both China and the United States, implying that future returns will be lower.

14.7. References

Aldy, J.E. and Zeckhauser, R.J. (2020). Three prongs for prudent climate policy. NBER working paper no. 26991.

Arnott, R., Cornell, B., Wu, L. (2021). Big market delusion: Electric vehicles. *Research Affiliates*, 1–9 [Online]. Available at: www.researchaffiliates.com/publications/articles/826-big-market-delusion-electric-vehicles.

Barber, B.M., Morse, A., Yasuda, A. (2021). Impact investing. *Journal of Financial Economics*, 139(1), 162–185.

Berg, F., Koelbel, J., Rigobon, R. (2019). Aggregate confusion: The divergence of ESG ratings [Online]. Available at: https://papers.ssrn.com/sol3/papers.cfm?abstract_id=3438533.

Business Roundtable (2019). Statement on the purpose of a corporation [Online]. Available at: https://opportunity.businessroundtable.org/wp-content/uploads/2019/08/BRT-Statement-on-the-Purpose-of-a-Corporation-with-Signatures.pdf.

Chatterji, A.K., Durand, R., Levine, D.I., Touboul, S. (2016). Do ratings of firms converge? Implications for managers, investors and strategy researchers. *Strategic Management Journal*, 38(8), 1597–1614.

Cornell, B. (2020a). ESG investing: Conceptual issues. *Journal of Wealth Management*, 23(3), 61–69.

Cornell, B. (2020b). ESG preferences, risk and return. *European Financial Management*, 27(1), 12–19.

Cornell, B. and Damodaran, A. (2020a). Valuing ESG: Doing good or sounding good. *Journal of Impact and ESG Investing*, 1(1), 76–93.

Cornell, B. and Damodaran, A. (2020b). Big market delusion: Valuation and investment implications. *Financial Analysts Journal*, 76(2), 15–25.

Dimson, E., Marsh, P., Staunton, M. (2015). Responsible investing: Does it pay to be bad? In *Global Investment Returns Yearbook*, Credit Suisse (ed.). Credit Suisse Research Institute, Zurich.

Dimson, E., Marsh, P., Staunton, M. (2020). ESG Investing. In *Global Investment Returns Yearbook*, Credit Suisse (ed.). Credit Suisse Research Institute, Zurich.

Dortfleitner, G., Halbritter, G., Nguyen, M. (2015). Measuring the level and risk of corporate responsibility – An empirical comparison of different ESG ratings approaches. *Journal of Asset Management*, 17(7), 450–466.

Engle, R., Giglio, S., Lee, H., Kelly, B., Stroebel, J. (2019). Hedging climate change news [Online]. Available at: http://pages.stern.nyu.edu/~jstroebe/PDF/EGKLS_ClimateRisk.pdf.

Fama, E.F. and French, K.R. (2007). Disagreement, tastes and asset prices. *Journal of Financial Economics*, 83(3), 667–689.

Fink, L. (2021). Letter to CEOs [Online]. Available at: https://www.blackrock.com/corporate/investor-relations/2020-larry-fink-ceo-letter.

Fink, L. (2022). Letter to CEOs [Online]. Available at: https://www.blackrock.com/corporate/investor-relations/larry-fink-ceo-letter#bodyWrapper.

Fish, A., Kim, D.H., Venkatraman, S. (2019). The ESG sacrifice [Online]. Available at: https://papers.ssrn.com/sol3/papers.cfm?abstract_id=3488475.

Friedman, M. (1970). The social responsibility of business is to increase its profitability. *The New York Times Magazine*, September 13.

Hong, H. and Kacperczyk, M. (2009). The price of sin: The effects of social norms on markets. *Journal of Financial Economics*, 93(1), 15–36.

Hsu, J., Liu, X., Shen, K., Viswanathan V., Zhao, Y. (2018). Outperformance through Investing in ESG in Need. *The Journal of Index Investing*, 9(2), 18–26.

Li, F. and Polychronopoulos, A. (2020). What a difference an ESG ratings provider makes. Research Affiliates Publication [Online]. Available at: https://www.researchaffiliates.com/documents/770-what-a-difference-an-esg-ratings-provider-makes.pdf.

Pastor, L., Stambaugh, R.F., Taylor, L.A. (2021). Dissecting green returns [Online]. Available at: https://papers.ssrn.com/sol3/papers.cfm?abstract_id=3864502.

Semenova, N. and Hassel, L.G. (2015). On the validity of environmental performance metrics. *Journal of Business Ethics*, 132(2), 249–258.

Welch, I. and Cornell, B. (2022). *Global Climate Change: The Pragmatist's Guide to Moving the Needle*. Bowker, New Providence, NJ.

List of Authors

Amir AMEL-ZADEH
Said Business School
University of Oxford
UK

Andrew ANG
BlackRock Inc.
USA

Monica BILLIO
Ca' Foscari University of Venice
Italy

Steffen BIXBY
AQR Capital Management LLC
USA

Marie BRIÈRE
Amundi Asset Management
and
Paris Dauphine University
France

Alfie BRIXTON
AQR Capital Management LLC
USA

Ryan M. BROWN
Allspring Global Investors
USA

Guillaume COQUERET
Emlyon Business School
France

Brad CORNELL
Professor Emeritus
UCLA Anderson
USA

Michele COSTOLA
Ca' Foscari University of Venice
Italy

Claes EKMAN
Andra AP-fonden
Sweden

Johannes FREY-SKÖTT
Independent
Sweden

Fiona FRICK
Ungestion
Switzerland

Andreas G.F. HOEPNER
UCD Michael Smurfit Graduate Business School
University College Dublin
Ireland

Jason C. HSU
Rayliant Global Advisors
UCLA Anderson
USA

Emmanuel JURCZENKO
EDHEC Business School
France

James KELLY
Connected Asset Management
UK

David W. KRIDER
Allspring Global Investors
USA

Sebastian LANCETTI
PanAgora Asset Management
USA

Théo LE GUENEDAL
Quantitative Research
Amundi Asset Management
France

Yin LUO
Wolfe Research LLC
USA

Peter MANNERBJÖRK
Andra AP-fonden
Sweden

Christian MORGENSTERN
Independent
UK

Tomas MORSING
Andra AP-fonden
Sweden

Martin NERLINGER
School of Finance
Swiss Institute of Banking and Finance
University of St. Gallen
Switzerland

Björn ÖSTERBERG
Independent
Sweden

Loriana PELIZZON
Ca' Foscari University of Venice
Italy
and
Leibniz Institute for Financial Research SAFE
Germany

Lukasz POMORSKI
AQR Capital Management LLC
USA

Francesco PORTIOLI
CRIF S.p.A.
Italy

Sébastien POUGET
Toulouse School of Economics
University of Toulouse 1 Capitole
TSM
France

Max RIEDEL
Ca' Foscari University of Venice
Italy
and
Leibniz Institute for Financial Research SAFE
Germany

Thierry RONCALLI
Quantitative Research
Amundi Asset Management
France

Martin SCHMALZ
Saïd Business School
University of Oxford
UK

Katharina SCHWAIGER
BlackRock Inc.
USA

Harindra DE SILVA
Allspring Global Investors
USA

Jim SNOW
BlackRock Inc.
USA

Sascha STIERNEGRIP
Independent
Sweden

Jérôme TEILETCHE
Unigestion
Switzerland

Loredana URECHE-RANGAU
LEFMI
University of Picardie Jules Verne
France

Daniele VERGARI
CRIF S.p.A.
Italy

Viktoria-Sophie WENDT
BlackRock Inc.
USA

Gabija ZDANCEVICIUTE
UCD Michael Smurfit Graduate Business School
University College Dublin
Ireland

Index

G, H, I

M, N, O

P, R

S, T, Z

Other titles from

iSTE

in

Science, Society and New Technologies

2022

AIT HADDOU Hassan, TOUBANOS Dimitri, VILLIEN Philippe
Ecological Transition in Education and Research

CARDON Alain
Information Organization of The Universe and Living Things: Generation of Space, Quantum and Molecular Elements, Coactive Generation of Living Organisms and Multiagent Model
(Digital Science Set – Volume 3)

CAULI Marie, FAVIER Laurence, JEANNAS Jean-Yves
Digital Dictionary

DAVERNE-BAILLY Carole, WITTORSKI Richard
Research Methodology in Education and Training: Postures, Practices and Forms
(Education Set – Volume 12)

ELAMÉ Esoh
Sustainable Intercultural Urbanism at the Service of the African City of Tomorrow
(Territory Development Set – Volume 1)

FLEURET Sébastien
A Back and Forth Between Tourism and Health: From Medical Tourism to Global Health
(Tourism and Mobility Systems Set – Volume 5)

KAMPELIS Nikos, KOLOKOTSA Denia
Smart Zero-energy Buildings and Communities for Smart Grids
(Engineering, Energy and Architecture Set – Volume 9)

2021

BARDIOT Clarisse
Performing Arts and Digital Humanities: From Traces to Data
(Traces Set – Volume 5)

BENSRHAIR Abdelaziz, BAPIN Thierry
From AI to Autonomous and Connected Vehicles: Advanced Driver-Assistance Systems (ADAS)
(Digital Science Set – Volume 2)

DOUAY Nicolas, MINJA Michael
Urban Planning for Transitions

GALINON-MÉLÉNEC Béatrice
The Trace Odyssey 1: A Journey Beyond Appearances
(Traces Set – Volume 4)

HENRY Antoine
Platform and Collective Intelligence: Digital Ecosystem of Organizations

LE LAY Stéphane, SAVIGNAC Emmanuelle, LÉNEL Pierre, FRANCES Jean
The Gamification of Society
(Research, Innovative Theories and Methods in SSH Set – Volume 2)

RADI Bouchaïb, EL HAMI Abdelkhalak
Optimizations and Programming: Linear, Non-linear, Dynamic, Stochastic and Applications with Matlab
(Digital Science Set – Volume 1)

2020

BARNOUIN Jacques
The World's Construction Mechanism: Trajectories, Imbalances and the Future of Societies
(Interdisciplinarity between Biological Sciences and Social Sciences Set – Volume 4)

ÇAĞLAR Nur, CURULLI Irene G., SIPAHIOĞLU Işıl Ruhi, MAVROMATIDIS Lazaros
Thresholds in Architectural Education (Engineering, Energy and Architecture Set – Volume 7)

DUBOIS Michel J.F.
Humans in the Making: In the Beginning was Technique
(Social Interdisciplinarity Set – Volume 4)

ETCHEVERRIA Olivier
The Restaurant, A Geographical Approach: From Invention to Gourmet Tourist Destinations
(Tourism and Mobility Systems Set – Volume 3)

GREFE GWENAËLLE, PEYRAT-GUILLARD DOMINIQUE
Shapes of Tourism Employment: HRM in the Worlds of Hotels and Air Transport (Tourism and Mobility Systems Set – Volume 4)

JEANNERET Yves
The Trace Factory
(Traces Set – Volume 3)

KATSAFADOS Petros, MAVROMATIDIS Elias, SPYROU Christos
Numerical Weather Prediction and Data Assimilation (Engineering, Energy and Architecture Set – Volume 6)

KOLOKOTSA Denia, KAMPELIS Nikos
Smart Buildings, Smart Communities and Demand Response (Engineering, Energy and Architecture Set – Volume 8)

MARTI Caroline
Cultural Mediations of Brands: Unadvertization and Quest for Authority (Communication Approaches to Commercial Mediation Set – Volume 1)

MAVROMATIDIS Lazaros E.
Climatic Heterotopias as Spaces of Inclusion: Sew Up the Urban Fabric (Research in Architectural Education Set – Volume 1)

MOURATIDOU Eleni
Re-presentation Policies of the Fashion Industry: Discourse, Apparatus and Power (Communication Approaches to Commercial Mediation Set – Volume 2)

SCHMITT Daniel, THÉBAULT Marine, BURCZYKOWSKI Ludovic
Image Beyond the Screen: Projection Mapping

VIOLIER Philippe, with the collaboration of TAUNAY Benjamin
The Tourist Places of the World (Tourism and Mobility Systems Set – Volume 2)

2019

BRIANÇON Muriel
The Meaning of Otherness in Education: Stakes, Forms, Process, Thoughts and Transfers (Education Set – Volume 3)

DESCHAMPS Jacqueline
Mediation: A Concept for Information and Communication Sciences (Concepts to Conceive 21st Century Society Set – Volume 1)

DOUSSET Laurent, PARK Sejin, GUILLE-ESCURET Georges
Kinship, Ecology and History: Renewal of Conjunctures (Interdisciplinarity between Biological Sciences and Social Sciences Set – Volume 3)

DUPONT Olivier
Power (Concepts to Conceive 21st Century Society Set – Volume 2)

FERRARATO Coline
Prospective Philosophy of Software: A Simondonian Study

GUAAYBESS Tourya
The Media in Arab Countries: From Development Theories to Cooperation Policies

HAGÈGE Hélène
Education for Responsibility
(Education Set – Volume 4)

LARDELLIER Pascal
The Ritual Institution of Society
(Traces Set – Volume 2)

LARROCHE Valérie
The Dispositif
(Concepts to Conceive 21st Century Society Set – Volume 3)

LATERRASSE Jean
Transport and Town Planning: The City in Search of Sustainable Development

LENOIR Virgil Cristian
Ethically Structured Processes
(Innovation and Responsibility Set – Volume 4)

LOPEZ Fanny, PELLEGRINO Margot, COUTARD Olivier
Local Energy Autonomy: Spaces, Scales, Politics
(Urban Engineering Set – Volume 1)

METZGER Jean-Paul
Discourse: A Concept for Information and Communication Sciences
(Concepts to Conceive 21st Century Society Set – Volume 4)

MICHA Irini, VAIOU Dina
Alternative Takes to the City
(Engineering, Energy and Architecture Set – Volume 5)

PÉLISSIER Chrysta
Learner Support in Online Learning Environments

PIETTE Albert
Theoretical Anthropology or How to Observe a Human Being
(Research, Innovative Theories and Methods in SSH Set – Volume 1)

PIRIOU Jérôme
The Tourist Region: A Co-Construction of Tourism Stakeholders
(Tourism and Mobility Systems Set – Volume 1)

PUMAIN Denise
Geographical Modeling: Cities and Territories
(Modeling Methodologies in Social Sciences Set – Volume 2)

WALDECK Roger
Methods and Interdisciplinarity
(Modeling Methodologies in Social Sciences Set – Volume 1)

2018

BARTHES Angela, CHAMPOLLION Pierre, ALPE Yves
Evolutions of the Complex Relationship Between Education and Territories
(Education Set – Volume 1)

BÉRANGER Jérôme
The Algorithmic Code of Ethics: Ethics at the Bedside of the Digital Revolution
(Technological Prospects and Social Applications Set – Volume 2)

DUGUÉ Bernard
Time, Emergences and Communications
(Engineering, Energy and Architecture Set – Volume 4)

GEORGANTOPOULOU Christina G., GEORGANTOPOULOS George A.
Fluid Mechanics in Channel, Pipe and Aerodynamic Design Geometries 1
(Engineering, Energy and Architecture Set – Volume 2)

GEORGANTOPOULOU Christina G., GEORGANTOPOULOS George A.
Fluid Mechanics in Channel, Pipe and Aerodynamic Design Geometries 2
(Engineering, Energy and Architecture Set – Volume 3)

GUILLE-ESCURET Georges
Social Structures and Natural Systems: Is a Scientific Assemblage Workable?
(Social Interdisciplinarity Set – Volume 2)

LARINI Michel, BARTHES Angela
Quantitative and Statistical Data in Education: From Data Collection to Data Processing
(Education Set – Volume 2)

LELEU-MERVIEL Sylvie
Informational Tracking
(Traces Set – Volume 1)

SALGUES Bruno
Society 5.0: Industry of the Future, Technologies, Methods and Tools
(Technological Prospects and Social Applications Set – Volume 1)

TRESTINI Marc
Modeling of Next Generation Digital Learning Environments: Complex Systems Theory

2017

ANICHINI Giulia, CARRARO Flavia, GESLIN Philippe,
GUILLE-ESCURET Georges
Technicity vs Scientificity – Complementarities and Rivalries
(Interdisciplinarity between Biological Sciences and Social Sciences Set – Volume 2)

DUGUÉ Bernard
Information and the World Stage – From Philosophy to Science, the World of Forms and Communications
(Engineering, Energy and Architecture Set – Volume 1)

GESLIN Philippe
Inside Anthropotechnology – User and Culture Centered Experience
(Social Interdisciplinarity Set – Volume 1)

GORIA Stéphane
Methods and Tools for Creative Competitive Intelligence

KEMBELLEC Gérald, BROUDOUS EVELYNE
Reading and Writing Knowledge in Scientific Communities: Digital Humanities and Knowledge Construction

MAESSCHALCK Marc
Reflexive Governance for Research and Innovative Knowledge (Responsible Research and Innovation Set - Volume 6)

PARK Sejin, GUILLE-ESCURET Georges
Sociobiology vs Socioecology: Consequences of an Unraveling Debate (Interdisciplinarity between Biological Sciences and Social Sciences Set – Volume 1)

PELLÉ Sophie
Business, Innovation and Responsibility (Responsible Research and Innovation Set – Volume 7)

2016

BRONNER Gérald
Belief and Misbelief Asymmetry on the Internet

EL FALLAH SEGHROUCHNI Amal, ISHIKAWA Fuyuki, HÉRAULT Laurent, TOKUDA Hideyuki
Enablers for Smart Cities

GIANNI Robert
Responsibility and Freedom (Responsible Research and Innovation Set – Volume 2)

GRUNWALD Armin
The Hermeneutic Side of Responsible Research and Innovation (Responsible Research and Innovation Set – Volume 5)

LAGRAÑA Fernando
E-mail and Behavioral Changes: Uses and Misuses of Electronic Communications

LENOIR Virgil Cristian
Ethical Efficiency: Responsibility and Contingency (Responsible Research and Innovation Set – Volume 1)

MAESSCHALCK Marc
Reflexive Governance for Research and Innovative Knowledge (Responsible Research and Innovation Set – Volume 6)

PELLÉ Sophie, REBER Bernard
From Ethical Review to Responsible Research and Innovation (Responsible Research and Innovation Set – Volume 3)

REBER Bernard
Precautionary Principle, Pluralism and Deliberation: Sciences and Ethics (Responsible Research and Innovation Set – Volume 4)

VENTRE Daniel
Information Warfare – 2nd edition

Printed and bound by CPI Group (UK) Ltd, Croydon, CR0 4YY

16/09/2026

14963076-0006